AF167720

Car IT Reloaded

Roman Mildner · Thomas Ziller ·
Franco Baiocchi

Car IT Reloaded

Disruption in the Car Industry

Roman Mildner
Bergisch Gladbach, Nordrhein-Westfalen,
Germany

Thomas Ziller
Besigheim, Germany

Franco Baiocchi
Dudelange, Luxembourg

ISBN 978-3-658-47690-8   ISBN 978-3-658-47691-5   (eBook)
https://doi.org/10.1007/978-3-658-47691-5

Translation from the German language edition: "Car IT kompakt Reloaded" by Roman
Mildner et al., © Springer Fachmedien Wiesbaden 2024. Published by Springer Vieweg. All Rights
Reserved.

© The Editor(s) (if applicable) and The Author(s), under exclusive license to Springer Fachmedien
Wiesbaden GmbH, part of Springer Nature 2025

This work is subject to copyright. All rights are solely and exclusively licensed by the Publisher, whether
the whole or part of the material is concerned, specifically the rights of translation, reprinting, reuse
of illustrations, recitation, broadcasting, reproduction on microfilms or in any other physical way, and
transmission or information storage and retrieval, electronic adaptation, computer software, or by similar
or dissimilar methodology now known or hereafter developed.
The use of general descriptive names, registered names, trademarks, service marks, etc. in this publication
does not imply, even in the absence of a specific statement, that such names are exempt from the relevant
protective laws and regulations and therefore free for general use.
The publisher, the authors and the editors are safe to assume that the advice and information in this book
are believed to be true and accurate at the date of publication. Neither the publisher nor the authors or
the editors give a warranty, expressed or implied, with respect to the material contained herein or for any
errors or omissions that may have been made. The publisher remains neutral with regard to jurisdictional
claims in published maps and institutional affiliations.

Planung/Lektorat: Petra Steinmüller

This Springer Vieweg imprint is published by the registered company Springer Fachmedien Wiesbaden
GmbH, part of Springer Nature.
The registered company address is: Abraham-Lincoln-Str. 46, 65189 Wiesbaden, Germany

If disposing of this product, please recycle the paper.

Preface

Since our first issue of *Car IT Compact* in 2015, the automotive industry has developed at a breathtaking pace. Electrification has become a technological driver of progress that seems unstoppable. Once seen as a crazy idea and often mocked, it has reached a dimension that hardly anyone could imagine. The unexpected success of Tesla Motors, a company once the subject of ridicule, has now made electric cars a serious contender in the automotive industry.

Since then, established car manufacturers have found themselves in a partly latent, partly open crisis: the sales figures for their EVs have been sobering. It seems as if the traditional car manufacturers have been caught on the wrong foot. Since then, we are still in a state of shock. How do we get out of it? We know there can be no patent concepts: the socio-political and economic situation is too complicated for that on both sides of the Atlantic. Therefore, we aim to offer a few suggestions that make a meaningful and constructive contribution to the global strategy debate.

Another reason for our book's innovative approach is the realization that new technologies, especially artificial intelligence (AI), have rendered books that offer simple inventories increasingly obsolete. In Car IT Reloaded, we have focused on findings and ideas that are innovative, unusual, or uncommon in the field, aiming to stimulate discussion and provide suggestions for the automotive industry's future.

Car IT Reloaded was born out of the realization that the automotive industry is changing so rapidly that we—who have grown up in this business—often struggle to keep up. We want to offer a view of the car industry that considers this rapid pace. As the automotive industry continues to open up new areas of technology—as is the case with artificial intelligence, for example—our book has grown considerably. While the previous book, *Car IT Compact*, presented a concise overview of the then-current manifestations of the increasingly digitalized world of the automotive industry, *Car IT Reloaded*, therefore, offers a more detailed perspective, not only of the vehicle industry but also of innovative ideas and some (self-)critical voices that expand this context.

However, even though *Car IT Reloaded* has grown significantly in volume compared to its predecessor, this book is not a comprehensive compendium but a synthesis of findings and advanced ideas to better understand and tackle current

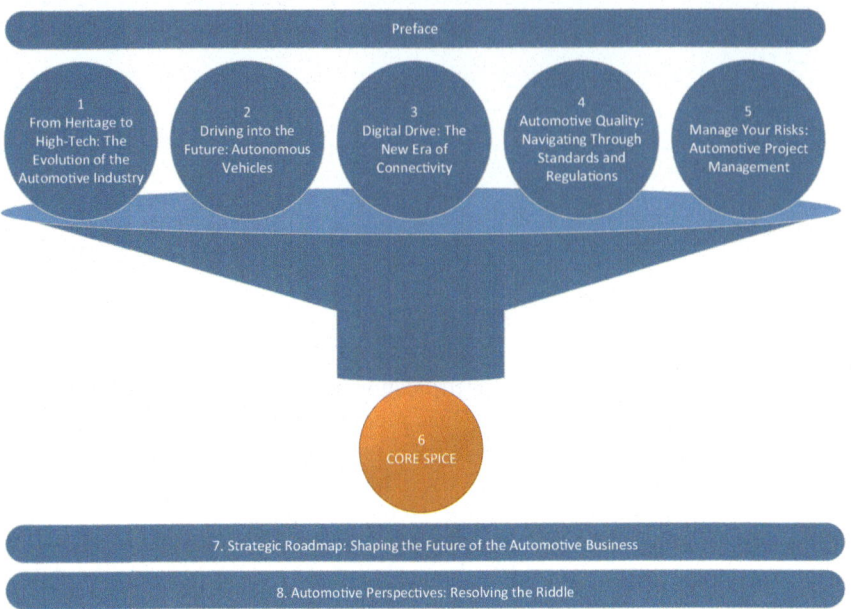

Fig. 1 Book concept of *Car IT Reloaded*

and future challenges in the car industry. We focus on practical ideas that can be applied in the short term, such as CORE SPICE.

Our book is based on a concept illustrated in Fig. 1.

The basic idea is based on a broader range of highly relevant topics in the vehicle industry. Starting with a historical outline covering particularly interesting subjects such as car propulsion technologies, electrification, vehicle safety, and autonomous driving, through to the historically outlined process improvement topics, the extensive Chap. 6 on CORE SPICE plays a special role. CORE SPICE supplements Automotive SPICE, widely used in the vehicle industry, with a practice-oriented, pragmatic component. While Automotive SPICE is an assessment model, CORE SPICE offers a pragmatic and practical project coaching approach. The benefits of systematic project implementation can be realized with less effort and in less time. This is particularly advantageous around safety-relevant projects (especially in functional safety). The book is rounded off in the concluding Chaps. 7 and 8, in which strategic topics relevant to the vehicle industry are discussed.

We are holding back on long-term forecasts this time. While electric vehicles continue to celebrate new successes, the situation with autonomous driving is more nuanced. The complexity of AI-based systems is significantly higher than anticipated. While we are looking forward to a fully autonomous car, no such vehicle is commercially available. Although we are convinced that self-driving cars are only a matter of time—the first vehicles, which still enable quite limited autonomous driving, are gradually becoming more popular—we realize it will be years before the automobile is so advanced that a steering wheel becomes superfluous.

Car IT Reloaded offers a wide-ranging outline of topics that affect today's vehicle industry. We focus on the critical issues we encounter daily as automotive project managers and technology advisors. However, the book is designed for a wide audience and will interest anyone working in the automotive industry or looking to broaden their horizons: project managers, consultants specializing in the automotive industry, quality managers, engineers, and anyone interested in automotive mass mobility. The result is a selection of topics that are both practical and entertaining. We hope you enjoy the book and find inspiration in it to utilize the solutions included and apply them to your practice.

Bergisch Gladbach, Germany   Roman Mildner
Besigheim, Germany   Thomas Ziller
Dudelange, Luxembourg   Franco Baiocchi

Acknowledgments *Car IT Reloaded* resulted from the intense collaboration, dedication, and support of many individuals, to whom we extend our deepest thanks.

Our special thanks go to Tobias Hummel for his invaluable advice and factual corrections, which greatly clarified the content.

A special thank you goes to Volker Johanning, whose groundbreaking work on the previous book, *Car IT*, inspired the creation of *Car IT Reloaded* and gave us the courage to tackle this new project.

We thank Alison Thompson of The Proof Fairy for her meticulous proofreading and editing. Her exceptional attention to detail and unwavering dedication have greatly enhanced the quality of this book.

We would also like to thank Springer-Verlag for supporting the formal design and encouraging us to conclude the manuscript successfully.

Lastly, we thank everyone who permitted us to use their sources and illustrations. Their contributions have greatly enriched this book.

We hope *Car IT Reloaded* will meaningfully contribute to the continued success of the car industry and to driving digital innovation. This work is dedicated to all who have been with us throughout this journey.

Cologne-Besigheim-Luxembourg   Roman Mildner
December 2024   Thomas Ziller
                        Franco Baiocchi

Competing Interests The authors have no competing interests to declare that are relevant to the content of this manuscript.

Contents

About the Authors

Roman Mildner is a project manager, automotive SPICE consultant, strategy consultant, leadership coach, and book author with over 25 years of experience in the software and systems industry. He holds a degree in computer science from RWTH Aachen University. He has made a name for himself through his work in the automotive industry. He is also successful and recognized in other industries, such as telecommunications, logistics, software houses, and life sciences. His international projects, which he has implemented in Germany, the USA, Australia, and Japan, demonstrate his global presence in the industry. His numerous clients include IBM Deutschland GmbH, Deutsche Telekom AG, Telefónica Germany GmbH & Co. OHG, T-Systems International GmbH, Harman International, Robert Bosch GmbH, Seeing Machines Limited, Aptiv Services Deutschland GmbH, Hitachi Astemo Ltd., ZF Friedrichshafen AG, Olympus Deutschland GmbH, CARIAD SE, Schaeffler AG, and many more. His practice-oriented approaches to project management and process improvement are particularly valued, especially concerning Automotive SPICE and safety-related projects. He has worked across the entire spectrum of the system development world, not only as a consultant but also as a practitioner, successfully fulfilling these roles as a project manager, quality manager, software architect, software developer, and test manager. This practical experience is the basis of his successful consulting work. His main goals are to establish effective project management and help clients overcome inefficient practices to achieve optimal results. His systemic approach leads to high customer satisfaction and timely project results, which he can proudly point out.

Thomas Ziller M.Sc. Dipl.-Ing. (FH), is self-employed and has over 20 years of professional experience in the automotive industry. He is the founder of martozi consulting and engineering. He supports companies in sales, business development, and engineering services in his consulting work. His focus is on team and strategy development as well as project management. He began his professional career with an apprenticeship as a communications electronics technician. After studying electronics and information technology, supplemented by a master's degree in signal processing and control, he first encountered the automotive industry at Valeo in the radar division. There, he worked in several areas, from engineering to sales and project management. He later moved to Carmeq (now CARIAD) as a team leader within the VW Group. He worked there for many years, successfully implementing challenging projects and building successful teams. Ziller found a new challenge at ASAP Engineering, where, using his technical expertise, he developed optimal solutions for established companies in complex projects. Today, in addition to project management and sales, he offers creative solutions for start-ups as a freelancer. As a Heilbronn University of Applied Sciences lecturer, he shares his knowledge to ensure that the next generation of engineers benefits from his insights—successes and failures—with practical "lessons learned."

Franco Baiocchi is an Intacs™-certified Competent Assessor and Project Management Professional (PMP) with over 30 years of experience in the automotive sector. He has been working as an independent consultant and freelancer since 2021. After his studies (FH Electrical Engineering), he started at Siemens in telecommunications sales and, after a short time, moved to Robert Bosch GmbH, where he held various positions in the development and application of gasoline injection systems, project management, sales, quality management, process management and consulting across multiple product areas. He led the piloting of a process management tool and developed the definition and introduction of a quality database application, including the interface to the project management tool to enable measurable quality reporting. In the consulting field, he has managed and completed a project in production to reduce zero-km failures for a product and advised on product areas in CMMI and ASPICE to achieve the corresponding maturity levels. He has also been leading assessments for almost 10 years.

From Heritage to High-Tech: The Evolution of the Automotive Industry

1

Abstract

This chapter traces the automotive industry's evolution from its early beginnings to today's high-tech era. The narrative covers key moments in automotive history, including the initial competition between electric, steam, and gasoline-powered vehicles, the eventual dominance of the internal combustion engine, and the industry's major safety and quality challenges. The chapter explores how the "Dieselgate" scandal contributed to the current electric vehicle renaissance and discusses Tesla's disruptive role in transforming the industry. It concludes with an analysis of current challenges in electrification, including battery technology, charging infrastructure, and the changing dynamics of the automotive market.

We do not intend this to be a comprehensive compendium on the world history of the automobile. Others have already done this in detail, and such an undertaking would go far beyond the scope of this book. However, we would like to place the following chapters in the automotive industry context, as it is a topic of great interest and importance. Our main aim is to provide a deeper understanding of this industry's current challenges. In recent years, the automotive industry has made enormous technical progress. This development has been multifaceted and so profound that many details have already been overlooked or forgotten. We will first outline the historical background—the beginning, significant highlights, and setbacks, as well as the current status—in order to deal with topics relevant to the future in later chapters. We can only shed light on a section of automotive history, but our examples represent the entire dynamic since the invention of the automobile. American and European history, especially German history, is, of course, close to our hearts. While we do not emphasize some important events from other countries, such as France, because of our geographical focus, our intent is to speak for the entire industry.

© The Author(s), under exclusive license to Springer Fachmedien Wiesbaden GmbH, part of Springer Nature 2025
R. Mildner et al., *Car IT Reloaded*, https://doi.org/10.1007/978-3-658-47691-5_1

We will now embark on a brief journey through time to look at aspects of automotive history.

1.1   The Dawn of the Automotive Industry

We know from general history that the wheel was invented in Mesopotamia around 3500 BC. It was invented simultaneously in other places, but no written records exist. It took around 300–500 years for the first carts to be developed. Horse-drawn carts were invented around 1900 BC. The situation mostly stayed the same until James Watt patented his practical steam engine in 1769. The new era of mobility began with the invention of the steam locomotive in 1804, which, for the first time, traveled without human or animal propulsion. In the following chapters of this book, we will outline how individual mechanical mobility developed in the years that followed.

Today, we take the automobile for granted. Many even feel tempted to grumble about what is sometimes called the "endless stream" of cars on our roads. We also like to use the argument that cars fundamentally harm the environment. Of course, it cannot be denied that cars hurt the environment; mass mobility has considerable consequences for nature. However, the environment is actually better with motor vehicles. Just imagine if we still had only horses as a means of transportation. Horses are beautiful animals, but their excrement is anything but harmless. In 1890, it was estimated that London's inhabitants would live almost three meters deep in horse excrement by 1950. Americans even feared that, by 1930, horse excrement in Manhattan would reach up to the third floor [RM-092]. One can smile at such extrapolations, but the global population has increased fivefold, from around 1.6 billion in 1900 to 8 billion today. If there were proportionally as many horses on our roads, our everyday life would smell very unpleasant and the environmental impact would be enormous.

In this sense, the automobile was a blessing; we didn't see it coming, but we wouldn't want to be without it today—and probably shouldn't.

1.1.1   Early Automobiles

As early as the fifteenth century, Leonardo da Vinci pursued the vision of a powered neither by humans nor animals. Over time, various and sometimes adventurous attempts were made to develop alternative methods, such as propulsion by air pressure. However, the steam engine was the first successful propulsion system. In 1769, the French Army captain Nicolas Cugnot developed the first verifiably functioning steam vehicle: a tricycle that could reach up to 3.6 km per hour.

The potential of these new machines was quickly recognized. Steam buses were already in operation in Paris at the beginning of the nineteenth century. Such buses were also experimented with in the USA and England, but the operation of these vehicles was inefficient, and most roads were unsuitable for them.

Moreover, mechanical vehicles also had an image problem at the time: they were generally regarded as strange, smelly, dangerous machines. For this reason, restrictive laws were passed in England. For example, a steam vehicle could only be operated if someone ran in front of it with a red flag to warn other road users. These restrictions were not lifted until the end of the nineteenth century, and such draconian laws hindered the progress of steam-powered mobility. Nevertheless, steam-powered vehicles became increasingly popular in Germany, the USA, France, and other countries at the beginning of the twentieth century. Some of these vehicles could also travel quite quickly. In 1906, the Stanley brothers, known for their steam-powered car races, reached the incredible speed of 205 km per hour with their *Locomobile.*

Progress—by its very nature—often encounters resistance—a common aspect of human nature. When early automobiles became popular, whole sections of society and some professions worried about the impact the new devices might have. Would horse-drawn coachmen now be out of work? What would happen to the whole "horse infrastructure": the stables, horse breeders, carriage manufacturers, farriers, horse feed suppliers, and so on? In the meantime, the new machines were by no means generally touted as "the future of mobility" at the time. Until 1914, automobiles were regarded as toys for the rich [RM-080]. They were status symbols—a thorn in the side of society, proof of modern decadence and rampant social injustice.

It was only when mass production of cars lowered the price to such an extent that "normal" people could afford them that this stigma disappeared and doubts about cars' usefulness fell silent.

Doubts about progress are inevitable. The latest developments in mobility, such as autonomous driving, are also under scrutiny. Self-driving cars have yet to become a reality, and the timeline for their widespread adoption remains uncertain. Nonetheless, we are confident that autonomous driving is simply a matter of time.

One thing seems certain: the automobile brought the masses prosperity, freedom, and joy. Despite environmental concerns and the many accident victims, the desire for individual mobility remains. The only question is: what will this mobility look like in the future?

The following chapters will revisit the automotive past to answer this question.

1.1.2 The First Electric Cars

Remarkably, one of the first electric vehicles was built as early as 1835 by Thomas Davenport, a blacksmith from Vermont, USA. He had already developed the first electric motor a year earlier. However, his patented electric vehicle did not bring him the hoped-for commercial success. Instead, his dream of electrically powered cars ruined him financially. When he died in 1851, electric driving was still far from being practical or popular. At that time, the steam engine was already widespread, diminishing interest in the new and still rather unreliable electric vehicles.

In 1884, Briton Thomas Parker developed and marketed one of the first commercially successful electric vehicles. Parker also developed various innovative solutions for the automobile, including a hydraulic all-wheel brake.

At the end of the nineteenth century, many manufacturers produced prototypes of electric vehicles. In 1899, Ludwig Lohner and Ferdinand Porsche (then employed by Lohner) developed the Porsche-Lohner "Toujours Contente" electric car, which was exhibited at the 1900 World Exhibition in Paris [RM-039].

Lohner was an environmentally conscious engineer.

> Leave us "world villagers" the last remnants of oxygen and clean air our wonderful society has given us. This air is ruthlessly polluted by the combustion products of the growing number of gasoline engines [RM-039].

Interest in electromobility also grew on the other side of the Atlantic. While most early electric vehicles used lead-acid batteries, the famous inventor Thomas Edison developed the nickel-ion battery used in Detroit Electric vehicles, for example. Henry Ford bought an electric vehicle from Detroit Electric for his wife Clara and gave Edison such a car for Christmas. Ford, who considered Edison's battery technology to be potentially revolutionary, thought about producing electric vehicles on a large scale. As early as 1914, he expressed the conviction that electric cars would be the next big thing; he was already planning a factory to manufacture them.

Electric vehicles became increasingly popular towards the end of the 19th and early twentieth centuries. By 1910, around 20 vehicle manufacturers were already vying for the favor of customers. The advantages included simple operation, uncomplicated maintenance, and low noise levels. However, like today, they suffered from an inadequate infrastructure. Charging stations where batteries could be changed were rare, and charging took many hours. In addition, electricity was not available everywhere at the time. By the 1920s, these obstacles seemed to have been overcome—at least in larger cities in the United States. The race to develop the best propulsion system was wide open.

1.1.3   The Triumph of the Internal Combustion Engine (ICE)

At the beginning of the twentieth century, three different types of driving were equally popular: steam-powered, electric, and petrol-powered vehicles. The latter were considered particularly unreliable and noisy. However, the first two engine types also had considerable disadvantages. Steam-powered vehicles required a cumbersome starting process and a longer preheating time before they were ready for operation. On the other hand, electric cars were slower and plagued by their limited range.

Like other types of propulsion, the gas engine was developed independently by various inventors. Frenchman Étienne Lenoir was already working on a petrol engine in 1862, and the Austrian Siegfried Marcus did the same in 1864. The German Nikolaus August Otto invented a fully developed four-cylinder engine in 1876; however, Frenchman Alphonse Beau de Rochas developed the underlying principle two years earlier. Otto founded an engine factory in Cologne, where he further perfected his internal combustion engines. One of his later employees was a certain Gottlieb Daimler.

Carl Benz patented the first gasoline-powered vehicle in 1886. It was a two-stroke engine on three wheels, marketed commercially for the first time.

After Gottlieb Daimler left his former boss, Otto, in 1885, he designed a vehicle with a four-cylinder engine. To this day, this type of engine remains very successful in automotive engineering worldwide.

However, the vehicles of the time were more like mechanical carriages than the cars we know today. Some terms in the automotive world, such as the "mudguard," first used on carriages and resembling a wing, come from this era when mobility was a rather dirty business. It was not until 1901 that a vehicle was designed and commercially produced that was more like today's cars in its essential features: the Mercedes 35 PS, developed by Wilhelm Maybach. Its 35-horsepower engine could reach up to 85 km/h [RM-040].

For more than a century, the gas engine has been predominant. Still, given the rapid pace of technological progress, it might appear surprising that the internal combustion engine was overshadowed by steam and electric vehicles for so long. The disadvantages of combustion engines continued to exist. Their costly maintenance and high noise level made them unpopular, especially among female drivers. Although gas engines offered clear advantages because of their higher energy density (and therefore greater range), there was a bigger problem: starting the engine required considerable physical effort. This obstacle was eventually solved through the invention of the electric engine starter by Cadillac engineer Charles Franklin Kettering in 1912. And thus, Kettering, a talented inventor with over 300 patents, paved the way for the dominance of the gas engine.

Coincidentally, the discovery of abundant and inexpensive oil wells in Texas at the beginning of the twentieth century made oil increasingly cheaper. In the 1920s, petrol was an absolute bargain, costing only around 5 cents per liter. This price trend and the constant further development of combustion engine technology meant other engine types were pushed into the background. Finally, the introduction of the assembly line by Henry Ford in 1914 drastically reduced the production costs of gasoline-powered cars, marking the end of the dominance of other car propulsion systems.

Since then, Ford models have rolled off the production line ever faster. Between 1908 and 1927, 15 million Model T cars were produced and delivered. The technology race seemed decided, and the internal combustion engine—often referred to in modern parlance as ICE, for *Internal Combustion Engine*—dominated the market. Electric cars gradually disappeared from the streets; by the mid-1930s, they were rarely seen. One of the last manufacturers of electric cars was Detroit

Electric. An interesting connection between Detroit Electric and the famous inventor Thomas Edison was that Edison, who owned a Detroit Electric automobile, was working on electric powertrains before the success of Ford's Model T overtook all other engine types. However, by the end of the 1930s, that time was only a vague memory. Detroit Electric had to file for bankruptcy during the Great Depression in 1929. It was bought out, and electric vehicles continued to be produced, but the race for a better powertrain had already been decided. In 1939, the factory closed its doors for good [RM-039].

It should be noted that the rapid demise of car brands did not only affect e-cars. Hundreds of new car manufacturers had been founded, particularly in the turbulent start-up phase of the early twentieth century. Over time, most of them went bankrupt or were taken over by successful competitors. The defunct American car manufacturers list includes over 1,600 names [RM-041]. Worldwide, the number could be around 3,000. With recent bankruptcies of electric car manufacturers such as Fisker, the list continues to grow. The automotive dream has always come at a high price.

1.2   Era of Automotive Turmoil

The automotive industry has experienced ups and downs over the decades, leaving deep scars on both the industry and motorists' minds. Growing safety concerns have led to technological advances and stricter standards. At the same time, the car part supplier industry and the entire supply chain have faced challenges. The "Diesel Gate" scandal shook confidence in the industry and accelerated its electrification.

This section addresses these challenges.

1.2.1   "Unsafe at Any Speed"

While post-war Europe lay in ruins, the 1950s was an exciting time of change for America, and industry experienced an impressive rise. Technological breakthroughs and earth-shattering inventions were announced at every turn: the transistor and integrated circuits and microprocessors, nuclear reactors, color television, the polio vaccine, new antibiotics, the commercial success of the Boeing 707, and much more—all unleashed on a society that, after the horror of the Great Depression, the shock of Pearl Harbor and the subsequent Second World War, was finally able to look to the future with confidence. It felt as if five decades had been compressed into fifteen years. America became an unprecedented consumer society—and the car became a cult object, celebrated above all in Detroit. The Detroit "Big Three" (General Motors, Chrysler, and Ford) dominated the global automotive industry. Detroit grew into a cosmopolitan, vibrant city, the equivalent of Silicon Valley at the turn of the millennium.

Belief in technology was the dominant spirit of the time, and no one dared oppose it. In the dawning era of space travel, the exterior design of vehicles often resembled rockets. Modern cars were often changed, like cell phones today: a new vehicle had to be bought every other year to avoid being ridiculed as outdated. In the 50s and 60s, car design became an expression of personality and individuality. Owning a modern car was a statement. The car had become an object of fashion, a way of expressing oneself.

The triumph of the Detroit carmakers seemed unstoppable: they dominated 90% of the global car market. Among the "Big Three," quality did not play a decisive role; the design was the driving force. Today, it is hard to imagine that defective cars used to be the norm. If it broke down, you just bought a new one. Looking back, it's clear this approach was unsustainable. Indeed, a scandal was only a matter of time.

In 1965, the book *Unsafe at Any Speed* [RM-036], written by a young lawyer from Washington, DC, became a bestseller. In his book, the then 31-year-old Ralph Nader described precarious safety problems that occurred in some cars popular at the time. He focused on the then-successful Chevrolet Corvair (General Motors) as a prime example of poor vehicle safety. The Corvair, advertised as a "compact" modern car, was equipped with a rear-engine drive that could produce up to 180 hp (special edition with a turbocharger). This allowed the six-cylinder engine to reach a top speed of 230 km per hour (there was no speed limit in the USA then). The Corvair was innovative in many respects; for example, it had independent suspension on all wheels, which increased ride comfort. Like the VW Beetle, which it was intended to compete with, it was air-cooled, had rear-wheel drive, and was quite affordable at around 2,000 US dollars.

Overall, the Corvair was a successful car that earned the "Car of the Year" award from the influential magazine *Motor Trend* in 1960.

But then problems and complaints began to accumulate. One of these problems was that the Corvair required significant differences in front versus rear tyre air pressure due to its different weighting. If this was not considered and the wheels were pumped up evenly, the vehicle could easily oversteer and roll over, becoming a death trap. Vehicle stability in sharp bends at high speed was poor; a Corvair would probably never pass today's "moose test." Additional stabilizers could be ordered to alleviate these problems, but they were not advertised and thus usually not ordered.

What followed reads like a thriller. While these problems could have been dismissed as driver error and 'not unusual at the time, 'GM management, unfortunately, chose to address the crisis with the character assassination and blackmail of Ralph Nader. His phone was tapped, and prostitutes were hired in an attempt to seduce the young lawyer and ruin his reputation. However, GM's efforts backfired. Ralph Nader sued them—and won. For GM's highly questionable "risk management" campaign, they had to pay Nader nearly half a million US dollars (around $4.5 million in today's dollars). Ralph Nader emerged as a champion of consumer rights.

As a result of the scandal, the NHTSA—the US National Highway Traffic Safety Administration, founded in 1970—launched a multi-year investigation into the Corvair incidents. In 1972, they announced that the Corvair was no less safe than other vehicles in its class. But it was too late—by 1969, GM had already discontinued production of the Corvair [RM-037].

The PR lessons learned from the Corvair debacle were more significant than the technical criticisms of an unqualified journalist with no engineering background. Instead of denying everything and taking dubious countermeasures, admitting the alleged problem and moving on would have been more effective. A well-thought-out PR strategy is crucial when a crisis occurs. In Chap. 7, we will take a closer look at the phenomenon of PR in the context of our industry.

While Ralph Nader's book startled the American industry, it was not its last scandal. At the end of the 1960s, Ford Motors developed the Pinto as a small, compact car to counter the increasingly successful competition from Japan and Europe, such as the Toyota Corolla and Volkswagen Beetle. Costing 2,000 US dollars, the Ford Pinto rolled off the production line for the first time in 1971. It was a successful compact car that was particularly popular with younger customers. One day in 1978, following a rear-end collision with another car, a Pinto exploded, and all three occupants were burned to death [RM-038]. Following this horrible accident, US authorities launched a large-scale investigation, which revealed that the Pinto had already been known to explode a year earlier. The Pinto's fuel tank was mounted too close to the rear bumper. It was particularly embarrassing that Ford engineers were already aware of the tank's design weakness at the time of the vehicle's development, including its tendency to explode. But nothing was done about it for years for cost reasons.

Fortunately for Ford's management, the driver who crashed into the Pinto was already known to the police for his careless driving. Therefore, the cause of the terrible accident lay with the driver and, at least from a formal point of view, not with Ford.

Such disasters—both human and in terms of public relations—caused the American car industry to lose favor with consumers. Japanese and European car manufacturers gradually became a threat to the "Big Three."

While European manufacturers, especially VW and its Beetle, were making waves, Japanese car manufacturers saw their chance to conquer the US market. However, although manufacturers such as Toyota and Nissan tried to gain a foothold in the USA in the 1960s, their success initially remained modest. Although they were able to improve quality steadily and at the same time offer their cars at comparatively low prices, the Detroit brands continued to dominate the American market thanks to their characteristic design. However, this changed abruptly with the first global oil crisis in 1973.

The problem with American cars was not solely related to the PR disasters. The decisive factor was that those brands earned the reputation of being less reliable, and the public eventually realized there were alternatives to poor-quality US cars. During the oil crisis, American cars' fuel consumption suddenly became another important purchasing factor. Gradually, smaller, cheaper, higher-quality cars became more in demand. As the first oil crisis slowly faded into oblivion and Detroit carmakers hoped for a return to the "good old automotive days," the second oil crisis began in 1979, leaving a lasting mark. The American consumer no longer wanted unreliable "rocket cars" with astronomical fuel consumption. As a result, buying "Made in America" cars became less importantl. Americans now wanted to drive from A to B as reliably and cheaply as possible.

In addition, Detroit carmakers faced increasing cost pressures due to growing scrutiny from the US trade union UAW (United Auto Workers).

Meanwhile, foreign manufacturers began to set up their plants in the USA. In 1978, Volkswagen AG, among others, opened its first American plant in Pennsylvania, with 2,500 employees assembling the Volkswagen Golf. Japanese manufacturers also continued to expand in the US. Honda opened its first plant in Ohio in 1982. In contrast to the German factories, however, the Honda plant represented a challenge of a different dimension to the "Big Three." Volkswagen opened the plant in Pennsylvania under the patronage of the UAW, and as a result, the quality of the VW Golf suffered from the same issues that Detroit manufacturers encountered. However, Honda chose a different path: the newly hired young American workers were sent to Japan, where they were taught the Japanese ways of quality manufacturing, such as *Total Quality Management* (TQM) and *Kaizen.*

This was one of many differences between Japanese and German working methods. German and American mentalities clashed sharply at the VW plant in Pennsylvania. The UAW employees also brought bad habits to the VW plant, similar to those of their Detroit colleagues: arrogance, unequal treatment at different employee levels, and a different way of dealing with mistakes and suggestions for improvement. VW finally capitulated to these problems and closed the plant in Pennsylvania in 1988 [RM-039].

While the VW managers failed with the help of the UAW, the Honda managers got off to a successful start with their efficiently functioning Japanese production methods. Initially, many feared that the cars produced in Ohio would be of poorer quality than the same vehicle types built in Japan. However, these concerns proved to be unfounded. Apparently, when treated with respect and led by effective managers, American employees could deliver the same quality as Japanese employees in Japan [RM-039]. The real triumph lay not only in the superior leadership of the Japanese management but, above all, in the UAW's inability to unionize the workers at the Honda plant. In a secret ballot conducted in 1986, the Honda plant workers—referred to as *"associates"* rather than *"employees"*—rejected membership of the UAW. Despite the UAW's subsequent attempt to force an open vote, which might have changed the outcome, Honda management stood firm. Thus, the Honda plant remained independent.

There is an interesting parallel here to the current situation in Tesla factories. There, too, the management has successfully resisted the spread of unions. There are other similarities, such as Honda's ability to build the first plant in Ohio in a record time of 18 months. The first Tesla plant in Germany was built in a similar time scale: it was completed in March 2022 after just 22 months of construction, despite political, bureaucratic, and environmental pressure.

Encouraged by Honda's success, other Japanese manufacturers, such as Toyota and Nissan, followed suit. The monopoly of the "Big Three" was soon broken. With 44 million vehicles sold worldwide, the Toyota Corolla became the most successful car ever, followed by the Volkswagen Beetle, while the Ford Model T was tantamount to a sentimental obituary.

Meanwhile, US car manufacturers were increasingly experiencing economic difficulties. As a result of the second oil crisis in 1979, Chrysler was the first of the "Big Three" to file for bankruptcy. It was the beginning of a long ordeal that reached its temporary climax with the insolvency of General Motors in 2009. One of the most successful and largest American companies, often inseparable from the "American Dream" and a jewel of America, was on its knees 101 years after it was founded. Although the company was "reanimated" after bankruptcy, it has remained a shadow of its former self. That same year, Chrysler also joined the ranks of bankrupt automakers. It was a frustrating time for American industry.

Germany, known as Europe's automotive stronghold, has experienced a roller-coaster ride of ups and downs over the decades. The 1990s, in particular, were a difficult period for the German automotive sector. The massive success of the VW Beetle was followed by a less clamorous period. Increasingly, front engines replaced rear engines, and competition in the global automotive market became ever more fierce.

With German reunification in 1990, German car manufacturers expected a boom in sales. They assumed the East Germans' newfound desire for freedom would be achieved primarily with German cars. But these high expectations were disappointing. Many people in East Germany and Eastern Europe did not have the necessary "small change" for such purchases. The resulting overproduction led to such an oversupply that, in 1992, there was even speculation about the possible insolvency of Volkswagen AG [RM-072]. The following year, Ferdinand Piëch succeeded Carl Hahn as Chairman of Volkswagen AG. At Audi, he had been known as a ruthless reorganizer. As the new CEO, he undertook several restructuring measures.

Firstly, he invited Japanese experts to teach VW the *Kaizen* strategy. The Japanese ideas of *lean management* and *just-in-time* were also copied. Secondly, he hired the notorious cost-cutter Ignacio López, about whom we write in Sect. 1.2.3 in the context of the supplier issues. Furthermore, Piëch consequently implemented the modular principle. This involved sharing vehicle components from several brands throughout the Group. In addition, under Piëch, the column dimension became a fetish: he knew the appearance of quality brands was essential to buyers and that this quality radiated to other, externally invisible aspects of a vehicle. Volkswagen also developed new, economical vehicles, some with unprecedented fuel efficiency. For example, the VW Lupo became famous for consuming 3 L of diesel per 100 km. Under Piëch, Volkswagen successfully negotiated a four-day week, which meant jobs were saved despite the slump in sales. The four-day week was accompanied by a 10% reduction in income for Volkswagen employees—a measure still in place today. The situation was different for top management in the Volkswagen Group; Piëch replaced almost all the members of the Board of Management [RM-072]. All these measures brought about a turnaround at Europe's largest car company: Volkswagen was able to return to a growth trajectory and later, in 2015, finally became the largest car company in the world.

Learning from each other is an essential skill for car manufacturers. Volkswagen adopted Ford's assembly line production methods in the 1930s and the *Kaizen* strategy from the Japanese in the 1990s. The Japanese, in turn, learned the TQM methodology from the Americans (in particular, from the famous quality expert William Deming).

Our industry's ability to learn is critical. This includes strategies known everywhere today, such as *lean management, just-in-time delivery, total quality management*, employee empowerment, continuous improvement (*Kaizen*), and much more.

The belief that Americans can always bounce back from devastating defeats is often considered a myth. But this vitality is by no means unique in the world. In fact, German and Japanese companies were also seemingly destroyed after the Second World War—and yet, amazingly, they returned to fame and fortune. This vitality of our industry gives us hope that other similar challenges can also be overcome.

In our industry, technological brilliance is not the only goal we have pursued. Vehicle safety has also steadily improved. Despite the PR and safety issues discussed in this chapter, the industry continues to learn from its mistakes and prioritize vehicle safety. In the next chapter, we will examine these efforts more closely.

1.2.2   Safety First

While vehicle design and the "coolness" of cars were often the main focus in the past, customer expectations have increasingly shifted towards safety in recent decades. The problem with the driving stability of the Audi TT at the end of the 1990s may serve as an example. The vehicle could oversteer dangerously at high speeds. Among the reasons experts gave for the car's behavior was a considerable aerodynamic lift at speed. In 1999, the German magazine *Der Spiegel* scrutinized a fatal accident involving an Audi TT. The issue was broadly discussed in German media [RM-035]. In the meantime, the physics of the sports car was admitted as a risk factor, and the cost pressure and vehicle design were questioned. The TT was meant to be an affordable, fast lifestyle car. The engineers' concerns about the vehicle were ignored. Audi chief engineer Werner Mischke is said to have said: 'We *wanted* this car to be *exactly* like this.' [RM-035] Many TT drivers accepted the technical risks. When driving such an unpredictable sports car, such risks were part of the fun for them.

But eventually, even many die-hard amateur racing car enthusiasts recognized that life holds a greater value than they might have acknowledged decades earlier. So, besides various safety improvements such as a new wishbone, the Audi TT also received a spoiler, although it was often criticized as "ugly."

An interesting parallel can be drawn between the case of the Audi TT and the scandalized Chevrolet Corvair 40 years earlier. Back then, engineers also insisted that vehicle design was no more dangerous or unstable than comparable vehicles in the same class. A safety-conscious driver would not have crashed in an Audi TT or a Chevrolet Corvair. Regardless of the purpose and design of modern cars, the driver's blind faith in technology was ultimately the biggest risk.

Today, car manufacturers boast that their vehicles have both passive and active safety features and are constantly being improved. In addition to the safety features we take for granted, such as airbags, ABS, ESP, reversing cameras, blind spot sensors, brake assist, and much more, a significant effort continues to be put into new safety features such as GSR-2, NCAP, etc. In the meantime, you could get the impression that drivers need to be protected from themselves—and we are getting better and better at it.

Yet progress in vehicle road safety is not always apparent. While the number of road deaths in Germany, for example, has continuously fallen, the situation is different in the USA (Figs. 1.1 and 1.2).

The number of road deaths is falling in almost every country in the world—but in the very country where owning a car has been a symbol of progress for over a century, the situation is developing negatively. There is speculation as to why, contrary to expectations, the relative number of road deaths has been rising for more than a decade in the US. It seems unlikely that the safety of US vehicles has suddenly deteriorated; the car market is so thoroughly globalized that similar safety standards apply everywhere. The increased number of traffic fatalities in America appears to be due to a structural change in road users. Many vehicles on American roads have become larger and heavier, partly due to the boom in trucks and SUVs that has been going on for years. This is a concerning development, especially for vulnerable road users such as pedestrians and cyclists. In contrast to many European countries, America has few cycle paths or traffic-calmed zones.

In such a traffic environment, road users who aren't protected by steel are particularly vulnerable. While car safety has steadily increased, it's hoped that US

Fig. 1.1 Traffic fatalities per 100,000 inhabitants in Germany. *Data source* OECD

Fig. 1.2 Traffic fatalities per 100,000 inhabitants in the USA. *Data source* OECD

authorities will also start considering the safety of other road users. Modern assistance systems will likely be able to recognize and protect vulnerable road users better and better. However, if this remains the only strategy for improving road safety, it will undoubtedly take decades before the relative number of road deaths in the USA decreases.

1.2.3 Issues with Suppliers

The vertical integration of manufacturing measures the extent to which a company outsources components for a product. Vertical integration in the German automotive industry has fallen for decades to less than 20% in 2015 [RM-042]. In other words, an OEM only carries out one in five manufacturing steps itself. The remaining steps are carried out by suppliers hired on a project basis. This can be seen as good supplier management—or as a major risk, as the OEMs ultimately have no direct access to suppliers if, for example, a component production deadline cannot be met.

Such a low level of vertical integration came at a price. Many European car manufacturers fell into a supplier crisis at the end of the 1980s and the beginning of the 1990s. The reason for this was the high cost of supplied vehicle parts. The quality of these components was also often not up to expectations. At the same time, pressure was increasing from Japanese manufacturers, whose TQM and extremely effective cost management were also becoming increasingly of a threat.

As a result of this development, OEMs began to increase the pressure on suppliers. All OEMs have established the institution of a *cost-cutter* as a critical corporate function. Ignacio López, who first worked at GM and from 1993 at Volkswagen, was one of the most successful. In the book *Das Schmarotzer-Prinzip—Wie Deutsche Automobilhersteller ihre Zulieferer Unbeaten* (*The Parasite Principle—How German Car Manufacturers Exploit their Suppliers*) [RM-043], author Volker Bauer describes how cost-cutters like Ignacio López whipped up systematic exploitation of their suppliers through various unfair tricks. These included:

- Insisting on unfair prices, threatening to terminate the contract, and announcing that already-developed concepts would be passed on to other suppliers.
- Simultaneous negotiation rounds: The cost cutter simultaneously negotiates with several suppliers in separate rooms, going from one table to the next and threatening that "the other" supplier will be awarded the contract unless the price is reduced.
- Uncoordinated changes: additional and costly features are suddenly taken for granted *after* a contract is agreed.
- Extended payment terms: enormous payment terms are imposed at the last minute, sometimes after a handshake agreement.
- Inconspicuous contractual clauses placing a financial burden on suppliers, such as stricter quality targets smuggled into contracts, which suppliers must then meet at their expense.
- Subsequent annual price reductions: Suppliers are pressured to reduce component prices annually, a tactic known as the "salami-slicing tactic."

As a result of such negotiating tactics, the development award is always granted very late. The logical consequence is enormous development pressure, and the supplier is forced against all reason into "simultaneous engineering." This means they must make do without sufficiently approved requirements and with an extremely "lean" (i.e., inadequate) system design. Such a project management approach "from hell" has dire consequences: project teams burn out under the subsequent deadline pressure, product quality fluctuates dangerously, and the additional costs for the resulting extra work naturally fall back on the supplier.

By systematically using such "parasitic" tricks, the cost pressure was shifted from the manufacturers to the suppliers while reducing vertical integration.

One consequence of such dubious practices was a wave of concentration in the supplier sector, which has been spreading since the end of the 1990s. Automotive suppliers became virtually uninvestable while car manufacturers continued using those "creative" practices. For example, Daimler demanded a 5% price reduction in 2001, followed by a further 10% [RM-044]. This is just one example; all other OEMs do the same. This "parasitic" strategy has unfortunately become the norm in our industry.

In doing this, automotive suppliers are taking considerable commercial and personal risks. The example of the insolvency of Japanese supplier Takata shows that even traditional and once-successful automotive suppliers are not spared when crises ensue. Founded in 1933 by Takezo Takada, the company initially manufactured parachutes. Later expansion into the vehicle safety belt business fitted well with this profile. From around 1988, Takata began building a new airbag business. However, unlike other airbag suppliers, after 1998, Takata used ammonium nitrate. Over time, it became apparent that ammonium nitrate can clump together, especially in high humidity. This led to a significantly faster combustion rate of the cartridge when an airbag-equipped with it was deployed. As a result, the controlled airbag explosion was much more violent. Parts of the exploded airbag came loose, flew through the passenger compartment at high speed, and injured the driver, sometimes dangerously or even fatally.

Takata, a listed company with a turnover of 4.5 billion euros and 60,000 employees [RM-045], filed for Chap. 11 bankruptcy in 2017 (Fig. 1.3).

Takata issued the largest recall of all time: more than 100 million airbags were replaced worldwide (67 million just in the USA [RM-046]). The recall continues

SRF 17136

Information to identify the case:

Debtor: TK HOLDINGS INC.   EIN: 13-3573416

United States Bankruptcy Court District of Delaware

Case Number: 17-11375 (BLS)   Date case filed for chapter 11: June 25, 2017

Official Form 309F (For Corporations or Partnerships)

Notice of Chapter 11 Bankruptcy Case

For the debtor listed above, a case has been filed under chapter 11 of the Bankruptcy Code. An order for relief has been entered. This notice has important information about the case for creditors, debtors, and trustees, including information about the meeting of creditors and deadlines. Read all pages carefully.

The filing of the case imposed an automatic stay against most collection activities. This means that creditors generally may not take action to collect debts from the debtor or the debtor's property. For example, while the stay is in effect, creditors cannot sue, assert a deficiency, repossess property, or otherwise try to collect from the debtor. Creditors cannot demand repayment from the debtor by mail, phone, or otherwise. Creditors who violate the stay can be required to pay actual and punitive damages and attorney's fees.

Confirmation of a chapter 11 plan may result in a discharge of debt. A creditor who wants to have a particular debt excepted from discharge may be required to file a complaint in the bankruptcy clerk's office within the deadline specified in this notice. (See line 11 below for more information.)

To protect your rights, consult an attorney. All documents filed in the case may be inspected at the bankruptcy clerk's office at the address listed below or through PACER (Public Access to Court Electronic Records at www.pacer.gov).

The staff of the bankruptcy clerk's office cannot give legal advice.

Do not file this notice with any proof of claim or other filing in the case.

Valid Picture ID is required for access to the J. Caleb Boggs Federal Building.

1. Debtor's full name
      TK Holdings Inc.

2. All other names used in the last 8 years: N/A

Jointly Administered Cases *[Other names, if any, used by the Debtor in the last 8 years appear in brackets and italics]*	Case No.	Tax ID.	Address
Takata Americas	17-11372	XX–XXX9766	2500 Takata Drive, Auburn Hills, MI 48326
TK Finance, LLC *[TK Finance Corporation]*	17-11373	XX–XXX2753	2500 Takata Drive, Auburn Hills, MI 48326
TK China, LLC *[TK China Inc.]*	17-11374	XX–XXX1312	2500 Takata Drive, Auburn Hills, MI 48326
Takata Protection Systems Inc. *[BAE Systems Safety Products Inc.] [Schroth Safety Products Corp.] [TKS Holdings, Inc.]*	17-11376	XX–XXX3881	1371 SW 8th Street, Suite 3 Pompano Beach, FL 33069
Interiors in Flight Inc. *[TPS Acquisition Inc.]*	17-11377	XX–XXX4046	5945 Hazeltine National Drive Orlando, FL 32822
TK Mexico Inc.	17-11378	XX–XXX8331	2500 Takata Drive, Auburn Hills, MI 48326
TK Mexico LLC	17-11379	XX–XXX9029	2500 Takata Drive, Auburn Hills, MI 48326
TK Holdings de Mexico, S. de R.L. de C.V.	17-11380	N/A	Carretera Santa Rosa Km. 3.5 Interior A Apodaca, Nuevo León 66600, Mexico

Fig. 1.3 Takata—official insolvency notice

to this day, and fatalities caused by the faulty airbags continue to be reported. In the USA alone, 26 deaths have been recorded so far. Billions of US dollars are being paid in compensation (one billion in 2017 [RM-047] resulted from the first lawsuit; others are currently pending).

Why did this terrible event happen?

It had been known since 2001 that Takata's airbags [RM-049] were defective. However, neither Takata nor the car manufacturers wanted to take action. One can only speculate why seven years passed before Honda initiated the first recall. It took another eight years for the NHTSA agency to demand a worldwide recall.

In our view, there were two main reasons for the hesitation. Firstly, the recall was costly, especially on such a scale. The other, more indirect reason was the pressure on suppliers—and, consequently, on car manufacturers—to reduce costs. Quite simply, ammonium nitrate was cheaper. Once it became clear that the substance in airbags was unstable, Takata would have had to use a safer and more expensive chemical. In an industry environment where cost-cutting was routinely passed on to suppliers, as described earlier, increasing costs would be unthinkable. For this reason, Takata used the cheaper chemical to the bitter end.

The relationship between car manufacturers and suppliers has been permanently damaged in recent decades. At the same time, quality and safety targets can only be achieved with the cooperation of everyone involved in the automotive ecosystem. However, many suppliers see their existence increasingly at risk, and some desperately fight for their survival—sometimes in vain. This is not the kind of "cooperation" needed for safe car design in our industry. Moreover, this constant tug-of-war is slowing down the drive for innovation. It's worth remembering that if the legacy OEMs don't manage to fix our automotive ecosystem, others eventually will. While the legacy OEMs persist in their entrenched supplier-customer roles, new OEMs have emerged unexpectedly in recent years, such as Tesla and Rivian, and they are taking previously unknown paths, such as different development methods, direct sales channels, insourcing versus outsourcing, and more. These newcomers follow very different methods and strategies from legacy OEMs, and some of them, like Tesla, have been quite successful. The industry needs to learn its lesson for the legacy OEMs to remain successful in the long term. More constructive cooperation would be appropriate instead of cultivating an antagonistic relationship between OEMs and suppliers. Only in this way can the industry develop its full innovative strength and remain globally competitive.

1.2.4 "Dieselgate" and the Consequences

In 2008, the then-CEO of Volkswagen, Martin Winterkorn, set the ambitious goal of VW becoming the global number one in the auto industry. In 2009, at the IAA (Internationale Automobil-Ausstellung—one of the world's largest and most prestigious automotive trade shows), he announced the goal of overtaking Toyota as the leading global car manufacturer [RM-050]. 'I know every screw in our cars, which is why I am involved in all the detailed decisions,' he said in an interview with the German newspaper *Die Welt* in 2006 [RM-052]. This proud attitude took hold. In 2012, VW AG completely took over its smaller competitor, Porsche. Just a few years earlier, that would have been unthinkable. 'The sky is the limit' was no empty promise. In the summer of 2015, the time had come: Volkswagen produced more cars than Toyota for the first time [RM-051]. German press celebrated this news extensively on the evening news. It was said that well-deserved German perfection had triumphed over the competition. The VW AG slogan, 'Volkswagen. The car.' summed up this pride in a nutshell.

Increasing environmental regulations and restrictions were of little concern to the German automotive industry. The diesel engine, which, thanks to its higher compression ratio, is more efficient and therefore has lower fuel consumption than the petrol engine, was seen as a *wunderwaffe*.

In 2015, just when it appeared that nothing could go wrong and the German industry was brimming with self-confidence, the worst possible catastrophe struck. It turned out that the diesel engine once celebrated as the unmatched marvel of German engineering, had only gained global acceptance through manipulated emissions tests. The problem was not the CO_2 emissions. The problem was the unacceptable emissions of highly harmful nitrogen oxides (NOx) suspected of causing cancer.

It initially appeared that a recall could fix the problem. Recalls are a sad and annoying ritual in our industry, but in this case, everything seemed to be under control [RM-052]. However, after a short time, the US Environmental Protection Agency (EPA) dropped a bombshell when they openly accused Volkswagen of installing illegal *defeat devices* to comply with the NOx limits. In other words, the EPA accused VW of outright lying [RM-055].

At first, it seemed it could be just another supplier issue. The automotive supplier Bosch was said to be involved in the scandal [RM-055], and for a while, it appeared as if it, once again, a case where a supplier had simply obeyed the demands of their customers. However, unlike the Takata case described above, Bosch was able to prove that both the supplier and the customer were in cahoots. It is hard to imagine what might have happened if this fact had not been revealed; Bosch might have suffered a similar fate to Takata.

What is particularly frustrating about the "Dieselgate" case is that *defeat devices* are not unusual. In 1995, General Motors was fined \$45 million for using a defeat device to avoid excessive carbon monoxide emission limits. Another scandal involved diesel trucks in the USA. Several manufacturers manipulated their engines to be more fuel efficient at a higher velocity, which increased emissions. This involved NOx emissions—the same emissions that were the focus of "Dieselgate." As a result, in 1998, several truck manufacturers were ordered to pay total fines of \$83.4 million, with a further billion US dollars spent on ECU conversions in the affected vehicles [RM-072].

In the EU, such fines were unheard of. The fact that manipulation of diesel emissions was not uncommon in Europe was also aided by the fact that such "tricks" were prohibited but generally had little or no legal consequences. Before the "Dieselgate" scandal, emissions manipulation was often dismissed as a minor offense [RM-072]. This would probably still be the case today if Volkswagen had not set itself the goal of conquering the USA with the "clean diesel." Diesel passenger cars were a rarity in the USA, and such vehicles were not as thoroughly tested as trucks. However, with the "clean diesel" offensive, diesel vehicles moved into the crosshairs of the US Environmental Protection Agency (EPA), which roused the attention of the authority's watchdogs.

Eventually, "Dieselgate" landed in US and German courts. So far, Volkswagen AG has been ordered to pay fines of 4.1 billion US dollars in the USA and 1 billion euros in Germany. It is estimated that Volkswagen AG's total losses due to the "Dieselgate" scandal amount to over 31 billion euros so far [RM-76]. Related lawsuits are ongoing, so we will not speculate further on the outcome.

What we can say with certainty, however, is that the scandal known as "Dieselgate" has reshuffled the cards in the automotive industry. In particular, the "clean diesel" initiative, promoted by Volkswagen as the future of clean mobility, turned out to be a dead-end. Diesel technology suffered such severe PR damage that no further improvement can be politically acceptable, no matter how sophisticated. Worse still, the combustion engine, in general, has become untenable. It has become apparent that the future of the car must be emission-free. It must be electric.

1.3   The Electric Renaissance

The combustion engine's dominance defined the twentieth century. The twenty-first century, however, will be dominated by the electric powertrain. This chapter sheds light on electric powertrains' development, challenges, and potential, from the first steps into the "electric winter" to the intermezzo of hybrid vehicles and the Tesla disruption. It also looks into the future of electrification.

1.3.1   The Long Electric Winter

If you search for "electric car" or "electric vehicle" on amazon.de, you will get dozens of electric toy cars. If you enter the same search query on the US site amazon.com, the result is an exhaustive list of publications dealing with developing and purchasing commercial electric cars. At first glance, this may seem like a funny anecdote. Still, it uncovers a hidden, more severe issue: for decades in Europe—and, in earlier times, in the USA, too—electric cars were not taken seriously.

Electromobility experienced a long "electric winter" when interest in electric vehicles was negligible and often ridiculed. The mobility infrastructure worldwide is geared towards fossil fuels, and the ever-improving combustion engines have made electric cars uneconomical and impractical. In the USA, there are around 145,000 gas stations; Germany has around 14,000 in operation. Since the end of the 1970s, the often lamented "death of gas stations" has led to a considerable reduction in the number in Germany from 45,000, but motorists have felt no cause for concern. It still feels like they can fill up "around every corner."

Combustion engines have long been a fixture in citizens' lives. If the infamous "Dieselgate" scandal hadn't occurred, we would still be buying and enjoying both petrol and diesel engines.

1.3.2   Hybrid Cars—An Intermezzo

As recently as the 1970s, some climate researchers believed a new ice age could soon be upon us. Various scientific books (for example, [RM-057, RM-058]) and scientific papers seemed to prove that the climate would cool down due to air pollution. Some of us still remember this curious period in climate science. But, since the 1980s, voices have been raised about fossil fuels leading to global warming. Computer-based calculations showed that burning fossil fuels could have catastrophic consequences due to increased carbon dioxide emissions. This discussion is highly politically charged. We are not politicians, so we will not pursue the scientific and political background here.

However, environmental policy was one of many decisive drivers behind the push for electrification. The second oil crisis in 1979 continued to raise demand for fuel-efficient cars, while further improvement in combustion engine efficiency hardly seemed possible. The combination of environmental and energy policy constraints made it clear that something had to change in our industry.

Japanese vehicle manufacturers were the first to recognize the need for innovative, more economical drive systems. In 1992, the then-Toyota boss Fujio Cho announced his commitment to a cleaner environment. "It depresses me when people tell me that Japan has done nothing [to protect the environment] in 100 years of automobile development. Toyota will do everything so that we can say that Japanese technology has contributed significantly to environmental protection" [RM-039]. This may sound like a common management spiel, but Toyota engineers actually developed the first hybrid powertrain, which was integrated into the first hybrid vehicle, the Toyota Prius.

Curiously enough, the Prius was a strictly internal project at Toyota. No automotive suppliers were involved in its development [RM-039]. This is an interesting parallel to the later Tesla strategy, in which Elon Musk internally developed several critical vehicle components and technologies.

The decision to develop hybrid technology contributed significantly to Toyota's commercial success. In 2008, Toyota overtook the then-market leader General Motors and became the largest vehicle manufacturer in the world. The Japanese company held this title until 2015 when Volkswagen overtook Toyota for the first time.

However, it would require more than increasingly economical engines with different engine variants and hybrid vehicle technology to meet the ever-stricter emission standards. "Dieselgate" was just the straw that broke the camel's back. By 2015, it was clear that mobility had to become completely climate-neutral, and this could only be achieved by using powertrains that do not require a single drop of petrol or diesel. One country after another announced it would ban the production of combustion engines altogether. Some countries are in a particular hurry; Norway and South Korea will be ready by 2025 (Table 1.1).

Table 1.1 List of countries and territories that prohibit the sale of internal combustion engines

Country/territory	Year	Details
Norway	2025	New vehicles
South Korea	2025	New vehicles
Belgium	2026	New company vehicles
Austria	2027	New cabs or car-sharing vehicles
Washington	2027	Government fleet
Slovenia	2030	New vehicles
Iceland	2030	New vehicles
Netherlands	2030	New vehicles
Denmark	2030	New vehicles
Ireland	2030	New vehicles
Israel	2030	New vehicles
Sweden	2030	New vehicles
India	2030	New vehicles
Germany	2030	New vehicles
United Kingdom	2030	New vehicles
Scotland	2030	New vehicles
Japan	2035	New vehicles
California	2035	New vehicles
China	2040	New vehicles
Singapore	2040	New vehicles
Sri Lanka	2040	New vehicles
Taiwan	2040	Bus (2030), motorcycle (2035), cars (2040)
Canada	2040	New vehicles
France	2040	New vehicles
Spain	2040	New vehicles
Portugal	2040	New vehicles
Egypt	2040	New vehicles
New Jersey	2040	New vehicles
District of Columbia	2045	Government and private fleet
Costa Rica	2050	New vehicles
Colorado	2050	New vehicles

Data source [RM-056]

In principle—at least as things stand now—it is expected that all countries except Saudi Arabia and Iran will soon ban combustion engines.

One has to wonder if we are about to throw the baby out with the bathwater, but even if some industry experts question the viability of these goals, "Dieselgate" has, to all appearances, heralded the final demise of combustion engines.

1.3.3 Tesla, the Disruptor

In 1990, California passed "zero-emission vehicle" legislation, significantly boosting electric vehicles. In 1996, General Motors developed the first EV of the new generation—the EV1. Toyota (RAV4-EV), Honda (Honda EV Plus), and Nissan (Altera) followed in 1997. Ford joined the ranks in 1998 with the Ranger EV.

Initially, it looked as if electric vehicles were gaining momentum. The EV fan community, many of whom leased GM's EV1, were thrilled. The car reached a speed of 129 km per hour—electronically limited. Without it, it could get an impressive top speed of over 290 km/h [RM-039]. The EV1 fan base grew steadily, but the model had its drawbacks. With all its shortcomings, the EV1 used a conventional lead-acid battery: it was heavy, took at least six hours to charge (though it took around 15 h when using a regular power socket), and there were capacity problems at low temperatures. The actual range of the EV1 under realistic conditions was only about 80 km. Nevertheless, none of this bothered the EV1 enthusiasts. They loved their vehicles and developed emotional attachments like some Tesla owners do today.

In 2003, General Motors suddenly withdrew from the EV business. No further leasing contracts could be concluded for the EV1, and GM also strictly refused to sell the EV1 after the leasing contracts had expired. Without exception, all EV1 vehicles were gradually scrapped. The outcry from the EV1 community was resounding. They had formed strong bonds with their cars and always said they were pleased with them. GM's decision was disastrous for their PR strategy: the EV1 was symbolically "buried" in a cemetery in California, to great publicity.

The reason for discontinuing production was simply the economic realization that the EV1 was not "scaling." After an investment of $1 billion, only around 800 EV1s were produced and leased [RM-039]. Building a meaningful business with the EVs was impossible with such measly numbers.

While GM made negative headlines with the demise of the EV1, the Japanese manufacturers did not give up. Although Toyota's EV RAV4 was also not commercially viable (Toyota sold around 1,000 units), they carried on. Similarly, Nissan's first EV with a lithium-ion battery was launched in the USA in 2010, selling only 22,000 units.

Eventually, GM executives realized it had been a mistake to pull out of the EV business altogether [RM-039], and they launched the Chevrolet Volt in 2010. Overall, however, it is safe to say that EVs remained a small niche in the car market. For EVs, the *"economy of scale"* was not achievable. It could have remained that way if an unknown start-up from California called Tesla had not entered the EV battle ring and an entrepreneur named Elon Musk had not invested in this start-up.

Musk is a colorful figure who has made headlines in the IT industry for over two decades. Born in South Africa, he made his first fortune in 1999 by selling the Internet service provider Zip2 to Compaq during the dotcom boom [RM-059]. He made $22 million from the sale and invested it in a financial services provider, X.com, which was subsequently merged with another service provider, Confinity. The resulting company was renamed PayPal. In 2002, eBay acquired PayPal for $1.5 billion. Musk earned 165 million dollars from this deal. In the same year, Musk founded Neuralink, which aimed to develop a computer-brain interface, and The Boring Company, which focuses on underground transportation solutions.

Tesla was founded in 2003 by Martin Eberhard and Marc Tarpenning. Elon Musk was not a co-founder. He invested in Tesla in 2004 and was initially not personally involved in developing the new electric vehicle [RM-059]. The Tesla Roadster, Tesla's first electric vehicle, was launched in 2008 using a clever PR strategy: the then Governor of California, Arnold Schwarzenegger, tested the Roadster and ordered it directly.

Tesla Roadster was not the first EV to use lithium-ion batteries. Nissan Altra did it back in 1997, but the EVs of the time, including the Altra, were designed to be an unobtrusive means of transportation. The motorized public did not find these vehicles particularly attractive. The Roadster, on the other hand, which was built on the chassis of the Lotus Elise, was designed as a lightweight sports car and was visually and technically "sexy." It had a range of around 200 km and could accelerate from 0 to 100 km/h in 4 s. That was an automobile that fit perfectly into Arnold Schwarzenegger's collection.

However, financial difficulties and the Great Recession caused the young company to falter. In 2008—the same year that sales of the Roadster began—Elon Musk became Tesla's CEO to stabilize the company's critical financial situation. Musk said Tesla was on the verge of bankruptcy [RM-057] and, at the last minute, he invested his money to keep the company alive.

As early as 2008, Tesla sold around 500 Roadsters (including reservations) at a list price of $109,000 [RM-060]. The Model S brought the first volume in 2012 [RM-061]. Further models followed: the 2015 Model X and 2017 Model Y and the large SUV Cybertruck in 2023.

The number of Tesla vehicles produced exceeded all expectations—with the possible exception of Musk himself, who has never suffered from a lack of self-confidence (Fig. 1.4).

Since 2017, it has become clear that Tesla was not a flash in the pan, as the competition often joked and secretly hoped. Tesla shares soared to astronomical heights. 2020 Tesla was worth more than Toyota, at least on paper. Elon Musk became the wealthiest person in the world for a time. It was a bold achievement, even if Tesla made much of its money selling environmental certificates; in 2022 alone, Tesla's revenue from the regulatory credits was 1.7 billion US dollars [RM-078] (Fig. 1.5).

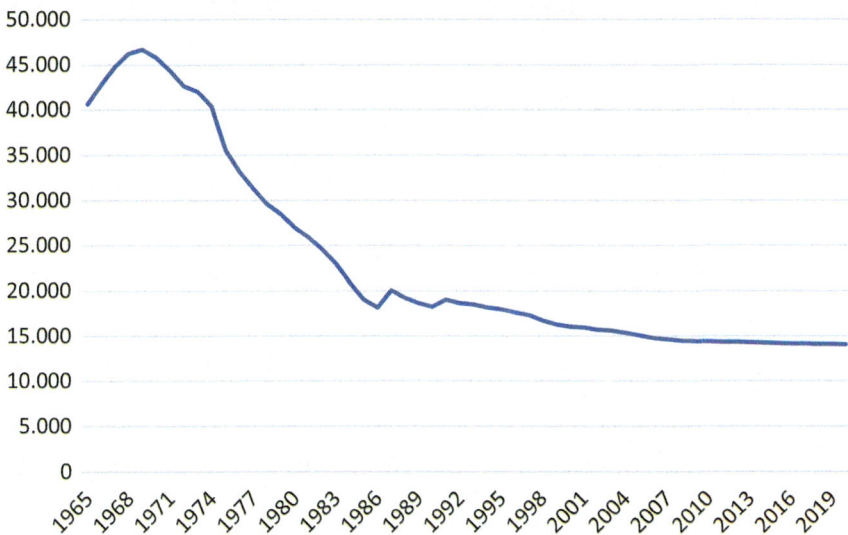

Fig. 1.4 Number of petrol stations in Germany. *Data source* ADAC

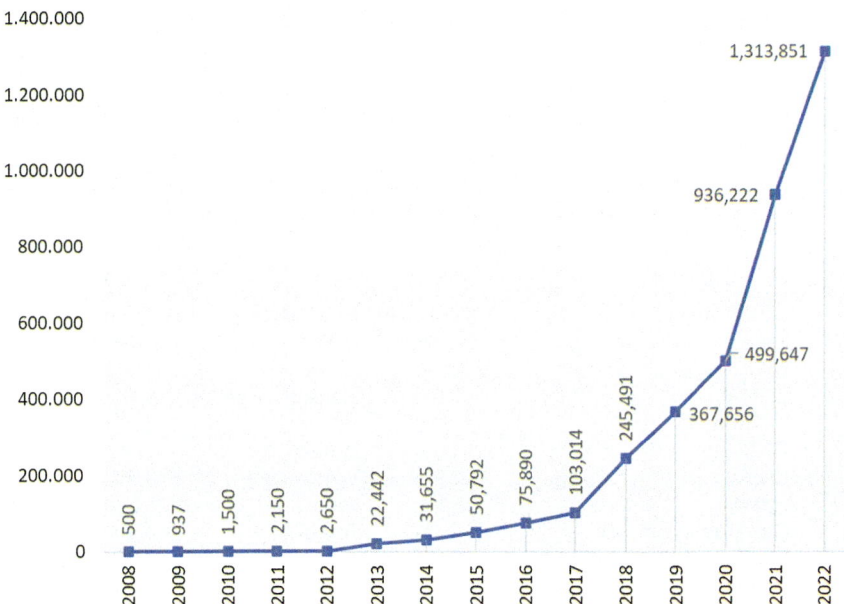

Fig. 1.5 Tesla sales figures

How did this happen? Musk, who had no prior experience in the car industry, quickly caught up with almost 130 years of automobile development. Most industry veterans dismissed Tesla's goals as ridiculously unrealistic.

One of the main reasons for Tesla's success was the realization that electric motors are less complicated than combustion engines. An EV does not require engine oil, complex fuel injection systems, or gearboxes. Replicating the rest of the mechanical design was relatively easy. Musk brought in enough experienced mechanical engineers to do this.

After the first shock, other manufacturers slowly followed suit. Chinese vehicle manufacturers, in particular, sensed an opportunity. So far, the legacy car manufacturers have been able to defend themselves against the Chinese "invasion." A particularly embarrassing incident happened to the Chinese car manufacturer Brilliance, who wanted to compete against the VW Golf with their car, quite unfortunately called BS4, with a price tag of 16,000 euros [RM-062]. However, the BS4 failed the 2009 EuroNCAP with a shameful zero stars. It meant the attempt to gain a foothold in Europe failed for the time being. The sigh of relief from the European vehicle manufacturers was almost audible in the German press [RM-063].

But in 2023, Chinese manufacturers celebrated a massive comeback with EVs and plug-in hybrids (Fig. 1.6).

Tesla has long dominated the market for "pure" EVs (BEV—Battery Electric Vehicles). However, the Chinese manufacturer BYD ("*Build Your Dreams*") has now conquered the market for plug-in electric vehicles (PHEV—*Plug-in Hybrid Electric Vehicles*) and overtook Tesla in the BEV segment in January 2024. However, PHEVs are not a long-term solution as they use fossil fuels and will not be permitted long-term.

For the time being, Tesla continues to dominate the BEV market. The question is, how much longer will it dominate this segment?

BEV Brand	BEV Share	PHEV Brand	PHEV Share
Tesla	16.6%	BYD	34.5%
BYD	11.5%	BMW	7.3%
SGMW	7.6%	Mercedes-Benz	6.7%
Volkswagen	4.2%	Volvo	5.1%
GAC Aion	3.5%	Li Auto	4.9%
Chery	2.9%	Volkswagen	3.5%
Hyundai	2.5%	Jeep	3.3%
Chang'an Automobile	2.4%	Toyota	3.0%
BMW	2.2%	Kia	2.8%
MG	2.1%	Lynk & Co	2.6%
Status: 2022. Source: TrendForce			

Fig. 1.6 Current EV/plug-in market shares

1.3.4   Battery Technology

In conventional ICE vehicles, the combustion engine often represents an innovation bottleneck. An engine's further development and optimization depends heavily on its design and innovative fuel systems. While particularly successful petrol and diesel engines dominated the market, exotic engines such as the Winkler remained irrelevant. Instead of pursuing alternative, disruptive motor concepts, developers gradually improved the established combustion engine variations, resulting in smaller improvements, such as turbochargers.

Electric motors have also hardly changed over the decades. Improving them seems unnecessary since they are very efficient and require little maintenance. Despite this, engineers still work on innovative concepts. For example, while most engines are radial motors, axial motors are used in small series, as they promise a much higher torque. Otherwise, an electric motor requires few extras, such as a gearshift, at least in its simplest form: the torque is available from the start and remains relatively constant over a wide speed range. Conventional electric motors appear sufficient for most EVs.

In contrast, batteries have become the innovation drivers in EVs. The constant desire for more range is prevalent. *"Range anxiety"*—the fear that an electric vehicle's battery will run out of charge before reaching a charging station—spurs engineers and chemists to solve this problem. Battery technology is the equivalent of the petrol engine in the era of combustion vehicles.

The lead-acid batteries used in earlier EVs were not powerful enough for mass use—as we saw with the GM EV1. Other battery types were, therefore, increasingly popular in the 1990s. The suitability of a battery for an EV is determined by several factors, such as:

– Availability of suitable raw materials
– Longevity (time until the battery is disposed of)
– Susceptibility to the "memory effect" (a phenomenon whereby the time to full discharge becomes shorter over time)
– Environmental sustainability
– Range of an EV with this battery technology—proportional to energy density
– Time required for charging
– Weight
– Fire hazards
– Sticker price
– Maintenance costs.

There is a whole host of different battery technologies. Some are unsuitable for EVs, while others are already used in car variants. An overview of various battery types can be found in Table 1.2.

The Li-ion battery is particularly popular and used for numerous applications. EVs are just one of many (Fig. 1.7, reproduced with permission from Fraunhofer Institute for Systems and Innovation Research IS).

Table 1.2 Overview of common battery technologies

Battery	Advantages	Disadvantages	Examples of use in EVs
Nickel–cadmium (NiCd)	Inexpensive, reliable	Strong memory effect, harmful to the environment	90s e.g. Peugeot 106 Electric
Nickel metal hydride (NiMH)	Greater capacity than NiCd, less memory effect	Heavy, still smaller capacity than lithium-ion (Li-ion)	90s Toyota RAV4 EV
Lithium-ion (Li-ion)	Higher energy density than its predecessors, lighter, longer service life	More expensive than the predecessors, risk of overheating	Tesla Model S
Lithium iron phosphate (LiFePO$_4$)	Thermally stable, longer service life	Lower energy density, higher weight	BYD e6
Lithium polymer (LiPo)	Lighter	More expensive, more vulnerable	Some prototypes and small series

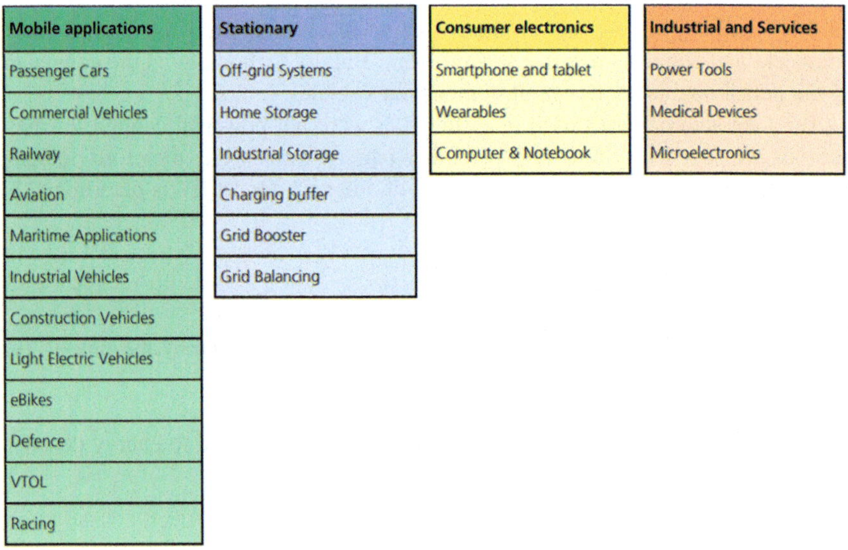

Mobile applications	Stationary	Consumer electronics	Industrial and Services
Passenger Cars	Off-grid Systems	Smartphone and tablet	Power Tools
Commercial Vehicles	Home Storage	Wearables	Medical Devices
Railway	Industrial Storage	Computer & Notebook	Microelectronics
Aviation	Charging buffer		
Maritime Applications	Grid Booster		
Industrial Vehicles	Grid Balancing		
Construction Vehicles			
Light Electric Vehicles			
eBikes			
Defence			
VTOL			
Racing			

Fig. 1.7 Applications of the lithium-ion battery. *Source* Fraunhofer Institute for Systems and Innovation Research ISI [RM-082]

Li-ion batteries have become the Swiss Army knife of energy storage. This universal cell technology can be found everywhere. Electric cars, laptops, cell phones—Li-ion batteries are ubiquitous.

They are also available in various shapes (see Fig. 1.8, reproduced with permission from Fraunhofer Institute for Systems and Innovation Research IS).

The different variants offer different advantages. The "stack" shape enables better space utilization, making batteries easily "stacked." The round shape ("cylindrical") is more robust and offers advantages in terms of heat dissipation. The flat coil variant allows cells to be mounted in different formats.

Li-ion batteries are also constantly being improved. For example, Elon Musk has implemented the idea of a *"tabless"* cell for Tesla. In the conventional method for cylindrical cells, strip-like *"tabs"* (small connecting tabs) are welded to the current collector foils. With the new tabless method, there is no separate tab. Instead, an uncoated section of the current collector foils is connected directly to the cell contacts (Fig. 1.9, reproduced with permission from Fraunhofer Institute for Systems and Innovation Research IS).

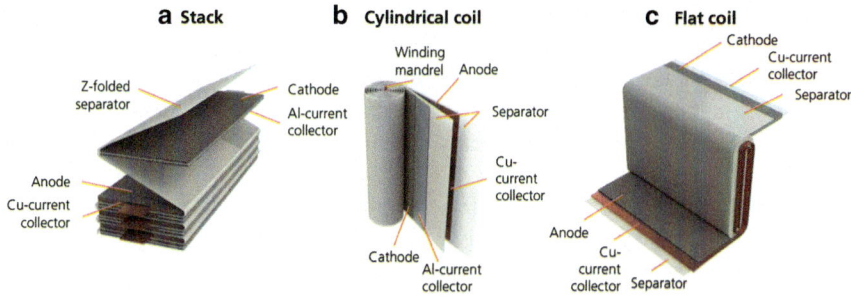

Fig. 1.8 Basic techniques of electrode construction. *Source* Fraunhofer Institute [RM-082]

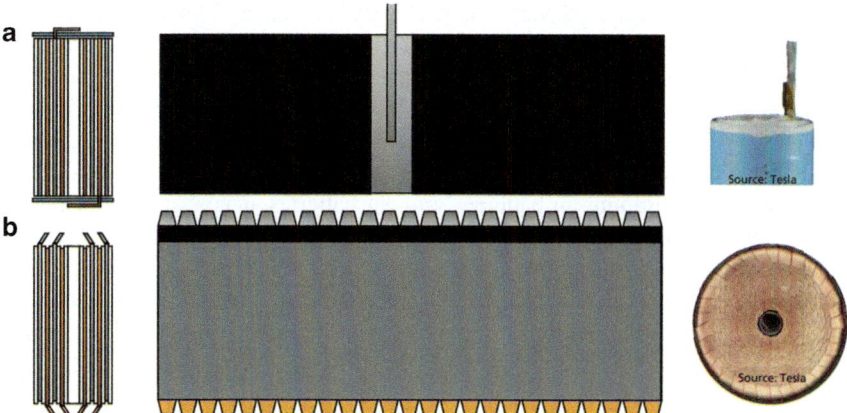

Fig. 1.9 Cells with tabs versus tabless (Fraunhofer Institute for Systems and Innovation Research ISI/Tesla) [RM-082]

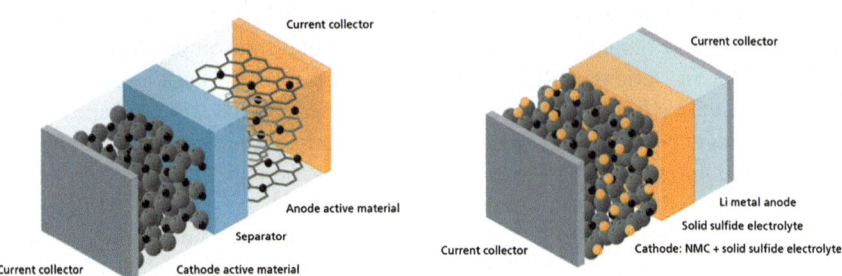

Fig. 1.10 Traditional lithium cells versus "solid-state" cells. *Source* Fraunhofer Institute for Systems and Innovation Research ISI [RM-082]

The advantage of tabless batteries is that the cells can be packed closer together without a separate wire. Tesla also increased the size of the cells. Previously, the so-called 2170 cells were used; 2170 stands for diameter (21 mm) and length (70 mm). The new cells are larger: 4680. According to Musk, the tabless batteries enable a 16% greater range and a 50% cost saving [RM-073].

Meanwhile, engineers keep working on entirely new battery technologies. For instance, solid-state technology seems promising (Fig. 1.10, reproduced with permission from Fraunhofer Institute for Systems and Innovation Research IS).

The term "solid-state battery" refers to a battery that uses a solid electrolyte instead of a liquid or gel-like electrolyte, as used in most of today's Li-ion batteries. The figure above shows that the "solid-state" cell (right) is much more densely packed than the conventional battery (left), enabling a higher energy density. Solid-state batteries promise higher operational safety (as the electrolytes are generally non-flammable), better range, extended lifespan, and faster charging times.

Solid-state batteries were originally expected to hit the market in 2035 [RM-082]. Surprisingly, in May 2023, Toyota announced a plan to deliver an EV with this new battery technology to end customers by 2027 [RM-084]. This demonstrates the immense pressure to advance in the field of electric batteries.

The pressure to innovate—and the resulting demand for electric batteries will have plenty of surprise end customers—is so intense that start-ups and scientists often feel tempted to make exciting announcements that later fail to materialize. These include lithium-air batteries, zinc-air batteries, magnesium-ion batteries, sodium-ion batteries, liquid-salt batteries, aluminum-air batteries, nanowire batteries, and many more, some promising an extraordinary range and other benefits. Perhaps these ideas will bear fruit one day, or utterly new energy storage concepts will be invented. Future batteries may achieve a range of several thousand miles. One thing is clear: there will be plenty of surprises in electric batteries for a long time.

1.3.5   Beyond Batteries: Alternative Energy Sources for EVs

Lithium-ion batteries are now prevalent, but alternative energy sources, like hydrogen, are still being discussed. These include:

- Hydrogen fuel cells
- Bio-fuels
- E-fuel.

We will now briefly examine these technologies, including the core concept and advantages and disadvantages of each alternative to the conventional battery drive.

Hydrogen fuel cells have been discussed for years as a promising technology. Hydrogen must be stored in a specially developed tank, while oxygen comes from the ambient air.

A platinum catalyst helps split hydrogen into atoms. However, the critical component is the intermediate membrane. Made from a special polymer (e.g., Nafion), it allows only protons to pass through, not electrons, which must detour through the motor. The reaction results in pure water. The hydrogen fuel cell drive is simply an electric motor.

Advantages of the hydrogen fuel cell drive (*FCEV—Fuel-Cell Electric Vehicles*)

- A quick refueling process (fill up and drive off).
- A range of over 800 km is possible.
- Lower weight than lithium-ion batteries.

Disadvantages of the hydrogen fuel cell drive

- Hydrogen production is energy-intensive and, therefore, expensive. Added to this is insufficient renewable production sources to cover demand.
- High transportation and storage costs.
- There are hardly any hydrogen filling stations. In 2017, there were 100 hydrogen filling stations in Japan, 14 in South Korea, 69 in the USA, 56 in Germany, and 15 in China [RM-065]. In 2021, there were around 550 worldwide [RM-066]. That is still a negligible number.
- Safety concerns when storing large quantities of liquid hydrogen under a pressure of 700 bar.
- Use of expensive rare earth minerals, such as platinum.
- More complex technology. For example, an additional battery is required for regenerative braking, as the fuel cells cannot store energy.
- Due to the small number of FCEVs sold to date, there is little experience with the durability and safety of this technology.

Overall, FCEVs remain a promising technology, particularly for specific purposes such as truck or bus transportation. Hydrogen is also considered a possible energy storage technology that can balance the energy mix.

Bio-fuels (biodiesel)

Combustion motors can use synthetically synthesized fossil fuels and remain potentially climate-neutral. These include:

- Bio-ethanol: alcohol produced by fermentation of carbohydrates such as sugar (e.g., cane sugar). In Germany, for example, a fuel mixture of 90% petrol and 10% ethanol is currently sold as "E10"; no modification of combustion engines is required in this case.
- Biogas: Gas (such as methane, known as "bio-methane") is produced from organic waste. Combustion engines can use this gas.
- Biodiesel: Like biogas, biodiesel is produced from certain (rapeseed), used cooking oil, etc. Biodiesel is also added as an additive, similar to E10 for gasoline (e.g., B7).

The basic idea behind these propulsion types is to use conventional combustion engines as a CO_2-neutral alternative while using the proven technology of combustion motors.

Advantages of biofuels

- Proven and climate-neutral drives.
- Use of existing infrastructure (in particular, filling stations and standard transport routes).

Disadvantages of biofuels

- Agricultural land is repurposed, which can lead to social tension.
- Environmental pollution due to soot and NOx emissions.
- Reduced biodiversity: large areas are used to grow the same crop.

Despite their potential climate neutrality, "Dieselgate" has made biodiesel a politically unacceptable option. Combustion engines are considerably more complicated than BEVs, so biofuels are likely to remain a niche technology.

E-fuel

In contrast to biodiesel, e-fuel (synthetic fuel) is not obtained from organic biomass but is produced using electrolysis. In this process, water is split into hydrogen and oxygen. A mixture of the hydrogen and CO_2 obtained is then chemically synthesized to create a petrol- or diesel-analog fuel. The e-fuels can then be used in conventional combustion engines.

Advantages of e-fuels

- E-fuels are 100% compatible with conventional combustion engines (ICE) but are climate-neutral.
- No additional investment is needed in the infrastructure; it can be maintained as it is.
- Jobs in the combustion engine industry are preserved. For comparison, BEVs require an estimated 30–40% fewer components.

Disadvantages of e-fuels

- There are no cost savings compared to BEVs.
- Substantial investment in the corresponding chemical plants is required.
- Climate neutrality cannot be achieved if electrolysis is not produced using renewable energies.
- Combustion engines are about to be banned. Suitable legislation would have to be drawn up, which is not politically unproblematic.

Some politically driven policies aim to support these technologies. In theory, e-fuels represent a socio-economic "magic bullet" that would save ICE-based production. However, this is now increasingly unrealistic. As previously stated, the combustion engine appears "burnt out." Therefore, it seems unlikely that e-fuels could be adapted as part of the "green revolution." A shift towards e-fuels would only occur if lithium-ion batteries were shown to be an ecological disaster, which is unlikely, regardless of the rationale.

1.3.6 Charging Infrastructure: The Future Cash Cow

The electrical infrastructure represents a unique challenge that some global players must take seriously at an early stage. Tesla currently offers the only global network of charging stations. The Supercharger network is constantly expanding in America, Europe, and Asia. In China, a key electrification market, there are already 1.8 million charging stations, of which 650,000 were installed in 2022 alone [RM-102].

Tesla's electric charging infrastructure stands out as a source of diverse business opportunities. It is challenging to differentiate products when they are homogeneous. Despite attempts to differentiate themselves with special, brand-specific fuels such as Shell V-Power or BP Ultimate, filling station networks have largely failed to build strong customer loyalty. In contrast, a charging station network can be leveraged in various ways to strengthen customer loyalty. Although electrical energy is even more homogeneous than fossil fuels, this is precisely where the opportunity lies, as charging stations can be more easily integrated into commercial infrastructure with environmental benefits. The term "charging stations" can be misleading, suggesting brief stops similar to traditional gas stations. Until a "flash charger" is invented that enables EV charging in 1 to 2 min, a different infrastructure will be needed compared to traditional filling stations.

The opportunities presented by the new EV charging infrastructure are manifold. The factors that can play a role in this include:

- The customer experience is changing, including shopping and entertainment options. Legoland provides an example of how this can work. However, car cities such as the VW Autostadt in Wolfsburg and The Henry Ford Museum in Michigan can also expand in many ways.
- Integration into the electric grid: intelligent grid technology cooperations with electricity providers can offer an excellent opportunity to expand customer loyalty.
- Personalization: charging stations can collect customer data, increasing customer retention and loyalty. Of course, legal regulations must be observed in this regard, but this is a detail (see Payback and similar loyalty concepts, for example).
- Scaling: besides service stations, charging stations can be integrated into other sectors. Examples: hotels, supermarkets, gyms, hair salons, cinemas, medical facilities, certain public buildings (where you usually have to wait a long time, for example, vehicle registration offices), car repair shops, temporary apartments (Airbnb), monitored rental blocks, parking lots, public and private underground garages, youth hostels, philharmonic halls, event halls, etc. One example is the partnership announced in October 2023 between the Hilton hotel chain and Tesla, under which Hilton plans to install 2,000 Tesla charging stations [RM-118].
- PR opportunities: the silent world power of public relations offers almost unlimited possibilities. They range from loyalty programs such as Payback to targeted, brand-specific campaigns that can promote brand building to drive long-term customer loyalty and positive public perception.

With the right strategy, traditional car manufacturers can connect with customers more deeply.

More convenient charging technologies, such as wireless charging stations, are also provided. Tesla has recognized this and is looking for innovative solutions to bring this technology to mass market maturity. As part of this strategy, Tesla bought the German company Wiferion in 2023, which offers this technology [RM-103]. Even though Tesla immediately sold the company (RM-125), the automaker kept the developers. It further indicates that electrification offers a very agile market potential in which quick decisions and a willingness to invest will determine future market leadership in the car business.

1.3.7   Barriers to BEV Adoption

BEVs with the usual lithium-ion batteries are not entirely unproblematic. Despite the enthusiasm of an army of Tesla fans and highly effective PR (to which Elon Musk has undoubtedly contributed significantly and continues to do so, including his breakdancing interludes and an incessant X/Twitter bombardment), numerous questions remain.

Range Anxiety
In conventional combustion, the only option is often vehicles (ICE); fuel consumption is easy to estimate. You are rarely surprised—the fuel gauge works quite reliably. Not so with BEVs. In advertising, the range is sometimes exaggerated beyond all measure. Depending on driving style, use of the air conditioning, outside temperatures, and battery age, it can happen that the enthusiastic BEV driver suddenly stops in the middle of nowhere. With an ICE vehicle, you can easily find fuel or quickly walk to a nearby gas station and keep driving. But if your BEV runs out of charge, often the only option is to call a tow truck.

It is particularly frustrating that much misinformation is being spread with "official" range figures for BEVs. The range is a critical selling point for some BEVs, but fact-based information on EVs is distorted by the desire of the EV makers to promise unrealistic ranges (Fig. 1.11).

The example of the Porsche Taycan is quite laudable: a measured value of up to 35% above the manufacturer's specifications should set an example for other BEVs. The motto should be "better to understate than to disappoint customers."

This test series does not include the Tesla Roadster, for which Tesla claims a range of around 1,000 km, but should serve as a reference for other manufacturers.

Extensive charging times
Charging an empty BEV to full should take just a few minutes. Unfortunately, even charging a Tesla at a proprietary Supercharger, for example, takes at least 20 min. The time it takes to charge a BEV depends mainly on the power connection type. Chargers are classified as follows:

– **Type 1**: This is the slowest option, for example, a regular household socket.
– **Type 2**: This type is usually available at public charging stations and some workplaces.
– **Type 3**: This is the fastest type of charging, usually only available at unique high-performance fast-charging stations.

A full charge at a regular socket (type 1) can take 20 h or more. BEVs remain at a significant disadvantage compared to ICEs (Table 1.3).

BEVs often require appropriate power connectors. In Germany, these are alternating current (AC) connections or a CSS connection for direct current (DC), enabling an output of up to 350 kW.

BEV	Price	EPA (km)	Measured	Difference
2022 Lucid Air Dream Edition Range	$169.000	836.86	804.67	-4%
2022 Mercedes EQS 450+	$102.310	563.27	635.67	13%
2022 BMW iX xDrive50 w/20" Wheels	$83.200	521.41	555.22	7%
2023 Cadillac Lyriq RWD w/20" Wheels	$62.990	502.09	531.08	6%
2021 Tesla Model 3 AWD	$48.990	567.99	498.9	-12%
2023 Porsche Taycan RWD 93 kWh Battery (New Software	$81.150	362.11	490.86	36%
2021 Tesla Model S Plaid w/21" Arachnid	$134.490	560.02	482.8	-12%
2021 Porsche Taycan RWD 93 kWh Battery	$85.470	362.11	471.78	30%
2019 Tesla Model 3 AWD	$47.990	518.18	466.71	-10%
2021 Ford Mustang Mach-E California Route 1 Edition	$50.400	490.39	461.64	-6%
2023 Ford Mustang Mach-E Premium AWD Extended Rang	$66.295	466.71	458.58	-2%
2020 Porsche Taycan 4S 93 kWh	$103.800	326.69	447.56	37%
2020 Tesla Model Y AWD	$49.990	508.26	444.5	-13%
2022 Ford Lightning Lariat Extended Range	$77.474	515.33	434.38	-16%
2022 Rivian R1T Large Pack. 20" all-terrain tires	$79.500	505.2	408.78	-19%
2021 Porsche Taycan 4 Cross Turismo 93 kWh Battery	$93.700	346.01	405.72	17%
2021 Porsche Taycan Turbo Cross Turismo 93 kWh Batter	$153.500	328.31	396.24	21%
2022 Kia EV6 GT-Line AWD 20" Wheels	$56.400	441.06	394.18	-11%
2022 BMW i4 M50 w/20" Wheels	$65.900	365.17	384.65	5%
2020 Hyundai Kona EV	$37.190	415.1	383.09	-8%
2021 Volkswagen ID.4 First Edition	$43.995	402.34	376.53	-6%
2022 Hyundai Ioniq 5 AWD SEL w/19" Wheels	$45.900	411.99	365.17	-11%
2021 Ford Mustang Mach-E AWD Std Range	$50.300	339.47	363.6	7%
2020 Chevrolet Bolt EV	$36.620	416.75	363.6	-13%
2021 Polestar Polestar 2	$59.990	375.01	363.6	-3%
2022 Hyundai Ioniq 5 AWD Limited w/20" Wheels	$54.500	411.99	313.82	-24%
2022 Jaguar I-Pace EV400 w/22" Wheels	$69.900	376.53	313.82	-17%
2020 Nissan LEAF SL +	$43.900	346.01	305.78	-12%
2022 Ford Lightning Pro Standard Range	$39.974	370.15	344.32	-7%
2019 Audi e-tron	$74.800	328.31	302.26	-8%
2020 Hyundai Ioniq EV	$33.045	273.59	275.2	1%

Fig. 1.11 Range of BEVs. *Data source* [RM-067]

In the USA, there are two different sockets:

– Tesla Supercharger (proprietary) is a direct current connection with an output of 250 kW.
– SAE J1772 (alternating current) delivers up to 100 kW.

Until recently, Tesla vehicles had a monopoly in this area; only Tesla vehicles could be charged with the Tesla Supercharger. However, that has recently changed, and Tesla now allows other electric vehicles to charge at the Supercharger [RM-132].

Battery life
Conventional ICEs can last several hundred thousand kilometers without engine replacement. For petrol engines, it is up to approx. 200,000; for diesel engines, approx. 400,000 km. There are rare examples of diesel cabs with over 1 million kilometers on the clock. The *Guinness Book of Records* gives the example of an American cab driver from Alaska who clocked up almost 5 million kilometers with his Volvo P-1800S [RM-097]. However, it's well-known that diesel and petrol

Table 1.3 Charging times for selected BEVs

Vehicle model	Charging time type 1 (h)	Charging time type 2 (h)	Charging time type 3 (min)
BMW i3	18	3	45
BMW i4	26	8	31
Fiat 500 Electric	15	2.5	35
Ford Mustang Mach-E	26.5	7	38
Hyundai Ioniq 5	31	6	40
Kia EV6	24/32	7	18
Mercedes EQA	26	6	30
MINI Cooper SE	14	3	45
Opel Corsa-e	22	7	30
Polestar 2	38	8	32
Porsche Taycan	35/41	8/9	41
Renault Zoe	27/34	5	50
Skoda Enyaq iV	28/39	6/8	40
Tesla Model 3	28	7	20
Tesla Model 3 Long Range	35	8	25
Tesla Model S	48	6.5	25
Tesla Model X	50	6.5	25
Tesla Model Y	28	6	20
Volvo XC40 Recharge	39	7.5	40
VW ID.3	34	7	25
VW ID.5	35	7.5	29

Source [RM-086]/Tesla

engines lose performance over time due to mechanical wear. As a result, a TDi with 200,000 km won't drive as nimbly as a brand-new car.

With EVs, the limiting factor is not the motor mileage but the battery life. The constant charging and discharging reduces the range with each charge. There are reports that after 10 years, the range can drop by up to 30%. Unsurprisingly, EV manufacturers limit their battery warranties (Table 1.4).

Replacing an EV battery can be more expensive than replacing a conventional combustion engine. Replacing an engine can cost 5,000–10,000 euros or more, but high-performance EV batteries are even more costly, depending on their capacity. Tesla's costs range between 15,000 and 20,000 euros [RM-100]. For other models, it can be considerably more costly (Table 1.5).

The good news is that battery prices have been falling for years.

Table 1.4 Warranty for EV batteries (examples)

Manufacturer	Model	Warranty period	Minimum capacity (%)
Audi	All	8 years/160,000 km	70
BMW	All	8 years/160,000 km	70
Lexus	UX300e	10 years/1 million km	70
Mercedes-Benz	EQC, EQA, EQB, EQC	8 years/180,000 km	70
Mercedes-Benz	EQA	10 years/250,000 km	70
Porsche	Taycan	8 years/60,000 km	80
Tesla	Model 3	8 years/200,000 km	70
VW	All	8 years/160,000 km	70

Data source ADAC [RM-098]

Table 1.5 Examples of EV battery replacement prices (as of 2021)

Car	Price (Euro)
Honda e	7.283
Mercedes EQA	15.209
Mercedes EQC	28.516
Mercedes EQV	27.230
Nissan Leaf	10.306
Smart EQ fortwo	6.537
VW ID.3/4	10.000–15.000
VW e-Golf	10.000–20.000

Source [RM-099]

However, the less good news is that lithium-ion battery prices rose for the first time in 2022 (Fig. 1.12). This means the battery price trend is not a one-way street, and the decline could slow down. The price of an EV battery is expected to remain a significant overall price factor for a long time.

Fire hazard

The lithium-ion batteries currently in use pose a fire hazard. Internal short circuits can cause the batteries to overheat, leading to fires—for example, in a car collision. Worse still, such fires are practically impossible to extinguish. The source of the fire is not the ambient oxygen, as is the case with conventional ICEs. Instead, it is a chemical reaction within the batteries. Under certain circumstances, this reaction can continue until it is complete, sometimes taking hours or even days. Worse still, these fires can flare up again unexpectedly, even days after the original fire event. In rare cases, a BEV battery can ignite spontaneously when a car is flooded with salt water ("thermal runaway") [RM-068].

Interestingly, these risks have not materialized in NCAP crash tests. So far, no BEVs have performed poorly in these tests, even regarding fire risk.

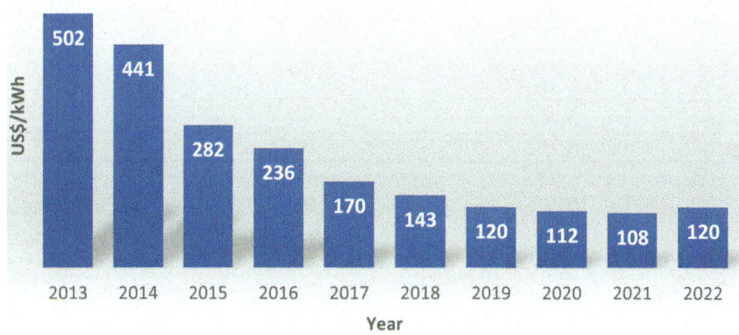

Fig. 1.12 Prices for a lithium-ion battery. *Data source* [RM-101]

Availability of charging stations

On paper, Germany has more than 92,000 charging points(!). However, the number is deceptive. Most of these are low-power chargers unsuitable for a power-hungry BEV. Also, one in ten chargers in Germany is out of service [RM-085].

Only the top two performance classes (see Table 1.6) are suitable for a BEV fast charger, resulting in 10,563 available charging stations. But even this figure is misleading: the access points to these charging stations are often not optimally located (for instance, several hundred meters away from a gas station), requiring you to search for them using a suitable app. Gas stations are also unprepared for a rush of long-waiting customers; you may be expected to have to wait hours before you can charge your vehicle.

The good news is that the number of suitable charging stations is steadily increasing. According to the German Federal Network Agency, around 5,700 charging stations were installed in Germany in the twelve months from January

Table 1.6 Charging stations in Germany

Number of charging points	1/2023	1/2022
0–3.7 kW	1,931	1,671
>3.7–15 kW	13,660	9,832
>15–22 kW	60,052	46,157
>22–49 kW	1,655	1,663
>49–59 kW	3,468	2,879
>59–149 kW	1,343	901
>149–299 kW	6,371	2,410
>299 kW	4,192	2,367
All performance classes	92,672	67,880

Data source Federal Network Agency

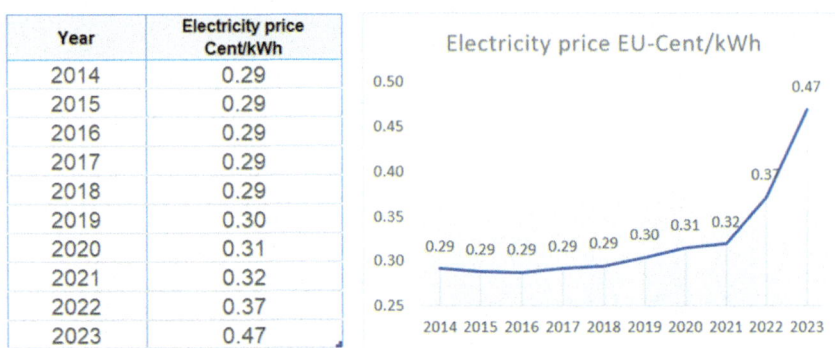

Year	Electricity price Cent/kWh
2014	0.29
2015	0.29
2016	0.29
2017	0.29
2018	0.29
2019	0.30
2020	0.31
2021	0.32
2022	0.37
2023	0.47

Fig. 1.13 Electricity price trends for private households in Germany. *Data source* [RM-069]

2022 to January 2023 alone. In addition, most EV customers charge their vehicles at home, namely 59% [RM-133]. At the same time, it remains to be seen how EV users will deal with the scarcity of available charging points if the density of EVs increases significantly in the future and no charging station is available near their homes.

Tesla was the first company to recognize the high demand for fast chargers. There are currently more than 50,000 Tesla Superchargers installed in the USA, yet this is far from enough, given the size of the country, even though the total number of charging stations is increasing rapidly. Other energy companies have also recognized that there is much money to be made from electrification and charging stations. BP, for example, announced that the company intends to create 20,000 "ultra-fast" charging stations in Germany by 2030, in addition to other "green" measures [RM-081]. The investment volume is enormous, yet it is just one example of many.

It can, therefore, be expected that the bottleneck caused by inadequate charging station coverage will be resolved globally within the next few years.

Electricity price
As electricity is the new "fuel" for EVs, it is hardly surprising that the demand for electricity has been increasing steadily. Electricity price curves are pointing upwards worldwide, especially in countries that continue to rely on the import of fossil fuels, as their conversion into electricity continues to be a significant power source. Wind and solar power prices are also constantly rising, not least due to the recent inflationary pressure. In addition, geopolitical developments influence global and local electricity prices (Fig. 1.13).

However, in everyday life, things look even gloomier. Unfortunately, prices between 45 and 79 cents per kilowatt hour are not uncommon. Electricity prices are always on the move, but the trend is always upward. Operating EVs could become more expensive than ICEs, leading to political turbulence.

Model	Propulsion	Base price (EUR)	Costs per KM incl. apportionment		
			10.000	20000 km	30000 km
Tesla Model 3 Performance AWD	BEV	58.560 €	105,5	63,7	50,6
Tesla Model Y	BEV	47.568 €	92,6	54,8	42,7
Audi Q4 40 e-tron 150	BEV	51.900 €	101,8	61,4	47,7
VW ID.3 Pro Performance Life	BEV	39.995 €	76,4	46,4	36,4
VW Golf 1.5 eTSI Life DSG	Gasoline	34.970 €	79,3	49,6	39,9
Golf 2.0 TDI SCR Life DSG	Diesel	37.945 €	86	51,5	40,5
Toyota RAV4 2.5 Hybrid Business Edition 160	Gasoline/Hybrid	43.790 €	89,2	55,4	44,2
BMW 320i Steptronic 135	Gasoline	50.700 €	109	65,6	52,1
BMW 320e Steptronic 150	Gasoline/Plug-In	54.000 €	107,3	63,8	50,3

Fig. 1.14 Operating costs (TCO)—examples. *Data source* ADAC [RM-047]

Power grid capacity

It's not just the incessant rise in electricity prices that worries EV advocates and potential EV buyers. The overall capacity of electric grids is also making it challenging to switch entirely from ICEs to BEVs. If everyone were to switch to EVs at once, the capacity of the German electricity grid would have to be **increased sixfold** [RM-070]. The taxpayer would have to bear the price for such a massive national investment.

BEV price

We discussed the EV prices in Fig. 1.11. BEVs are still quite expensive compared to conventional ICEs. However, prices will likely drop significantly due to the economy of scale law. Furthermore, the price is variable because many countries subsidize BEVs, sometimes considerably. As of the completion of this manuscript, the subsidy in the USA for a new BEV was USD 7,500.

The sticker price is, however, not the only decision factor when buying a car. A more insightful measure is the TCO—Total Cost of Ownership. TCO calculates all operating costs, such as purchase price, depreciation, maintenance, and energy expenses. The ADAC recently analyzed the cost for several vehicles, including the type of propulsion. The results were revealing (Fig. 1.14).

At the beginning of the EV era, some EV buyers were happy about the special tax breaks and the opportunity to charge their EVs at work. This non-representative survey makes it clear that the bargain era is over. If the TCO is the measure, EVs appear neither cheap nor convenient. Though some EV manufacturers have slightly reduced prices, we still need a genuinely affordable electric car.

1.3.8 Electric Vehicle Outlook

Who would have thought the closure of the Detroit Electric plant in 1939 would not be the final chapter in the history of electric mobility? EVs are suddenly "in," and combustion engines are "out"—at least, that's what it looks like for now.

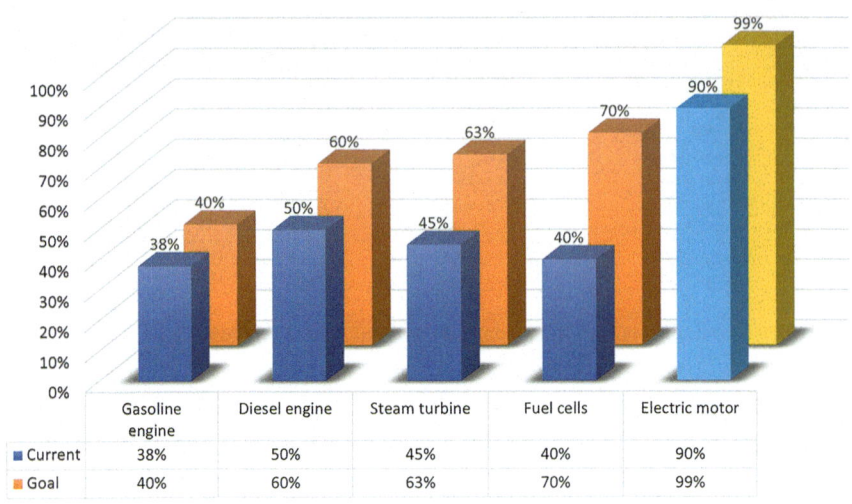

	Gasoline engine	Diesel engine	Steam turbine	Fuel cells	Electric motor
■ Current	38%	50%	45%	40%	90%
■ Goal	40%	60%	63%	70%	99%

Fig. 1.15 The efficiency of different drive types ([RM-088,089,090,091])

From a scientific perspective, it's surprising that this has taken so long. The search for the best vehicle propulsion system should also be a quest for the most efficient one, and electric motors have always been the most efficient (Fig. 1.15).

How can it be explained that for over 130 years, we viewed an inefficient type of drive as the best and only option? This must have been mainly due to the underdevelopment of battery technology. But maybe we weren't looking hard enough for an alternative to the lead-acid battery.

Regardless of the reasons, it has become clear that the future belongs to efficient, electric EV motors. The consulting firm Ernst & Young provided an interesting forecast in January 2023, predicting the future of EV development:

– EV sales in the USA, China, and Europe are expected to exceed all other motor types' sales by 2030—three years earlier than previously assumed.
– In 2040, sales of ICEs will account for less than 1% of global vehicle sales.
– Sales of electric vehicles in Europe will exceed those of other drive systems by 2027.

Therefore, consumer demand, government measures, tax incentives, and aggressive electrification targets set by politicians and vehicle manufacturers should accelerate the EV revolution [RM-071].

It appears safe to assume that a simple extrapolation of BEV development may need to be revised. For example, according to some predictions, there might not be enough lithium available for EV production, and lithium-ion battery technology may reach its limits. However, this goes against patterns observed in past technological progress. Many things were invented because the inventor "didn't know it was impossible." Bold, innovative inventions often border on madness. A few centuries ago, the idea of a self-propelled vehicle without the need for horses or similar animals would have been enough to land someone in an insane asylum for life. Battery technology will likely follow a similar path; other battery solutions will emerge if lithium isn't sufficient.

General Motors' EV1 was based on lead-acid batteries—the same technology used in early EVs like those from Detroit Motors. For a long time, little changed in the realm of EVs. But now, the time has come to show what engineers are capable of. It is almost certain that further battery technologies will be developed that will be cheaper and more environmentally friendly. Who knows, maybe they won't even use lithium at all.

While the future of mobility seems certain to be electric, the exact path to this future is still uncertain. However, as the need for electromobility grows, the "electrification" in our industry will continue for the foreseeable future.

1.4 The Automobile Time Machine

In our book, we have emphasized the milestones in the automotive industry, from the invention of the first automobile to the rise of the combustion engine and the advent of EV. The development from the first car to today's electric vehicle spans several centuries. Examining automotive history's successes, inventions, and crises is truly captivating. Below, we present a selection that reflects the most significant inventions and events directly related to the automotive industry and plays a role in the broader context of world history (Table 1.7).

Table 1.7 Selected key events in automotive construction, the automotive industry, and worldwide

Year	Inventions	Industrial events	Special events
1769	Nicolas-Joseph Cugnot develops the first steam-powered vehicle (F)	–	James Watt patents his improved steam engine (Scotland)
1776	Gasoline engine (D)	–	American Declaration of Independence
1800	Invention of the electric battery (V)	–	Napoleonic Wars begin
1834	First practical electric motor (D)	–	–
1835	First electric vehicle (US)	–	Invention of Morse code
1837	Electric motor is patented (US)	–	–

(continued)

Table 1.7 (continued)

Year	Inventions	Industrial events	Special events
1860	First two-stroke engine (F)	–	The first railroad in Great Britain is opened
1865	–	A hydrogen gas-powered vehicle Lenoir's "Hippomobile" travels 180 km within 3 h (F)	"Locomotive Act" restricts the use of steam vehicles (UK)
1859	Invention of the lead-acid battery (F)	–	Opening of the Suez Canal (F)
1870	Internal combustion engine (DE)	–	The telephone is invented by Alexander Graham Bell
1884	The first commercially viable EV (GB)	–	–
1886	The first usable vehicle from Carl Benz (DE)	–	Opening of the Statue of Liberty (US)
1890	First two-stroke V2 engine (D)	Gottlieb Daimler and Wilhelm Maybach found Daimler Motoren Gesellschaft (D)	–
1892	Invention of the diesel engine (D)	–	–
1894	–	First motorsport event from Paris to Rouen (FR)	Olympic Games are revived (GR)
1896	–	Henry Ford builds his first automobile prototype	The first speeding ticket (GB)
1897	Tricycle from Carl Benz (DE)	The Automobile Club of Great Britain (RAC) is founded (UK)	–
1898	–	Renault is founded (F)	–
1899	–	Fiat is founded (I)	
1900	–	First motor show in New York (US); Mercedes is manufactured by Daimler (DE)	–
1901	Curved dashboard (US)	Mercedes 35 HP is presented (D)	–
1902	Speedometer (D)	AAA (American Automobile Association) is founded (US)	–
1903	Windshield wiper (US)	Ford Motor Company is founded (US) ADAC is founded (D)	The first road trip from the West Coast to the East Coast of the USA (US)
1904	Snow chain (US)	–	First driving school (D)

(continued)

Table 1.7 (continued)

Year	Inventions	Industrial events	Special events
1906	–	Foundation of Rolls-Royce (GB) Record speed of 205.4 km/h was achieved with a steam car (US)	Motor vehicle tax law is introduced (D)
1907	–	Beijing-Paris car race (distance of 15,000 km) (F/CN)	Great financial crisis (*Financial Panic*) (US)
1908	–	Henry Ford begins production of the Model T, called Tin Lizzie (US)	–
1908	–	Foundation of General Motors (US)	–
1910	Electrical lighting and signaling system (DE)	Alfa Romeo is founded (IT)	–
1912	Charles F. Kettering invents the electric starter (Cadillac) (US)	–	The first traffic light (US)
1913	–	Henry Ford starts mass production (US)	–
1914	–	Dodge Brothers Company is founded (US) First car wash (US)	Start of the First World War
1915	Tires with inner tubes (US)	–	–
1916	–	Bayerische Motoren Werke (BMW) is founded	–
1917	Hydraulic brake (US)	–	–
1920	Four-cylinder engine (DE)	–	The "Roaring Twenties" era begins (US)
1922	V8 engine (US)	Lincoln is taken over by Ford (US)	–
1924	Car radio (US)		
1925	–	Chrysler is founded (US)	Publication of F. Scott Fitzgerald's "The Great Gatsby" (US)
1927	–	Ford Model A is introduced (US)	First flight across the Atlantic by Charles Lindbergh (US)
1929	–	–	Great Depression
1931	Differential (DE)	Bugatti Royale is presented (FR)	–
1933	–	Foundation of Toyota Motor Corporation (JP)	Adolf Hitler comes to power

(continued)

Table 1.7 (continued)

Year	Inventions	Industrial events	Special events
1934	Streamlined body (DE)	Citroën presents the Traction Avant (FR)	–
1936	–	First diesel car (Mercedes-Benz 260 D Pullman, D)	–
1937	–	The first Volkswagen Beetle (D)	–
1938	Hydramatic automatic transmission (US)	Volkswagen begins mass production of the KdF (later known as the Volkswagen Beetle) (D)	–
1939	Car air conditioning (US)	Last electric car for sale (Detroit Motors, US)	Start of the Second World War
1940	Modern automatic gearshift (US) First all-wheel drive (US)	–	–
1945	–	–	End of the Second World War
1948	Citroën 2CV is launched (FR)	–	–
1950	First gasoline direct injection (DE)	–	Start of the Korean War
1952	Airbag (US)	–	–
1954	300 SL from Mercedes-Benz, first car with gullwing doors (DE)	–	–
1955	–	–	Polio vaccination is released
1957	–	Introduction of the Fiat 500 (IT)	First space probe "Sputnik" (USSR)
1959	Seat belt (SWE)	Introduction of the Austin Mini Cooper (GB)	–
1960	First all-wheel steering (JP)	–	–
1964	–	Porsche 911 is presented (D) Ford presents the Mustang (US)	–
1965	–	Volkswagen acquires Auto Union from Daimler (D)	–
1966	–	Toyota starts selling the Toyota Corolla in the USA (JP)	"Cultural Revolution" in China (CN)
1967	–	Chevrolet presents the Camaro (US)	–

(continued)

Table 1.7   (continued)

Year	Inventions	Industrial events	Special events
1969	–	First man on the moon (US)	Woodstock music festival
1970	Catalytic converter (US)	–	–
1971	Modern ABS (anti-lock braking system) (IT)	–	First pocket calculator (US)
1972	–	VW Beetle overtakes Ford Model T as the best-selling car (D) Volkswagen Passat presented (D)	–
1973	–	Introduction of the catalytic converter	The first oil crisis First cell phone (Motorola, US)
1974	–	The first VW Golf The last VW Beetle produced in Germany (D)	Richard Nixon resigns as US President
1975	Lamborghini Countach, known for its wedge shape and gullwing doors, is launched (IT)	–	End of the Vietnam War (US) First personal computer (Altair 8800, US)
1976	–	Honda Accord is launched (JP)	–
1977	–	Debut of the Porsche 928 (DE)	First commercial use of GPS (US)
1978	–	Introduction of the VW Golf Diesel, first mass-produced vehicle with turbo diesel (DE)	–
1980	–	–	Outbreak of the Iran-Iraq war
1981	Navigation system (JP)	–	–
1983	–	Introduction of the minivan by Chrysler (US)	First cell phones for consumers (Motorola, US)
1985	–	–	Perestroika in the Soviet Union
1986	–	Volkswagen acquires SEAT (D)	The space shuttle Challenger explodes (US)
1987	Airbag becomes standard in the EU	ABS becomes suitable for mass production (D)	Black Monday, stock market crash

(continued)

Table 1.7 (continued)

Year	Inventions	Industrial events	Special events
1989	–	Mazda MX-5 (JP) First Lexus (LS 400) as a Toyota brand (JP)	Fall of the Berlin Wall (D)
1990	–	Clean Air Act Amendments passed in California: the strictest emissions standard in the world (US)	Germany's reunification (D) The Second Gulf War (US)
1991	Electronic Stability Program (ESP) (DE)	–	Disintegration of the Soviet Union
1992	–	Introduction of the Dodge Viper (US)	–
1993	–	Ferdinand Piëch becomes Chairman of Volkswagen—VW turnaround (D)	–
1994	Smart Airbag (DE)	First Mercedes-Benz A-Class (DE)	–
1995	–	Volkswagen takes over Skoda (D)	First MP3 player (DE)
1997	–	The first vehicle with hybrid drive, the Toyota Prius (JP)	–
1998	First adaptive cruise control system (JP)	Daimler-Benz merges with Chrysler (US/DE)	–
1998	–	Volkswagen acquires Bugatti, Lamborghini, and Bentley (D)	–
1998	–	Daimler-Chrysler merger (D)	–
1998	–	BMW takes over Rolls-Royce (D)	–
2000	Electronic parking brake (DE)	First BMW X5, BMW's market entry into the SUV segment (DE)	Surprisingly, the 2000 bug has no significant effect
2001	–	–	September 11 terrorist attack (US)
2002	–	Introduction of the Mini Cooper by BMW (GB/DE)	–
2003	–	–	Iraq War
2004	–	Ford presents the GT, a tribute to the GT40 racing car (US)	–

(continued)

Table 1.7 (continued)

Year	Inventions	Industrial events	Special events
2006	–	First Bugatti Veyron is sold, with a top speed of over 400 km/h (FR)	–
2007	–	Tesla presents the Roadster model (US)	Start of the global financial crisis (*Great Recession*)
2007	–	Daimler-Chrysler merger falls apart	–
2008	–	Tata Motors takes over Jaguar Land Rover (IN)	–
2010	First electric cars with a usable range, e.g. Nissan Leaf (JP)	–	–
2012	–	Tesla presents the Model S (US)	–
2013	–	Launch of the Tesla Model S, an electric car with a long range (US)	–
2013	–	BMW presents the i3 (BEV) (DE)	–
2015	Progress in autonomous driving (US, DE, JP)	"Dieselgate" scandal concerns Volkswagen (DE)	Refugee crisis in Europe
2016	–	Chevrolet Bolt, an affordable electric car with a long range, is launched (US)	Brexit referendum
2016	–	Nissan takes over Mitsubishi (JP)	–
2017	–	Peugeot SA takes over Opel/Vauxhall (F)	–
2017	–	Takata files for insolvency following the airbag scandal (JP)	
2018	–	Jaguar I-Pace, the first all-electric SUV from a traditional car manufacturer (UK)	–
2019	–	Presentation of the VW ID.3 (DE)	–
2020	–	The first all-electric Porsche Taycan (DE)	COVID-19 pandemic
2020	–	Tesla sells 1 million BEVs	–

(continued)

Table 1.7 (continued)

Year	Inventions	Industrial events	Special events
2021	First certification as a Level 3 autonomous vehicle (Honda/JP)	BEV Lucid Air offers a range of 840 km (US)	Home manager delivery bottlenecks
2021	–	Foundation of Stellantis as a merger of Fiat, Chrysler, and PSA	–
2022	–	–	Outbreak of war in Ukraine
2023	–	BYD becomes the largest vehicle manufacturer (cumulative BEVs and PHEVs) (CN)	–
2023	–	Tesla's Model Y becomes the best-selling vehicle in the world, Toyota Corolla comes second	–

References

[RM-035] Cult mobile in trouble, https://www.spiegel.de/wissenschaft/kultmobil-in-der-klemme-a-43dd6788-0002-0001-0000-000014937286, accessed 03.09.2023

[RM-036] Unsafe at Any Speed: The Designed-in Dangers of the American Automobile, Ralf Nadler, ISBN 10: 0670741590

[RM-037] Panel Evaluation of the NHTSA Approach to the 1960–1963 Corvair Handling and Stability, https://www.corvair.org/images/attachments/DOT_HS-800_676.pdf, accessed 09/04/2023

[RM-038] Crash Course: The American Automobile Industry's Road to Bankruptcy and Bailout-and Beyond, Paul Ingrassia, ISBN-10: ? 9780812980752

[RM-039] History of Electric Cars, Nigel Burton, ISBN-10 1847974619

[RM-040] The automobile and its history, Günter Barnickel, ISBN:9783750457171

[RM-041] List of defunct automobile manufacturers in the United States, https://en.wikipedia.org/wiki/List_of_defunct_automobile_manufacturers_of_the_United_States, accessed on 06.09.2023.

[RM-042] Car IT compakt, Volker Johanning/Roman Mildner, ISBN: 978-3-658-09967-1

[RM-043] Das Schmarotzer-Prinzip—Wie deutsche Automobilhersteller ihre Zulieferer ausbeuten: 40 unsaubere Tricks und Strategien, Volker Bauer, ISBN?: ? 3940445843

[RM-044] Consolidation continues, FAZ, https://www.faz.net/aktuell/wirtschaft/automobilzulieferer-die-konsolidierung-geht-weiter-130401.html, accessed 06-Sep-2023

[RM-045] The family crash, https://www.handelsblatt.com/unternehmen/mittelstand/familienunternehmer/shigehisa-takada-der-familiencrash/19980148.html, accessed 07-Sep-2023

[RM-046] NHTSA/Takata Recall Spotlight, https://www.nhtsa.gov/equipment/takata-recall-spotlight, accessed 07-Sep-2023

[RM-047] Takata Corporation Agrees to Plead Guilty and Pay $1 Billion in Criminal Penalties for Airbag Scheme, https://www.justice.gov/opa/pr/takata-corporation-agrees-plead-guilty-and-pay-1-billion-criminal-penalties-airbag-scheme, accessed 07.09.2023

[RM-049] What is the history of the Takata airbag recall? https://getjustice.com/faq/takata-airbag-recall/what-is-the-history-of-the-takata-air-bag-recall/ , accessed 07.09.2023

[RM-050] VW very self-confident—Porsche submits , https://www.dw.com/de/automesse-im-zeichen-der-krise/a-4694554, accessed 08.09.2023

[RM-051] Volkswagen overtakes Toyota, https://www.dw.com/de/vw-verkauft-mehr-autos-als-toyota/a-18611169, accessed 08.09.2023

[RM-052] Chronicle of a scandal , https://www.spiegel.de/auto/aktuell/vw-abgasskandal-chronik-eines-skandals-a-1122730.html

[RM-055] "Dieselgate"—A timeline of the car emissions fraud scandal in Germany, https://www.cleanenergywire.org/factsheets/dieselgate-timeline-car-emissions-fraud-scandal-germany, accessed 08.09.2023

[RM-056] 31 Countries and U.S. States with Gas Car Bans, https://www.chargedfuture.com/countries-and-states-with-gas-car-bans, accessed 08.09.2023

[RM-057] The Cooling: Has the Next Ice Age Already Begun?, Lowell Pointe, ISBN:? 013172312X

[RM-058] The new ice age (An Impact book), Henry Gilfond, ISBN:? 0531014584

[RM-059] Elon Musk, Tesla, SpaceX, and the Quest for Fantastic Future, Ashlee Vance, ISBN: 0008279659

[RM-060] 2008 Tesla Roadster, https://www.sportscarmarket.com/profile/2008-tesla-roadster, accessed 10.09.2023

[RM-061] Tesla Growth and Production Statistics: How Many Vehicles Are Sold Across the Globe? https://www.investing.com/academy/statistics/tesla-facts, accessed 10.09.2023

[RM-062] Brilliance BS2 will undercut Golf, https://europe.autonews.com/article/20080901/ANE09/808309895/brilliance-touts-bs4-de-emphasizes-bs6, accessed 10.09.2023

[RM-063] Zero stars for Brilliance BS4, https://www.n-tv.de/auto/Null-Sterne-fuer-Brilliance-BS4-article293481.html, accessed 10.09.2023

[RM-065] Key Technologies on New Energy Vehicles, Shichun Yang/Xinhua Liu/Shen Li and Cheng Zhang, ISBN: ISSN 2662–2920

[RM-066] Almost 150 new hydrogen refueling stations opened worldwide in 2021, https://www.sustainabletruckvan.com/hydrogen-refueling-stations-worldwide-2021, accessed 11.09.2023

[RM-067] What's The Real World Highway Range Of Today's Electric Cars? We Test To Find Out, https://insideevs.com/reviews/443791/ev-range-test-results, accessed 12.09.2023

[RM-068] Hurricane Idalia floodwaters cause Tesla to combust: What to know about flooded EV fires, https://eu.usatoday.com/story/news/nation/2023/09/01/tesla-fire-hurricane-idalia-flooding-ev-combustion-issue/70738027007, accessed 12.09.2023

[RM-069] Electricity Report: Electricity price development, https://strom-report.com/strompreise/strompreisentwicklung/#strompreisentwicklung-2023-details, accessed on 12.09.2023

[RM-070] Federal Ministry for the Environment: Brief information on electromobility regarding electricity and resource requirements , https://www.bmuv.de/fileadmin/Daten_BMU/Download_PDF/Verkehr/emob_strom_ressourcen_bf.pdf, accessed on 12.09.2023 43.

[RM-071] Electric vehicles continue charge towards sales dominance—EY analysis https://www.ey.com/en_gl/news/2023/01/electric-vehicles-continue-charge-toward-sales-dominance-ey-analysis, accessed on 12.09.2023

[RM-072] Faster, Higher, Farther, Jack Ewing, ISBNs 9780593077269

[RM-073] A road map for Europe's automotive industry, McKinsey, https://www.mckinsey.com/industries/automotive-and-assembly/our-insights/a-road-map-for-europes-automotive-industry, accessed 17.09.2023

[RM-078] Tesla Form 10k, https://www.sec.gov/Archives/edgar/data/1318605/000095017023001409/tsla-20221231.htm, accessed 21-Sep-2023

[RM-080] The Life of the Automobile, ISBN: 9781848877078

[RM-081] BP's gigantic electrical plan for Germany, https://www.welt.de/wirtschaft/article247428116/BPs-gigantischer-Elektro-Plan-mit-Aral-fuer-Deutschland.html, accessed 24.09.2023

[RM-082] Steffen Link, Christoph Neef, Tim Wicke, Tim Hettesheimer, Marcel Diehl, Oliver Krätzig, Florian Degen, Franziska Klein, Patrik Fanz, Matthias Burgard, Ricardo Kleinert, Development perspectives for lithium-ion battery cell formats, Fraunhofer Institute for Systems and Innovation Research ISI, 2022

[RM-084] Toyota says solid-state battery breakthrough can halve cost and size, https://www.ft.com/content/87cb8e92-8e82-4755-8fc3-2943f8f63e1d, accessed 25.09.2023

[RM-085] Every tenth charging station is out of service, https://www.electrive.net/2022/11/04/jede-zehnte-ladesaeule-ist-ausser-betrieb, accessed on 25.09.2023

[RM-086] Electric car charging time: charging times for all current models, https://www.carwow.de/ratgeber/elektroauto/wie-lange-laedt-ein-elektroauto-ladezeiten-aller-aktuellen-modelle, accessed 26.09.2023

[RM-092] The Wheel Infentions & Reinventions, Richard W. Bulliet, ISBN 978-0-231-17338-4

[RM-097] Guinness Book of Records—Highest vehicle mileage, https://www.guinnessworldrecords.com/world-records/highest-vehicle-mileage, accessed 02.10.2023 60

[RM-098] Electric car battery: service life, warranty, repair, https://www.adac.de/rund-ums-fahrzeug/elektromobilitaet/info/elektroauto-batterie, accessed 02.10.2023

[RM-099] The cost of a new battery for an electric car, https://www.t-online.de/auto/elektromobilitaet/id_89687372/e-auto-batterie-das-kostet-ein-neuer-akku-fuers-elektroauto.html, accessed 02.10.2023

[RM-100] How Often Do Tesla Batteries Need To Be Replaced?, https://www.jdpower.com/cars/shopping-guides/how-often-do-tesla-batteries-need-to-be-replaced, accessed on 02.10.2023

[RM-101] Lithium-ion Battery Pack Prices Rise for First Time to an Average of $151/kWh https://about.bnef.com/blog/lithium-ion-battery-pack-prices-rise-for-first-time-to-an-average-of-151-kwh, accessed on 02.10.2023

[RM-102] Branded charging networks for electric vehicles—Insights from China (Roland Berger), https://content.rolandberger.com/hubfs/Roland_Berger_Branded_charging_networks.pdf, accessed on 03.10.2023

[RM-103] Tesla Acquires a Wireless Charging Startup, https://insideevs.com/news/680401/tesla-acquires-wiferion, accessed 10/03/2023

[RM-118] MUSICIANS WHO SING THE MOST ABOUT CARS, https://www.goldeagle.com/tips-tools/musicians-sing-most-about-cars, accessed 10/08/2023

[RM-132] https://www.forbes.com/sites/alistaircharlton/2021/11/01/tesla-superchargers-can-now-be-used-by-other-electric-cars-here-is-how-it-works, Forbes, accessed on 20.10.2023.

[RM-133] https://emobilitaet.online/news/forschungsprojekte/7677-anteil-oekostrom-elektroautos, emobilitaet.online, accessed on 20.10.2023

Driving Into the Future: Autonomous Vehicles

2

Abstract

This chapter explores the rapidly evolving field of autonomous vehicles and their technological foundations. It details the six levels of autonomous driving (0–5), examining each level's technical and social challenges. The chapter provides an in-depth look at artificial intelligence's role in autonomous driving, including neural networks, machine learning, and sensor technologies. It also addresses critical aspects such as safety systems, regulatory frameworks, and public acceptance. The discussion concludes by analyzing how autonomous vehicles might reshape future mobility concepts and transportation systems.

Autonomous driving frequently makes headlines and is a hot topic in both work and personal conversations. Opinions range from 'I can't wait to use it' to 'I'll never trust it.' Development has been remarkable, with individual advanced driver assistance systems (ADAS) becoming ubiquitous and the associated networking capabilities ("connected cars") on the rise. Modern onboard systems assist drivers, with advanced features like highway pilots taking control of vehicles. Unlike past academic experiments, these solutions are now available for purchase.

In Germany, autonomous shuttles drive on selected routes but are still being tested at lower speeds only. Meanwhile, the United States is ahead, with self-driving cars already part of everyday traffic in cities like San Francisco.

Autonomous driving is now closely associated with emerging artificial intelligence technologies. There are several approaches to improve the functions of this technology, although challenges remain regarding the safety of such systems. It is a sensitive issue, as vehicle safety remains a high priority in our industry. Autonomous test vehicles cover millions of kilometers to collect suitable data. Nevertheless, new testing and validation concepts are required, as it is physically impossible to carry out this gigantic number of test drives under all possible conditions with sufficient safety margins in an acceptable amount of time. Although

© The Author(s), under exclusive license to Springer Fachmedien Wiesbaden GmbH, part of Springer Nature 2025
R. Mildner et al., *Car IT Reloaded*, https://doi.org/10.1007/978-3-658-47691-5_2

artificial intelligence achieves impressive results in development, its predictive ability often remains limited. Therefore, comprehensive safety mechanisms are crucial throughout the development lifecycle.

Achieving autonomous driving is a significant technical hurdle. Various legislative measures are required to make self-driving cars safe and acceptable to customers. These aspects must be successfully addressed to make autonomous driving technically and socially feasible. This chapter outlines the current state of development, summarizing the challenges and opportunities ahead.

2.1   Autonomous Driving—Levels 0 to 5

The idea of a self-driving car is many decades old. Even in the era of horse-drawn carriages, some people dreamed of it, but driverless carriages were expressly forbidden. In 1958, General Motors (GM) published details of a driverless vehicle prototype using metal strips inserted into the road surface. Of course, this concept was a vision that was not practical at the time and had little in common with today's idea of autonomous mobility [TZ-201].

An example of a milestone in the history of autonomous driving was the electronically-controlled driverless vehicle from Continental, which was tested on September 11th, 1968, on the Contador test track (Fig. 2.1, reproduced with permission from Continental AG).

The self-driving car concept offers more than a convenient way to travel from point A to point B. It also promises to use our roads, particularly the highways,

Fig. 2.1 First electronically-controlled driverless vehicle from Continental. *Image* Continental

more efficiently. Regardless of how driver behavior evolves, nobody drives error-free. Only a fully autonomous vehicle can truly be safe and maximize the use of our infrastructure. This is just one of the many advantages that future self-driving cars could bring.

The first major automotive OEM to test an autonomous car was Mercedes-They started pioneering work on autonomous vehicles through the EUREKA Prometheus Project, which ran from 1987 to 1995. This project involved developing and testing autonomous cars, with one of their prototypes achieving autonomous driving on European highways by the early 1990s.

Published in January 2014, years after initial development began, the SAE J3016 standard defines the various levels of vehicle automation, categorized into six levels split into two sections. In levels 0–2, drivers remain fully responsible for the car, though assistance systems provide support in certain situations. From 3 to 5, responsibility gradually shifts from the driver to the automated systems at higher levels.

Level 0—No Driving Automation
At this level, the driver is fully responsible for all driving, with existing systems supporting the driver in certain situations. Examples are ABS (Anti-lock Brake System) and ESP (Electronic Stability Program).

Level 1—Driver Assistance
In some driving modes, an assistance system takes over deceleration, acceleration, and steering, using sensors to consider and evaluate the surroundings. The driver still has a complete overview and is responsible for the vehicle's behavior. A well-known example, one of the first to fall into this level, is the cruise control system.

Level 2—Partial Driving Automation
This level has now found its way into many vehicles. It involves a system taking over lateral and longitudinal control in specific driving situations. The vehicle observes its surroundings more extensively than with Level 1. The driver has the option to intervene at any time and is warned if they do not have their hands on the steering wheel. Examples are the "IQ.DRIVE Travel Assist" function from Volkswagen and "BlueCruise" from Ford. The term "Level 2+" is sometimes used, whereby several Level 2 systems are combined to further increase driver comfort and safety. Level 2/2+ offer the same functions as Level 3 systems; however, as the driver is still actively involved, the question of liability is regulated and responsibility always lies with the driver.

Level 3—Conditional Driving Automation
The functions at Level 3 do not differ significantly from those of Level 2 or 2+. However, at Level 3, responsibility shifts to the vehicle or system; therefore, continuous monitoring by the driver is no longer necessary. The decisive factor here is that the systems can perform tasks autonomously, i.e., without human supervision. Nevertheless, should intervention be required, the driver must be able to regain

control within a specified time (e.g., 5–10 s). Mercedes-Benz was the first manufacturer in the USA and Germany to receive approval for such a system. BMW has been offering a correspondingly approved Level 3 system in the 5 Series since 2023. At Mercedes-Benz, this function is provided as "Drive Pilot" in the EQS models and the S-Class and is approved in Germany for speeds of up to 60 km per hour on certain multi-lane freeway sections.

Level 4—High Driving Automation

At this level, vehicles operate entirely autonomously, meaning the driver no longer monitors the surroundings while the system is active. This allows them to devote themselves to other activities, including the possibility of sleeping. The legal requirements for Level 4 still need to be fully clarified. A draft bill dating back to 2021 was rejected due to the unresolved issue of data protection, and only in 2022 did the EU approve the legislature.

No Level 4 car is available (at the time of writing). Prototypes from some manufacturers, suppliers, and research companies are intended to enable highly automated driving. However, these systems are only functional in a defined geographical area (they are "geo-fenced"), such as the A9 federal highway in Bavaria, Germany.

Level 5—Full Driving Automation

At this level, the vehicle takes over all driving areas, and the driver no longer has the option of intervening in the driving process during the operation. Data the vehicle receives from sensors and external sources, such as the cloud or infrastructure, must be exact and reliable. Introducing this level aims to overcome the challenge of "the last-mile" transport, creating a new customer experience. This experience will also be conditioned by the fact that the vehicle interior will likely change significantly. Steering wheels and pedals will no longer be necessary and may be replaced by screens, for example. Depending on driver preferences, the passenger compartment will become a living room, office, or fitness studio.

Level 5 driving also represents a quantum leap towards independent and autonomous mobility for people who cannot or are not allowed to drive due to their age or disability. It may sound like a platitude, but our industry is about to improve our lives and standards of living—once again.

2.2   The Challenge of Self-driving Cars

Engineers have long been fascinated by the idea of self-driving cars, yet, despite years of international development work, realizing this idea remains an unachieved goal due to various challenges, including legal regulations and technical complexities.

Legal challenges

The legal framework for autonomous driving is complex and needs to be fully defined. Various countries are actively working on creating regulations for autonomous driving. The legislative process is country-specific and often lengthy. It is clear that this new technology opens up a wide range of possibilities, and many aspects still need to be clarified. Experts and politicians are trying to determine the legal conditions for autonomous driving, but balancing safety and progress is challenging. Countries closely linked to the automotive industry often take a pioneering role in shaping the relevant legislation. The following examples illustrate this fact.

The United States

In the United States, legislation varies significantly depending on the state. Each state can enact its laws and regulations, leading to a complex legal landscape with many rules. The chart [TZ-202] shows the status of legislation in individual states as of June 2022. California is the leader in legislation for the passenger car sector. More than 40 companies with around 1,500 vehicles are registered there for testing and are conducting trials. The companies Waymo and Cruise have received approval for commercial operation.

Truck legislation is already further along in Florida, Texas, and Arizona, and tests required for developing autonomous trucks are being carried out there. California wants to amend its truck legislation by the beginning of 2024. A new regulation is planned here, according to which there is to be a safety driver—this would be a step backward for passenger car legislation. Discussions on this law are still ongoing.

Europe

Similar to the situation in America, the legal framework for autonomous driving in Europe is not centrally controlled. Within the European Union, sovereignty for legal regulations on autonomous driving lies with the individual member states. A critical step in this area was taken in Germany in 2021 when the legislature published a law on autonomous driving [TZ-203]. In Germany, self-driving vehicles can participate in public transport in designated areas. An early step was previously taken in Germany in 2017 when the legislation was revised to adapt the rights and obligations of drivers regarding the use of autonomous vehicles. This was the cut-off date from which it was permitted, under certain conditions, to allow the systems to drive the car. A prerequisite of this legislation is that the driver must be able to intervene at any time if necessary.

Mercedes-Benz celebrated significant success in 2023 [TZ-204] when the company received approval for a SAR Level 3 system in the US state of Nevada, with the "Drive Pilot" function being certified. In Germany, the SAE Level 3 system was approved for up to 130 km per hour on freeways [TZ-205]. Recently, BMW has also had such a system certified for the 5 Series.

The Autonomous Vehicle Approval and Operation Ordinance (AFGBV) was published in 2022. Germany is experimenting with autonomous driving at Level 4, including pilot projects with shuttles that have Level 4 capabilities.

Progress in other European countries varies, however. While the UK currently allows Level 3 and plans to introduce more far-reaching legislation by 2025, France has had a regulation in place since 2022 similar to the German one. On the other hand, Italy still needs to pass specific legislation for autonomous vehicles, and currently, it only permits the operation of cars up to Level 2. Spain is a little further along in this respect. New legislation, expected to follow the laws of other European countries, is being worked on there that will allow vehicles up to Level 4.

Asia

Three Asian countries are particularly advanced in developing and implementing autonomous driving systems: Singapore, Japan, and China [TZ-206]. Since 2017, these technologies have been permitted on public roads in the city-state of Singapore. The Transport Authority of Singapore has tested and approved over 40 driverless vehicles for operation in a limited geographical area.

The process has taken longer in Japan, but in April 2022, a new law was passed to regulate the conditions for self-driving vehicles on the road up to Level 4. However, the legal situation in Japan is only one building block for getting autonomous cars on the road. The law is implemented locally, and companies must obtain a permit and define the application area. For now, legislators are concerned about using minibusses in rural areas. Although the legislation was still incomplete, Honda launched its first Level 3 system with a traffic jam assistant in Japan in 2021.

The People's Republic of China has made autonomous driving a key technological priority, with Chinese companies benefiting from state funding and supportive legislation. Regulations for testing autonomous vehicles were introduced in the summer of 2021, requiring test vehicles to have a real-time remote monitoring system and a device that records relevant data for at least 90s before an accident.

While some insurance guidelines concerning liability in traffic accidents are still absent, additional steps are needed for continuous, regular operation. If an accident occurs with such systems active, the companies must prove that the system was developed according to state-of-the-art standards. The law also requires the storage of relevant data in the vehicle, as outlined in the Autonomous Driving Act of July 12, 2021, which amended the Road Traffic and Compulsory Insurance Acts.

Technical challenges

Another crucial aspect concerns technical challenges in the development of self-driving cars. A key concern is that these vehicles operate in unpredictable traffic conditions, unlike the less critical settings of agricultural terrains or highly controlled airspace. One example of such dynamic challenges is children running into the road unexpectedly.

Mapping the real world

Unlike a theoretical model, the traffic world is dynamically complex. Unpredictable events can occur anytime, cannot be learned in advance, and require a short reaction time. Therefore, the existing sensors in autonomous vehicles must record the environment in real time. This detection must function reliably under all weather conditions and be able to process complex scenarios, such as many vehicles, cyclists, different road markings, pedestrians, and animals in dense traffic environments.

The subject of road markings is particularly tricky. An autonomous system is designed not to cross solid lines. However, markings differ from country to country, and the nature of the lines that a camera must recognize also plays a significant role. Markings often need to be clarified, especially in roadworks. If errors occur in the road markings, the vehicle may draw the wrong conclusions, whereas a human driver can usually react appropriately in such situations. These examples show that no system can currently guarantee 100% safety.

Protection

Validating ADAS (Advanced Driver Assistance Systems) functions becomes increasingly complex with each SAE level, partly because interfaces are added and various suppliers are involved. The validation occurs in two phases: test drives on proving grounds and public roads.

In recent years, advanced simulations have become essential to the testing and validation strategy since not all scenarios can be tested in real-world conditions, and some validation cases are too hazardous for road traffic. The total estimated test kilometers to date exceeds 14 billion, with the complexity of test scenarios increasing 104-fold from Level 1 to Level 5 [TZ-207]. The following methods are used for simulations:

- Model in the loop
- Software in the loop
- Hardware in the loop
- Vehicle in the loop.

These validation procedures enable efficient testing of autonomous vehicle systems in various realistic scenarios. Development begins with the smallest units and extends to the entire vehicle for comprehensive system testing. A key element is the generation of high-quality data for the simulations, for example, through high-resolution sensor technology such as cameras, Lidar, and radar, to create a representation of the world that is as close to reality as possible. The usability of this data depends on how well it is prepared for the respective systems and is enriched with meta-information, such as the location of traffic signs or the classification of vehicle types.

Due to the effort required, extensive test drives are only practical at some stages of development. Therefore, advanced simulations will likely increase as they can automatically generate complex or rare scenarios, thus significantly improving test coverage.

Including artificial intelligence (AI) in safety verification leads to a further increase in complexity, as the decisions of AI-supported systems are only sometimes predictable. In our industry, where lives are on the line, extensive research activities are required to achieve higher autonomy levels. This research is often conducted with scientific institutes and facilities [TZ-208]. In contrast to less critical industries, we in the automotive sector must spare no expense or effort to ensure the safety of autonomous driving systems.

Connectivity

Networking the vehicle with the cloud, other road users, and the infrastructure is essential to the self-driving car concept. It is possible to make collected data available to other cars and receive information about the current environment, such as road conditions, road closures, etc. That is the key to holistic environment monitoring. However, there are still a few hurdles to overcome. For example, there needs to be more consistent standardization in this area. Manufacturers are currently working with different connectivity concepts, and data is seen as an essential competitive advantage, so such data is not shared across car brands.

Data privacy concerns

Data about the vehicle and the driver is collected, including driving style, sensor data, and energy consumption. Tesla's example illustrates the extent of this data collection. Tesla vehicles continuously collect enormous amounts of data, including video data, consolidated wirelessly over the air to the control center. There is no consistent transparency in data protection for Tesla owners. No specific data collection is legally mandated, but various data protection measures are in place. The German Drivers Association, ADAC, demands and supports a uniform EU law on data protection.

At the same time, private companies are offering specific data monitoring options as an opportunity to cut costs. For example, some insurance companies already provide telematics tariffs so drivers with less driving experience can get cheaper rates.

Safety requirements

Data security and safety are two distinct but equally important aspects of the automotive industry. The critical question is: what measures must be implemented during development to ensure proper protection? At least three standards relate to this challenge:

- ISO 26262—Functional Safety (FuSa),
- ISO 21434—Road Vehicles for Cybersecurity Engineering, and
- ISO 21448—Safety of The Intended Functionality (SOTIF).

Security is about protecting the system and data from external influences. Cybersecurity is a relatively new field of expertise. It aims to ensure that all vehicle data is stored and transmitted securely and that there is no unauthorized access from outside. The UNECE WP.29 regulations have been mandatory in the EU for newly homologated models since summer 2022 and must be implemented in all models delivered from mid-2024 [TZ-245].

Infrastructure

The infrastructure for automated driving is a tricky topic. Theoretically, it would be ideal to technically equip every person and object—such as pedestrians, cats, dogs, or traffic lights—so that they and their three-dimensional position could be identified. That is, of course, not feasible. Currently, external infrastructure only exists at selected locations, such as highways. The data that is passed on comes exclusively from other vehicles. Vehicle manufacturers aim to find an optimal configuration that does not require external infrastructure.

Electrical/electronic architecture

The large amount of data received by various sensors in the vehicle while driving poses an architectural challenge. For example, cars from EV manufacturer NIO collect eight gigabytes per second [TZ-242]. The amount of data depends mainly on the number and type of sensors used and where the data is processed in the system. The vehicle architecture must, therefore, be carefully designed. Attention must be paid to where the data is collected, the quality in which it is transmitted to the central control unit, and the capacity of the required bus lines. This is closely linked to the complexity and scope of the various ADAS solutions, which range from simple parking assistance to fully automated driving.

Zone architecture is a modern vehicle design approach that divides the vehicle's electronics into architectural sections or "zones", each responsible for managing specific functions. Zone architecture extends traditional system design with a single control unit per system. It is characterized by a distribution of functions across several zones, each managed by its control unit, with a central control unit responsible for overall coordination. The sensors in the vehicle are either connected directly to the central control units or to special gateways within the individual zones (zone ECUs), whereby the data can be pre-processed if necessary. Current bus systems such as CAN and CAN-FD are no longer sufficient; Ethernet, the industry standard, is replacing them. This technology was introduced by a new standard, IEEE 100BaseT1, specifically for the requirements of the automotive industry.

The development of new architecture is also driven by car manufacturers' efforts to purchase specialized software components from various suppliers while integrating their developments. Unlike some current systems, no single supplier delivers the complete system.

Hardware in the embedded sector will also evolve from smaller microcontrollers to larger computing units with multiple cores. These cores can then run different operating systems, such as AUTOSAR Adaptive or Classic or, in the case of infotainment systems, a variation of the Linux open-source operating system.

Separation of hardware and software enables the zone architecture to allow for the particularly flexible scalability of vehicle platforms. Zone systems can also run on different platforms or be ported with relatively little effort. The challenge lies in enabling the integration of new software and hardware concepts while protecting the intellectual property (IP) of the companies involved. The critical task for vehicle manufacturers will be to precisely define the specifications for all components to ensure smooth integration.

Ethics

In 2017, an ethics committee in Germany published a final report [TZ-209] to establish guidelines for developing and selling automated and autonomous vehicles. The primary objective of these vehicle systems must generally be to avoid accidents. The ethics discussion mainly revolves around how automated systems can and should make decisions regarding the well-being of human life and which life is prioritized. These systems must give priority in their decisions to the protection of human life and the avoidance of dangerous situations.

Public perception and acceptance

Autonomous driving has yet to gain the necessary acceptance among the general public. Acceptance of autonomous driving systems falls as respondents' age increases [TZ-210]. However, a feeling of uncertainty prevails across all age groups. In general, the new technology is viewed as untrustworthy. Car makers will undoubtedly have to work on this acceptance for some time. It will be crucial that autonomous vehicle systems work reliably so that (future) users of autonomous functions feel safe.

Euro NCAP will also play a role here. Technical development is reflected in awarding stars for the vehicles, and the requirements for five stars will be adjusted over the years. With new driver assistance systems, such as emergency braking assistants, vehicles will receive the highest rating if the safety of those systems can be demonstrated. Euro NCAP is also widely used by car buyers to help them make informed decisions.

Vehicle safety remains crucial. New technologies in cars are often viewed with skepticism. It is hard to imagine what would happen if an autonomous vehicle unexpectedly caused a risk to life and limb. While we have already become accustomed to the fact that careless or even negligent road users can cause life-threatening injuries to themselves or others, a single fatality involving an autonomous vehicle could be a reason to ban an entire vehicle technology, at least temporarily.

In a broader context, skepticism regarding autonomous driving poses a particular national risk in Europe. Because this skepticism is linked to the age of respondents, countries with younger populations, like China, can introduce autonomous driving and gain market share more swiftly. Legislators in European countries where many people are older must take conscious countermeasures. However, only time will tell whether this will happen.

2.3   Artificial Intelligence

Artificial intelligence (AI) has become the megatrend of our time. One indication of this is the widespread popularity of AI-related terms, as recorded by Google trend data. Figure 2.2 shows the trend over recent years [TZ-246].

AI has become indispensable as a vital component of autonomous driving. AI's academic and technological foundations are vast, but here, we'll focus on the aspects relevant to autonomous driving.

2.3.1   Introduction

The success of OpenAI's ChatGPT triggered a focus on everything related to artificial intelligence. It demonstrated how advancements have evolved over the past few years. AI-based tools and services increasingly take over tasks in the background (such as the automatic analysis of unstructured mass data) or the foreground, as with ChatGPT—an AI service with over 100 million daily users.

AI-powered features are increasingly popular, and ChatGPT and similar AI solutions are only the beginning. Whether it's chatbots or AI-generated images, AI is being used everywhere by experts and casual users alike. Initially an esoteric software for experts, it is now a tool for the masses. The potential of this technology is estimated at over 554 billion US dollars in 2024 [TZ-211], although we consider this estimate to be highly conservative.

The history of artificial intelligence goes back to the eighteenth century. Julien La Mettrie's book *L'homme Machine* (*Man as Machine*), published in 1748, was a pioneering work. La Mettrie was a French physician and philosopher who shaped the philosophical direction of materialism. In his conception, man resembles a machine—a mechanical device. This radical idea met with equally radical resistance: La Mettrie was expelled from the country and died in Berlin at the age of just 42. Despite La Mettrie's tragedy, this materialistic view of a human remains omnipresent in the age of modern semiconductors and software systems.

Modern AI traces its roots to 1936, when Alan Turing demonstrated that computing machines could execute processes. This is made possible by breaking down processes into sub-processes run by individual algorithms. At the time, Turing's foresight was something from science fiction; today, this knowledge is an integral part of the first semester of computer science.

Fig. 2.2 Development of interest in AI at Google over time

In 1956, John McCarthy invented the term "artificial intelligence" during the "Summer Research Project on Artificial Intelligence" discussion at the Dartmouth Conference in the USA [TZ-212]. Participants concluded that thought processes are also possibly independent of the human brain, establishing the scientific discipline of "artificial intelligence." The first program, "The Logic Theorist," was developed to prove mathematical theorems. McCarthy, a logician, computer scientist, and author, also created the LISP programming language. LISP became central to AI, and many of its concepts have influenced modern programming languages.

The dream of an AI conversation partner first became a reality in 1966 with the first chatbot. Scripts and basic options were programmed to simulate conversation, but those scripts were primitive compared to modern AI-powered chatbots. Today, millions of users worldwide use chatbots such as ChatGPT every day.

The logical ability of a machine equipped with artificial intelligence has long been anticipated. In 1945, Alan Turing predicted that computers would one day play chess better than humans. This vision became a reality 50 years later with IBM's "Deep Blue." "Deep Blue" achieved this success by calculating all sensible moves; however, the system determined the optimum values required. Today, modern AI systems are trained through an intensive learning process; a deterministic component, even in chess, is becoming less relevant.

In 1965, Gordon Moore, co-founder of Intel, formulated what became known as Moore's Law. It states that the number of transistors in a chip doubles approximately every two years, which leads to an exponential increase in computing power. This law has so far remained valid; processors, memory, and graphics cards are becoming more powerful at a geometric pace. This phenomenon has helped today's AI systems become suitable for everyday use. At the same time, the development of artificial intelligence in the software sector has made decisive progress. AI has become a dedicated discipline taught at universities and specialized institutes.

Given these developments, it was inevitable that artificial intelligence would increasingly find its way into our everyday lives. The possibilities of AI seem almost inexhaustible, so the automotive industry has also turned its attention to this technology, not least to tackle the still unresolved issues in automated driving. The Prometheus research project [TZ-213] was the first automated vehicle project that delivered tangible and promising results.

However, before we address the AI-relevant topics specific to the automotive industry, we will examine fundamental AI concepts, especially neural networks (NN).

2.3.2 Neural Networks

A neural network is a software program that mimics natural neurons, the fundamental units of the brain. In abstract terms, natural neurons can process and store information and transmit it to other neurons.

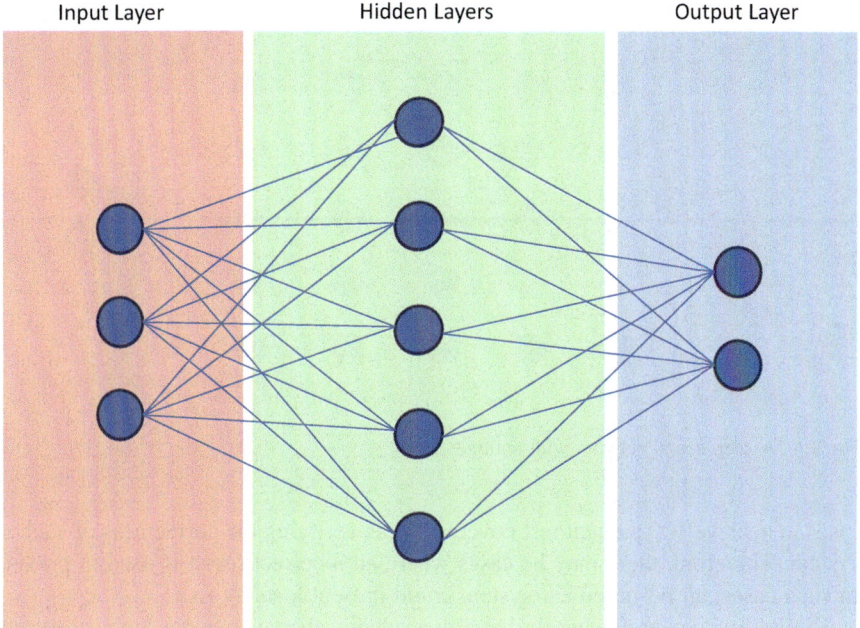

Fig. 2.3 Structure of a simple neural network

Neurons consist of three main parts: the cell body, the dendrites, and the axon. The cell body is the central part. The dendrites are the "receivers" of the neurons. Axons act as transmission lines, passing on signals to other neurons. Natural neurons communicate with each other via electrical signals.

Inspired by the natural brain, Warren McCulloch and Walter Pitts introduced the idea of machine-based neural networks in the 1940s. Research into neural networks intensified in the 1950s and 1960s, but at that time, they were not yet suitable for complex tasks. Simple neural networks are one layer and can hardly be used in practice.

Figure 2.3 shows a trivial neural network. It contains an input layer, an output layer, and a hidden layer. During the training phase, weights are assigned to the individual layers' nodes, which decide the output layer. The input layer passes the input data with a trained weighting to the hidden layer, as the outer layers cannot see the information. The information and weighting process takes place in this hidden layer, which can consist of one or more layers/levels. The output layer is the final layer, making the results available to the application.

Interestingly, there is no rational explanation for why the concept of neural networks works. This is why evaluating or validating these algorithms regarding their functional safety is tricky. Neural networks are a prime example of a heuristic— a solution that often works, but the results are not always optimal. Due to large neural networks' high complexity, it is impossible to generate all conceivable test

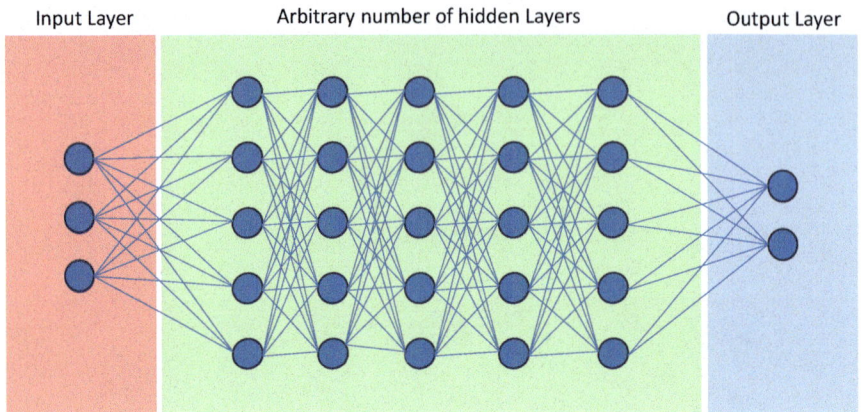

Fig. 2.4 A deep neural network with multiple layers

cases to achieve 100 percent test coverage, thus verifying the correctness of such a system. Therefore, there may be cases where an incorrect result is used in praxis. In such cases, an AI-based car system could present a safety risk.

While simple neural networks have an academic character, it was long suspected that better neural networks would be possible with several hidden layers.

Figure 2.4 shows an example of a neural network with four hidden layers.

However, while it was suspected that more layers could improve the NN results, for decades, despite various variations, network topologies, and numerous experiments with different approaches to developing practically applicable multilayer neural networks, increasing the number of hidden layers did not lead to consistently better results. It was not until 2012 that Geoffrey Hinton, a British-Canadian scientist celebrated as the "father of artificial intelligence," achieved a breakthrough. His technology, known as deep learning, made it possible to implement complex heuristics used in visual pattern recognition or AI solutions such as ChatGPT. It was only with this technology that neural networks, which until then could hardly be used in any meaningful way, earned the name "artificial intelligence."

2.3.3   Type of Artificial Intelligence

The various terms in artificial intelligence, both in the literature and general, often need clarification and appropriate interpretation. We classify these terms hierarchically below.

Artificial intelligence is a generic term originating from computer science. It represents a conceptual bracket for topics from machine learning (ML) and deep learning to the latest development—generative pre-trained transformers (GPT). The hierarchy is shown in Fig. 2.5.

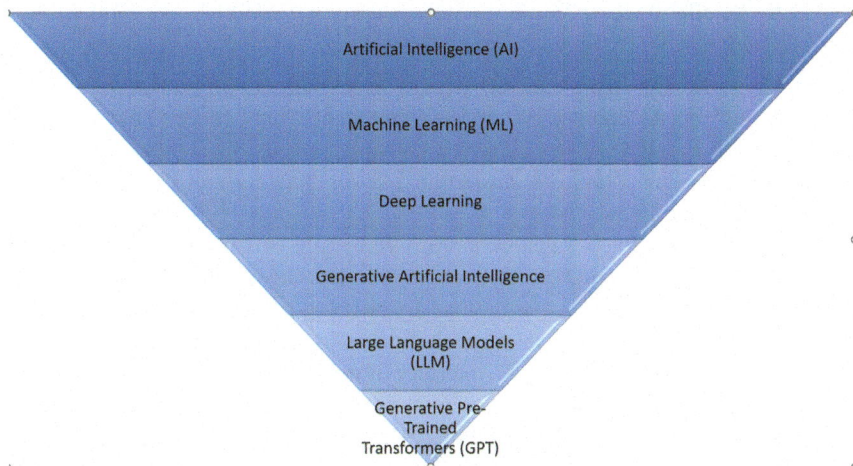

Fig. 2.5   Core concepts in the field of artificial intelligence

Artificial Intelligence: Artificial intelligence is the generic term that summarizes all the developments in the various areas of AI.

Machine Learning: Machine learning enables AI systems to learn from data and make decisions.

Deep Learning: Deep learning uses several processing layers, allowing complex patterns to be recognized.

Generative Artificial Intelligence (Generative AI): Systems capable of generating new textual and media content.

Large Language Models (LLM): Language models with an enormous number of parameters that are trained using large amounts of data.

Generative Pre-Trained Transformers (GPT): GPTs specialize in generating natural and consistent texts and are trained with large amounts of data (especially text data).

AI computer systems can solve complex tasks that previously could not be solved in an acceptable amount of time, even with the most efficient machines. Where deterministic algorithms fail, AI specializing in pattern recognition is an advantage. This AI-based search does not always provide the best solutions but usually offers "sufficiently good" ones.

AI algorithms can be divided into "weak" and "strong" AI.

Weak AI is trained for a specific task. It is used in many areas, including automotive. The aim is to make predefined solutions/decisions, such as classifying vehicles. Machine learning and deep learning are used in various algorithms in the automotive industry. ADAS often uses weak AI methods to handle subtasks such as traffic sign recognition, traffic light recognition, vehicle classification, free space recognition, etc.

Strong AI has human-like characteristics, as in science fiction books and movies. Examples include AI characters such as Lieutenant Commander Data from *Star Trek* or C3PO and R2-D2 from *Star Wars*. Such systems have the potential not only to solve various complex tasks but also to simulate emotions [TZ-214]. Also known as superintelligence, strong AI is a much more powerful, universally applicable adaptive technology. Such technology can apply existing knowledge to new situations and thus reason, plan, or even be creative. Strong AI is still sometimes classified as science fiction, even though the next big step has already been taken with ChatGPT, which appears deceptively human-like at times.

ChatGPT was originally a text interface used in dialogs with the user. The training data is extracted from the Internet, meaning a broad spectrum of knowledge is available in the different versions. The technological bases are generative pre-trained transformers with integrated large language models. A more detailed technology description can be found in [TZ-215].

An interesting observation is that although ChatGPT is categorized as "artificial intelligence," it is not very "intelligent." The functional principle of ChatGPT can be explained relatively simply. The underlying large language model is a pattern recognition system at its core. It interprets text passages as patterns, whereby the sequence of semantic structures is regarded as a series of patterns. The system identifies the most likely patterns from the learned data and generates answers based on these. This continuous generation of text passages is done by selecting patterns from an extensive collection with the highest relevance regarding a scalar "quality function." It is thus essential to understand that ChatGPT does not represent "intelligence" in the human sense but is merely an advanced pattern recognition system. Unlike earlier systems, such as those based on Prolog, which were difficult to use, ChatGPT does not base its operation on systematic reasoning. This approach also means that LLMs such as ChatGPT sometimes generate erratic, illogical, or irrelevant answers—a phenomenon often described as "hallucination." They may provide correct answers from recognized patterns but are occasionally inconsistent.

Combining LLMs with other technologies (such as fact-checking or reasoning systems) to increase the accuracy and relevance of answers could further enhance their development.

Machine learning
A key element in machine learning is the availability of comprehensive, high-quality, and correctly labeled data for the respective purpose. "Labeled" means each detail can be assigned a unique marker so the associated object can be distinctly classified—for example, objects like "road," "car," "tree," "house," etc. Machine learning (ML) independently learns patterns and correlations from the training data. The data is interpreted, and a suitable software code is created. The

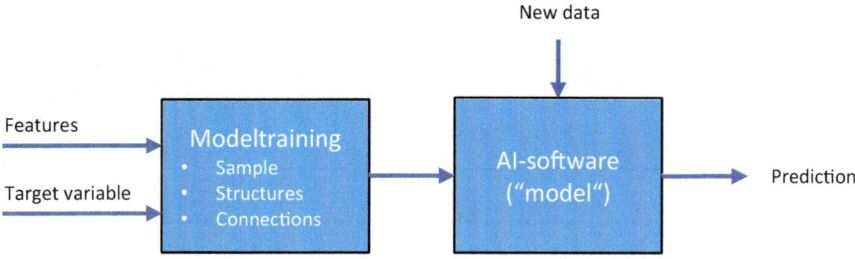

Fig. 2.6 Explanation of machine learning

data must be structured, and the features extracted so that ML can classify using the learning methods; thus, the network must be trained effectively.

Figure 2.6 shows the typical machine learning process, which is the same for every application in the field. Correct feature labeling and a well-defined objective are critical in model training.

Model training for machine learning requires careful preparation. The available data is often unstructured. This raw data must be carefully filtered and prepared to complete the training successfully. Data must be analyzed to work out patterns, dependencies, and structures. The AI software can then be created using the processed data. The training process and the AI software run independently of each other. The system is trained once and then fed new, unknown data during operation. The results are then provided iteratively.

The computing power required for training can be considerable. Training extensive AI models is an excessively energy-intensive process that can consume thousands of kilowatt hours of electricity, depending on the model complexity and training duration. Once the model's training has been completed, the quality of its predictions remains unchanged, as the software does not continue to develop independently. This means no further significant energy input is required to use neural networks that have already been trained.

The table Fig. 2.7 shows the three main types of machine learning. The suitability of each type depends on the problem to be solved.

The main types of machine learning are summarized in Fig. 2.7 and are discussed in the following sections.

Supervised learning
Supervised learning uses pre-processed and carefully labeled data. Data labeling is a critical step in supervised learning. Labeling is shown in Fig. 2.8, where vehicles, in this case, have been recorded (labeled).

In most cases, the learning process is carried out manually. Having the correct input data for the task is critical to obtaining the right results from the network. Error minimization methods, such as gradient or backpropagation, are used in the algorithms. The backpropagation method is a central approach for most neural networks. It is a method by which errors from the output are passed back through

Machine Learning	Unsupervised Learning	Dimensionality Reduction	Big Data Visualization
			Structure Discovery
			Feature Elicitation
			Meaningful Compression
		Clustering	Recommender Systems
			Target Marketing
			Customer Segmentation
	Supervised Learning	Classification	Identity Fraud Detection
			Image Classification
			Customer Retention
			Diagnostics
		Regression	Population Growth Prediction
			Estimating Life Expectancy
			Market Forecasting
			Weather Forecasting
			Advertising Popularity Prediction
	Reinforcement Learning	Real-time Decisions	
		Game AI	
		Robot Navigation	
		Learning Tasks	
		Skill Acquisition	

Fig. 2.7 Overview of the types of machine learning

Fig. 2.8 Marking of vehicles
for network training

the network (hence the name) to update the weights of the individual neurons in the neural network (NN). The example in Fig. 2.8 shows how this technology is used in object recognition. In automated driving, many objects, such as cars, buses, motorcycles, pedestrians, etc., must be reliably recognized in road traffic.

Unsupervised learning

In unsupervised learning, the goal of the neural network is to independently identify structures and patterns in the data without relying on previously classified or labeled data. The network learns to recognize relationships between the data points and to learn from them. This learning process requires no predetermined answers or labels and enables the model to discover complex data structures without explicit human guidance.

As there is no direct human feedback in unsupervised training, the training process in this mode is generally more time-consuming. It requires the data to be prepared carefully and correctly. One area of application for unsupervised learning is, for example, securities trading, where models are trained to recognize patterns and strategies for buying and selling. However, the results must be treated with caution; they must always be validated extensively, as the direction of the predictions may be incorrect without human supervision. Human experts must manually check NN-based results for certainty despite the learning process's autonomy.

Reinforcement learning

Reinforcement learning is the third primary learning method in machine learning. The neural network (NN) aims to learn complex or very complex questions by trial and error. There is no initial data for this learning method. The neural network learns through reward or feedback from the solution created. One of the best-known examples is the DeepMind product AlphaGo. In 2017, the software beat the world champion in the game of Go. It was trained using the reinforcement learning method.

Other unique forms of learning are hybrids, such as semi-supervised and self-supervised learning. In semi-supervised learning, labeled and unlabeled data are used to optimize learning further.

Deep learning

Deep learning is a unique form of machine learning. The foundations were laid in the 1990s by Yann LeCun, a French researcher who used early forms of backpropagation in his dissertation. Today, he works at Meta (formerly known as Facebook) and researches the topic of human understanding of AI. Earlier in his career, he designed the first convolutional neural network to recognize handwritten numbers. The method shows its strengths when identifying objects in images.

In 2012, Geoffrey Hinton, the scientist mentioned in the previous section, achieved a breakthrough by halving the error rate in the Large Scale Image Recognition Challenge with his new model [TZ-216]. The Large Scale Image Recognition Challenge is a competition in the field of machine vision that tests the ability of algorithms to correctly classify a large number of images and correctly recognize objects within these images. This achievement was made possible by algorithms that trained the network on new graphics cards. The training speed could be increased by 1,000 in this process.

The Convolutional Neural Network (CNN)—a deep learning architecture—is used in image and audio file processing. To solve increasingly complex problems, more and more hidden layers have been introduced. There are several advantages to integrating additional layers:

– The ability to learn complex patterns increases.
– A better representation of data for the respective task is possible.
– Broader generalization becomes possible so that environmental data can be responded to.

– Automatic feature extraction is possible so that additional features can be automatically extracted from the data.
– Complex tasks can be scaled to realize even more complex tasks.

With each layer added, the learning process becomes more complex and slower. It is performed sequentially and is always dependent on the last layer. With the introduction of backpropagation, the calculation starts from the output and works its way forward, layer by layer. Backpropagation is a learning algorithm that requires little computing power.

In addition to convolutional neural networks, there are other forms of deep learning, which are listed below:

– Perceptron
– Feed Forward Neural Networks
– Recurrent Neural Networks.

Each technology has its advantages in a specific application.

Development of Large Language Models (LLM)
A significant milestone in 2017 was the development of the Transformer architecture, a crucial advancement in artificial neural networks for large language models. This architecture, primarily used for natural language processing (NLP), has paved the way for developing models such as GPT, Gopher, PaLM, BERT, and others.

Several frameworks have become established in the field of deep learning. They provide all the critical components and enable a solution to be implemented quickly. Some leading examples are:

– TensorFlow—widespread use, open source; focus: automatic image recognition or natural language processing
– Keras—open source; focus: dynamic business environments
– PyTorch—open source; focus: use on mobile platforms such as Android or iOS.

These developments in large language models and transformer architecture, along with the emergence of powerful deep learning frameworks such as TensorFlow, Keras, and PyTorch, have revolutionized the artificial intelligence landscape and now enable advanced applications in a variety of fields, from automatic image recognition to natural language processing.

2.3.4   AI Application in a Vehicle

The automobile is currently undergoing a phase of intense change. Originally designed purely as a means of transportation, it has constantly evolved, and today, expectations are increasing, from automated driving to electric drives and connectivity. In addition, the car is progressively transforming into a mobile living room and multimedia platform [TZ-217]. This transformation poses challenges and offers new opportunities for the automotive industry. Car manufacturers see the opportunity to expand their value creation through innovative functions and new business models. At the same time, new players with expertise in big data, IT, and cloud technology are entering the market, including companies such as Google, Amazon, and Mobileye. In the development phase, weighing up the advantages and disadvantages and clarifying where and how these new functions will be integrated is necessary. For instance, can an automotive solution be embedded in the vehicle systems, or should it be implemented in the cloud? Hybrid solutions are emerging. In the case of voice control, for example, the primary function could be implemented in the vehicle. At the same time, full functionality is made possible by cloud connection and vehicle networking.

One example of an AI-supported solution is the valet parking feature in Fig. 2.9. The driver hands over their vehicle in a predefined handover zone. The parking area has cameras, sensors, and computers, forming an AI-supported infrastructure. It constantly communicates with the car and guides it independently to the assigned parking position. When the driver requests that the vehicle be returned, it is automatically returned to the handover zone. The start-up Kopernikus Automotive GmbH, in collaboration with Volkswagen subsidiary Cariad, provides such a solution [TZ-218]. As a further example, Mercedes-Benz has launched such a system on the market together with the supplier Bosch and the parking garage operator Apcoa [TZ-219]. It is the first Level 4 parking system to go into series production.

The valet parking feature is a rather complex function realized in interaction with the vehicle, the cloud, the infrastructure, and an app. However, this scenario does not have mixed operation, which means autonomous vehicles are not used together with manually driven cars. Mixed operation is a far more complex scenario (Fig. 2.9, reproduced with permission from Continental AG) (Fig. 2.10).

An important building block for automated driving in this more complex environment is the classification of pedestrians. The next important step is to recognize

Fig. 2.9 A parking garage in which the valet function is used. *Image* Continental

Fig. 2.10 Classification of pedestrian intentions and gestures using neural networks. *Image* Continental

pedestrians' intentions, gestures, and hand signals to enable interaction with the vehicle.

Artificial intelligence opens a wide range of potential applications and business models in the automotive industry, such as:

- Production: optimization of workflows through data analysis
- Human Machine Interface (HMI): use of speech recognition technologies
- Service: vehicles that recognize their service requirements
- Automated driving: object recognition to support driving
- Business models: activation of additional driver assistance systems
- Sales: determination of sales probabilities
- Mobility platforms: initiatives such as Waymo/Geely, Pony.ai, etc..

Artificial intelligence applications in the automotive industry constantly expand with new ideas and ongoing optimization. The dynamic nature of the field is a testament to the profound changes that AI technology is bringing. However, the extent to which customers are prepared to accept these new technologies remains critical. Unanswered questions revolve around data transparency, data security, AI regulation, functional safety, and the ownership of data collected in and around the vehicle.

2.3.5   Image Processing

Vehicles increasingly rely on visual data from cameras. These images provide a rich data source to be extracted and analyzed, making image processing a central component in advancing AI and ADAS.

In computer science, there are two primary approaches to image processing: image recognition and pattern recognition. Both utilize distinct algorithms to process the content of images, making them accessible to specially developed algorithms. This process enables the identification of structures, objects, and people. Pattern recognition relies on simple operations such as normalized grey value correlation, artificial intelligence, and deep learning, which have mainly gained traction in image recognition. These technologies serve different purposes, but both are integral to image processing.

Image recognition uses classification, marking, object recognition, and segmentation methods to recognize objects, places, people, vehicles, or text in images. The precision of localization increases from classification to segmentation until localization down to the last pixel is finally possible.

Classification defines the class or category to which the image belongs.

Marking increases the recognition accuracy of several objects in one image.

Recognition identifies the position of an object within the image.

Segmentation localizes the elements to the next pixel, used in autonomous driving.

During segmentation, the image is broken down into parts to achieve correlation with objects in the real world [TZ-220]. For each pixel, a decision must be made as to whether it belongs to an object in the image and, if so, to which object. Two types of segmentation are used: complete segmentation and partial segmentation. Here, you will find a list of different segmentation algorithms:

- Complete segmentation
- Partial segmentation
- Edge-oriented processes
- Region-oriented procedures
- Point-oriented procedures.

The image Fig. 2.11 (reproduced with permission from Continental AG) shows how each pixel of the captured image was assigned to a category via AI. This exact segmentation is critical as a reliable input value for sensor fusion (see Sect. 2.4.7), which is processed with further information, such as radar sensors or cloud data. The individual applications in the vehicle use the generated maps as the basis for the respective function.

TensorFlow's open-source library has become widely established in image processing and pixel-perfect segmentation. It is described as an end-to-end platform for machine learning [TZ-221]. In this context, "end-to-end" means the entire process is covered, from data preparation to productive code development.

Mobileye has proven itself over time as a leader in the field of Advanced Driver Assistance Systems (ADAS). Owned by Intel since 2017, it is a permanent fixture in the automotive industry. The company was founded in Israel in 1999, but its added value came to fruition in 2004 when the EyeQ1® chip was launched on the market. Mobileye's strategy was to offer integrated solutions from a single source in conjunction with software and hardware for ADAS platforms. Automobile manufacturers working with Mobileye found individual ADAS functions' performance superior during test drives. Other partnerships between Mobileye and

Fig. 2.11 Segmentation illustration. *Image* Continental

OEMs have been announced, such as with Porsche [TZ-243], who, in May 2023, partnered strategically with Mobileye on premium driver assistance system solutions [TZ-249]. It is interesting to note that Mobileye remains true to its core business. Its software and hardware are integrated into ECUs from various Tier 1 suppliers. Mobileye is now developing advanced functions that are implemented using artificial intelligence. Other competitors are entering an exciting and large market. Qualcomm, for example, has landed a significant coup and will replace Mobileye in BMW's driver assistance functions from 2025 [TZ-250].

In general, there are only a few Tier 1 suppliers left who develop their image recognition systems. Mobileye has often been chosen as a sub-supplier by vehicle manufacturers through a partnership in function development. Some OEMs are obliged to integrate Mobileye hardware and software into their control units. Bosch is one of the few suppliers that is driving forward its development in this field.

2.3.6 Voice Processing

The human–machine interface in the vehicle is an important selling point for owners. Voice recognition is increasingly important, but it is different for every vehicle brand. AI has also found its way into this arena, providing the driver more comfort during operation. In the first instance, we will look at how a voice model is defined.

A natural language model is a computer program that understands human language and uses statistical models to define speech and text data based on patterns. The patterns are used to predict new tests and speech data. Compared to big data, which is usually structured and clearly defined, speech processing presents more difficulty in analyzing and evaluating data. The major challenge in language processing is the order of the words in a text or sentence, which must be recorded correctly. Similarly, language evolves, so words cannot permanently be assigned a meaning.

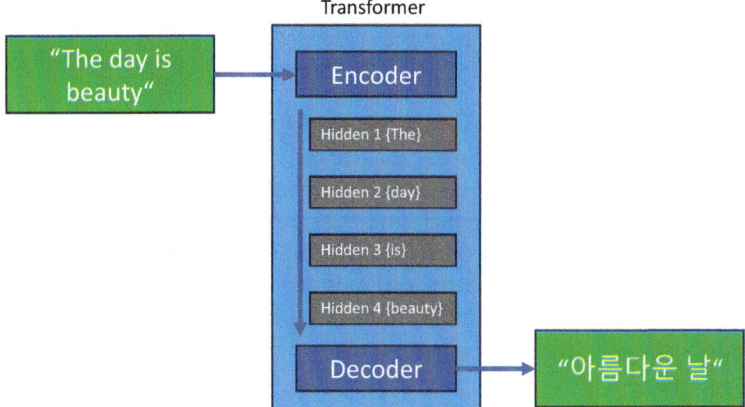

Fig. 2.12 Display transformer

First and foremost, the driver must be able to interact with the language model. Artificial intelligence has been presented with the challenging task of analyzing and interpreting speech and text, so the technology has been specially adapted and optimized for this use. The first models for machine speech processing were implemented with recurrent neural networks, which offer a limited text length and do not allow parallelization. Artificial neural networks with Transformer architecture form the basis for large language models; these were introduced as an innovation by Google researchers in 2017 (Fig. 2.12).

Positional encoding and self-attention, two transformative innovations, are integrated into the Transformer structure. Positional encoding and decoding assign a unique numerical value to each word's position in a sentence. The encoder and decoder are trained to ensure the input matches the output. This process forms multidimensional vectors from words, making them more amenable to mathematical algorithms. Self-attention, the attention mechanism, contextualizes words and captures their interdependencies within the text. Integrating several self-attention layers into the Transformer model further enhances the positional encoding architecture.

These two innovations have been incorporated into the structure of the Transformer and have significantly advanced the technology.

2.3.7 Energy Consumption

ChatGPT, a powerful language model, has been rigorously trained using deep learning and implemented as an artificial neural network [TZ-222]. The latest version, ChatGPT 4.0, has undergone intensive testing to demonstrate its capability to pass an intermediate-level test [TZ-223]. This extensive testing instills confidence in the model's performance.

ChatGPT's outstanding results are primarily due to the continuous increase in the scope and accuracy of its training data. More up-to-date data has significantly improved its performance, though the underlying algorithms have remained unchanged.

However, improvements in the accuracy of LLMs come at a cost. The length of the training phase is excessive; for version 4.0, training took around six months. The energy consumption was staggering; GPT-3 used 1,287 gigawatt hours [TZ-224]. ChatGPT is trained with data from a defined data baseline, and due to the nature of the training process, subsequent training of the neural network is impossible. Newer data cannot be used until the next full release of the data set. There is, therefore, no choice but to repeat the training process in its entirety, which means even more energy consumption and time as the total amount of data increases.

The energy demands of AI applications are expected to rise steadily in the coming years. Political pressure to rely on renewable energy is intensifying the pursuit for alternative energy solutions. This shift explains the renewed interest in nuclear power and the exploration of futuristic fusion reactors, which offer a potentially limitless energy supply.

2.3.8   KI-Hardware

The algorithms for AI development are extremely computationally intensive. For this reason, early considerations were given to practical implementation, which led to the development of graphics processing units (GPUs) and specific ASICs, such as Google's Tensor Processing Units (TPUs). Hardware manufacturer Nvidia is a leader in blockchain technology and AI development, particularly in the automotive sector. Founded in 1993 by Jensen Huang, Chris Malachowsky, and Curtis Priem, Nvidia launched its first graphics card in 1995 and introduced the first graphics processing unit (GPU), the GeForce 256, in 1999. With GPUs' enormous computing capacity, Nvidia became involved in deep learning as early as 2015. The technology is now also used as a near-standard in embedded systems in the automotive industry. While the development of microprocessors is progressing slowly due to the physical limits of semiconductor technology, GPU development is still in its infancy. Nvidia forecasts a possible thousand-fold acceleration in GPU performance by 2025 [TZ-228]. Qualcomm, AMD, Xilinx, and Intel are other renowned companies in the AI hardware sector.

2.3.9   Risks Associated with the Use of Artificial Intelligence

One major disadvantage of solutions like ChatGPT is the need for more user security. Misuse cannot be completely ruled out, and there is limited control over issues such as spam generation, fraud, or deliberate disinformation. Additionally, AI hallucinations—where systems like ChatGPT provide incorrect answers—can occur, often due to insufficient or vague data. A notable case involved a US lawyer who used ChatGPT to cite precedents that turned out to be entirely fabricated [TZ-225]. Another concern is AI's automatic code generation. While solutions like ChatGPT-4 can generate code effectively, corporate users must ensure the code does not violate copyrights, as the training data often comes from unverified online sources. Platforms like GitLab now offer AI-based code suggestions, further highlighting this growing trend [TZ-226].

Elon Musk's involvement in AI has fueled advancements in this field. In early 2023, he founded xAI, hiring top AI researchers, including Geoffrey Hinton, Yoshua Bengio, and Yann LeCun, to drive AI development. The growing importance of AI language models will likely lead to further breakthroughs in the coming years.

While AI applications' recent success has been astonishing, the long-term impact remains uncertain. Some experts have called for a pause in AI development to set appropriate regulations. At the start of 2023, a group of professors, AI entrepreneurs, and scientists published an open letter warning of the risks posed by AI advancements [TZ-229]. The European Union has also recognized this issue and is working on the first worldwide AI regulation, classifying AI applications into risk categories with corresponding legal requirements [TZ-230].

The first significant risk involves AI systems' learning processes. Humans must review the AI's output to ensure accuracy, but personal biases, cultural backgrounds, or differing interpretations can lead to misleading results. For example, Amazon's 2014 case, where AI discriminated against female candidates, shows how past hiring practices influenced AI decisions [TZ-231]. These lessons must be applied across industries to prevent bias from contaminating AI training data.

In vehicle development, especially autonomous driving, there is a critical risk of incorrect data or behavior being learned from flawed scenarios. For example, if test drivers who collect training data disregard traffic rules, the system might learn and replicate this behavior. Additionally, AI's non-deterministic nature means that slight changes in input can lead to an entirely different and sometimes wrong output, posing potential safety risks.

A broader concern is humanity's future alongside AI. Will the reliance on AI stifle human creativity and know-how, leading to technological stagnation? While AI can bring convenience, such as autonomous driving, society must continue cultivating human creativity and innovation.

Every new technology brings risks that must be understood and managed. Thus, systematic AI risk management must be integrated at every level of development and production.

2.4   Sensors

2.4.1   Introduction

The dream of self-driving cars is finally nearing reality. Many building blocks needed for autonomous driving are already in place. These include solutions such as:

– Adaptive Cruise Control
– Track stability systems
– Pre-Crash-System
– Automatic parking systems
– Line-Keeping-System
– Anti-lock system (ABS)
– Traction control system (ASR)
– Navigation
– Stereovision and radar systems.

Further functions can be implemented through clever cooperation between these systems. This interaction is known under the collective term "sensor fusion".

Sensor fusion is a sub-aspect of Advanced Driver Assistance Systems. The Society of Automotive Engineers (SAE) introduced the term in the early 1990s. However, the history of ADAS goes back to the 1970s, when Bosch introduced the ABS (anti-lock braking system).

An early, complex example of this development is the parking aid introduced by BMW in 1991, which used ultrasonic technology. Sensors of this type are a central element of ADAS. Sensor technology has developed rapidly since the 1990s and sensor-based systems are used in virtually every modern vehicle. For example, the previously mentioned parking aid is now a standard feature. Parking aid systems have evolved from a simple parking aid to fully automated parking and support for other driver assistance systems.

Numerous types of sensors are used in modern vehicles for different applications. An overview of the technologies and sensors is shown in Fig. 2.13.

The data from sensor fusion is used to create an environment model. This model collects all the information from the sensors, such as vehicles, obstacles, and open spaces. Decisions or strategies for the individual driver assistance functions can be derived based on the environment model. Only by using several sensor types can a system generate a comprehensive image of the vehicle's surroundings and thus increase accuracy.

Below, we present the most essential sensor types used in the ADAS sector.

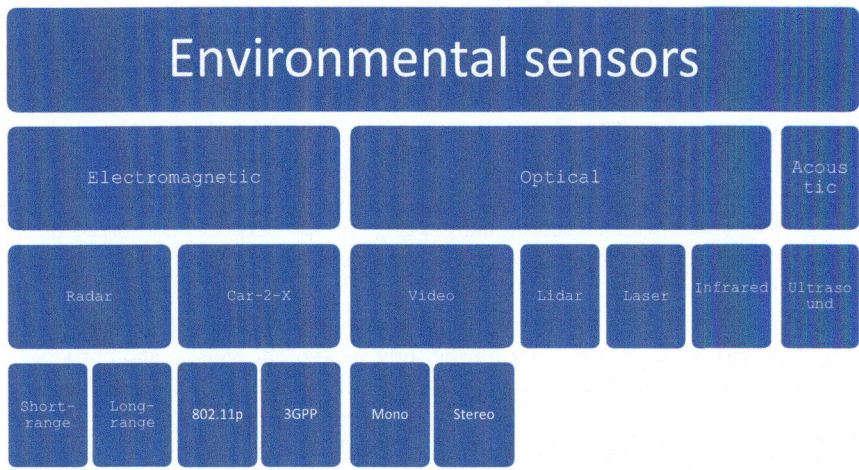

Fig. 2.13   Overview of sensors

2.4.2   Ultrasound Sensors

Ultrasonic sensors emit acoustic waves that are reflected by objects and returned to the receiver. Frequencies from 20 kilohertz to 1 gigahertz are generally called "ultrasound." For a vehicle sensor, frequencies between 43 and 60 kilohertz are used. This technology is sometimes referred to in specialist literature as "bat technology." The distance from the object to the sensor can be determined by measuring the transition and response times. The sensor principle is straightforward and can be implemented cost-effectively.

The ultrasonic sensor has become indispensable for automatic low-range maneuvering along with other sensors. Its advantages are obvious, and thanks to the large quantities produced, costs are falling steadily.

Examples of solutions based on ultrasonic sensors are:

– Parking aid
– Blind spot warning system
– Remote-controlled parking assistant
– Cross-traffic warning behind the vehicle
– Proximity detection when maneuvering
– Support for lane change maneuvers
– Monitoring of the area in front of and behind the vehicle in stop-and-go traffic
– Distance warning when approaching objects at the front or rear of the vehicle.

These sensors are often combined with others, such as the radar sensor.

2.4.3   Radar, Radio Detection, and Ranging

Another ADAS technology introduced by Mitsubishi in 2012 is the radar sensor with adaptive cruise. The system is known as ACC (Adaptive Cruise Control). It was initially installed in premium-segment vehicles. With further tightening of the NCAP requirements and the specification that the automatic emergency braking system (Automotive Emergency Braking) became mandatory, the sensor has increasingly become the standard, and the function has become widespread in the lower vehicle segment. The range of the sensors in the latest generations is now over 300 m.

The sensor principle is identical to that of ultrasonic technology. However, the frequency here is 76–77 gigahertz for the front radars. The side radars, introduced later, started at 24 gigahertz, but this frequency has not been used for vehicles since 2022; instead, the frequency used has been extended over the 77 gigahertz range. The distance resolution has improved by 20 thanks to the switch to 77 gigahertz. This distance resolution is possible because the pulse width increases with higher frequency if all other parameters remain identical. It is also possible to determine the relative speed of the detected objects via the phase difference between the transmitted and received signals.

Side radars have added further functions for the driver, such as blind spot detection. This technology has dramatically improved vehicle safety at higher speeds.

For many vehicle manufacturers, the "radar cocoon" is an essential factor in the move toward automated driving. The term refers to a system of radar sensors mounted around the entire vehicle that provides 360° protection. This solution, combined with other vehicle sensors, is intended to provide data redundancy and increase functional safety.

Examples of radar functions are:

– Distance control (including ACC)
– Traffic jam assistant
– Blind spot detection
– Exit assistant
– Rear Cross Traffic Alert
– Pre-crash systems.

The radar function is combined with other sensors, such as the camera.

2.4.4   Camera

The use of cameras began with the reversing camera, which developed from an exclusive optional extra to a standard function. Numerous cameras are now used in vehicles. For example, up to nine cameras are installed in a Tesla. Two camera variants are now used: the mono camera and the stereo camera. The stereo camera is used primarily as a front camera to support the radar sensor in measuring distance and detecting potholes, which enables the chassis to react adaptively. Cameras are now installed around the vehicle on the windshield, exterior mirrors, B-pillar, and trunk lid. Software is used to merge the images from the various cameras into an overall picture, which is displayed to the driver in the infotainment system, depending on the driving situation.

A unique feature is the "night vision" function, implemented using an IR camera (IR stands for infrared). However, this feature can only be found in a few vehicles.

The range of the mono camera now covers over 150 m, and the resolution has improved to over two megapixels. Various approaches are used in the software development of the camera, including conventional algorithms, optical flow, structure from motion, and artificial intelligence [TZ-247]. This multi-layered approach allows objects to be differentiated and road characteristics to be recognized. Thanks to learning methods in neural networks, semantic segmentation is possible, whereby, for example, a distinction can be made between the roadway and the edge of the roadway.

Examples of the camera function are:

- Rearview camera
- Surround-view
- Interior monitoring
- Adaptive cruise control
- Blind spot detection
- Traffic sign recognition
- Driver monitoring (drowsiness sensor)
- Lane Keeping Assist System
- Autonomous Emergency Braking.

2.4.5 Lidar

Lidar (Light Detection and Ranging) is the latest sensor technology currently mainly used in luxury and experimental (e.g., robotaxi) vehicles. Lidar technology can achieve a range of up to 150 m for smaller objects. The measurement in this process is carried out using a laser, which, like radar, can measure both speed and distance. The data is taken from the measured 3D data. There are two main categories of lidar systems: scanning and flash. Like a scan, the scanning method guides the light beam over the scenario. Mirror systems are used in the sensor for this purpose. On the other hand, the flash lidar method illuminates the entire scene in one go, and the measurement is then taken. This technology avoids moving parts entirely, increasing the system's robustness. Wavelengths of 850 and 905 nm are used in the sensor. This technology enables high-resolution depth information about the vehicle's surroundings and additional attributes to be added to the camera data.

This technology cannot be assigned an individual function, as it supports other sensors. This makes the current functions more robust, for example, in terms of reliability, or it makes other sensors redundant. Therefore, we do not provide separate examples for the lidar function.

2.4.6   Car-2-X

Car-2-X refers to a process in which a vehicle interacts with another communication partner, for example, another car or the infrastructure, such as traffic lights (in this case, it is referred to as "Car-2-Infrastructure"). This technology, which has been developing for over ten years, enables the vehicle to communicate with its environment. Rather than video data, sensor data is transmitted, which keeps the data volume manageable. Data can be transmitted via WLAN for vehicles (WLANp/pWLAN) or via mobile radio. This additional data contributes to road safety, warning of danger spots, roadworks, or dangerous weather conditions. The vehicle fleet sends the data to the backend, which is forwarded anonymously to the other connected road users.

Currently, this technology cannot fully exploit its potential, as only a few vehicles are equipped. Although some manufacturers are beginning to equip their vehicle fleets with the Car-2-Car function, no standard has yet been defined, making cross-manufacturer communication virtually impossible.

2.4.7   Sensor Fusion

Communication with infrastructure, such as traffic lights or pedestrians, is only just being implemented and researched in pilot projects. However, with the help of optimized, AI-supported algorithms, for example, an optimal route can be calculated for each vehicle in complex environments based on the information from the infrastructure data.

Figure 2.14 shows the principle of sensor fusion, including the integrated sensor data fusion.

The vehicle environment is detected using various sensors. The sensors can differ in the following properties: sampling frequency, latency time, transmission duration, and detection range. The data is forwarded to the sensor fusion after an appropriate adjustment, and a time stamp is added. In the sensor fusion, they are synchronized and linked together using time stamps. The applications then access this fused data.

The potential of sensor fusion can be seen in increased reliability, increased probability of detection, improved accuracy, and better visibility of objects. In addition, ambiguities are reduced through the use of different sensors. All these advantages contribute to optimizing vehicle functions and thus promote driver acceptance.

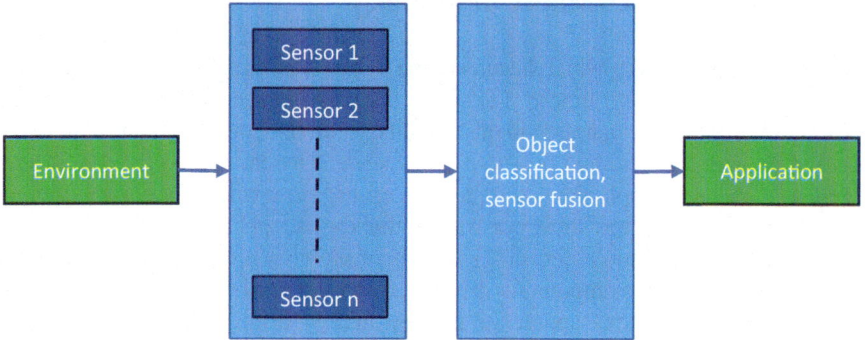

Fig. 2.14 Sensor fusion

The data is linked together using various procedures and methods. These include:

– Classification procedure
– Statistics
– Neural networks
– Stochastic methods
– Kalman filters
– Logical operations
– Methods based on rules.

The importance of sensor fusion has increased considerably in recent years as part of the development of ADAS systems and automated driving. Implementation without sensor fusion seems unthinkable, making it a core element of vehicle concepts.

Depending on the specific application, the user should develop sensor fusion based on the following criteria:

– Functionality
– Levels of sensor data fusion
– Functional architecture
– Fusion of algorithms
– Homogeneous and heterogeneous arrangements
– Sensor network (static/dynamic).

A suitable selection and integration of these criteria are essential to ensure that the developed sensor fusion delivers the desired results for the corresponding application.

2.5   Evolving Concepts of Mobility

The Volkswagen Beetle, which began selling in 1948, is regarded as the symbol of the German "economic miracle" [TZ-244]. The automotive industry in Germany experienced a boom, and individual mobility with one's own vehicle has been associated with freedom and prosperity ever since. The car has firmly established itself as a status symbol. Now, the question arises: How will autonomous driving shape the future, and how must our way of thinking adapt?

Thanks to new technologies, many aspects of mobility are changing. For example, the shift towards autonomous driving is accompanied by a change in urban planning [TZ-232]. Many appreciate getting directly from home to their desired destination without changing trains. This is why public transportation is less popular than some legislators or particularly environmentally conscious citizens would like it to be.

On average, a private vehicle is only driven for 43 min per day, with an average of two trips of 15 km each. Only 1 percent of all journeys exceed 100 km [TZ-233]. This raises the question of why many people still own a car. Why do they invest so much capital in this potentially extra "luxury"? Cars are becoming increasingly expensive the average price of a new vehicle rose to 42,790 euros in 2022 [TZ-234] and the trend is still growing. At the same time, cars are increasingly being pushed out of city centers to improve city dwellers' quality of life.

While the car as a status symbol was firmly anchored in our consciousness in the past, it is gradually losing its significance. This is particularly the case among younger people. Today, freedom is increasingly associated with an internet connection, which makes it possible to work from any desired location, for example. The mobility of the future is, therefore, likely to be different. It will be a mix of other modes of transportation, and we will rely on intelligent transport systems. For example, the center of our mobility activities will be an app linked to individual and shared calendars and other social networks. The "last mile," i.e., the last few kilometers from the train station or airport to our destination, will be dominated by bike or car-sharing services. Owning a car as a luxury item will likely fade into the background. Instead, as is already the case in other areas (e.g., music with Spotify or Apple Music), we will rely on rental or subscription-based services whenever necessary. It isn't easy to imagine this scenario becoming a reality in rural areas, as alternative mobility services are not yet as advanced there. Still, it is only a matter of time before this gap is closed.

A central component of future mobility is autonomous vehicles. Although autonomous cars and their technology have advantages and disadvantages, a study by the US research institute Rand found they offer significant added value to society [TZ-235]. From an environmental point of view, the advantages of this technology are clear: vehicles can continuously be operated optimally and, for example, drive significantly faster and more efficiently in a convoy [TZ-235]. This opens up new opportunities for older people, in particular, to remain mobile, as they can call a vehicle and be driven to their desired location. In addition, the tiresome search for a parking space could be eliminated, as cars can be requested when needed. The interiors of vehicles may transform into personal retreats [TZ-236]. Customers could choose vehicles with office equipment, fitness equipment, or even a movie theater depending on their needs. This makes it possible to use the journey time productively or to pursue hobbies. Networking within the vehicle will be necessary for autonomous driving and help keep passengers occupied during the journey.

What negative consequences could this social development have? Who can still afford traditional mobility (especially with their vehicle)? Will adapted concepts be developed for rural regions? These questions still need to be answered. But one thing is sure: work profiles will change. A current example is that in Berlin, fewer and fewer young people are getting their driver's licenses [TZ-237]. Driving instructors could become largely redundant in the future. Cab/Uber drivers and truck drivers could also disappear from the streetscape. Instead, more jobs could be created in the low-wage segment, as vehicles have to be cleaned after each use. The fear that many people will become unemployed is still partly unfounded. There currently need to be more than 100,000 truck drivers in Germany, for example [TZ-238]. It is, therefore, necessary to drive technology forward to ensure general prosperity and the security of supply chains.

The change in mobility behavior could also have other unexpected effects. For example, an increase in individual transportation with autonomous functions could lead to less use of public transport. This could, in turn, increase the volume of traffic—an undesirable scenario from an environmental perspective.

Fully comprehensive and binding standards and legislation will be required to manage the logistical challenge of autonomous vehicle transportation. For example, one risk could be that inadequacies or errors in the most widely-used networked mobility platforms could have far-reaching consequences when they scale up and are used by many vehicles. Theoretically, this could lead to total gridlock. The most minor errors could have a massive impact on the entire fleet and even result in serious accidents.

With the increasing complexity of modern vehicles, there is a growing risk of technical faults, which, although primarily non-critical, could be significant in quantity. It can already be observed today that both the frequency of recalls and the number of vehicles affected are increasing [TZ-239].

Another factor is the increasing connectivity of vehicles. Advancing digitalization means vehicles are constantly online, a situation we are already familiar with from smartphones. There are numerous vehicle access points, including keyless technology and infotainment systems. The risks exacerbated by these points of attack range from malware infiltration to car theft. Cybersecurity (see Chap. 4 for more information) will become a serious challenge. A case in which a Tesla 2021 was hacked wirelessly using a drone, allowing the vehicle doors to be opened remotely [TZ-240], is still a relatively harmless example. It is hard to imagine what could happen if Level 5 vehicles, whose occupants no longer have the option of manually intervening in the driving process, are manipulated.

Another example is the dependence of autonomous vehicles on the mobile infrastructure. In San Francisco, where a trial operation with robotaxis is being run, there were disruptions to the mobile network in the summer of 2023 caused by a large number of visitors to a festival. As a result, many of the operator Cruise's vehicles were left stranded [TZ-241]. Further details about the operation of Cruise's robotaxis have since become known. The service was discontinued in the fall of 2023, and it was reported that up to 1.5 employees per robotaxi were needed to ensure its operation [TZ-248].

Regardless of the technical challenges in vehicle autonomy development, the social component of the new technology must be noticed. While enthusiasts are happy about any progress, others have serious concerns. In San Francisco, for example, an active protest group ("Safe Street Rebel") is positioning itself against certain technology companies. Their criticisms include accidents caused by these technologies, blocking roads by autonomous vehicles, and increasing surveillance in the city associated with the new technology. Sometimes, protesters place traffic cones on the hoods of cars. As these are not programmed to react to such obstacles, they remain in place, and the companies then have to send employees to remove the cones. Such protests accompany the spread of robotaxis and extending their operating licenses to 24 h daily.

Given this rapid development, the question remains about when the fully autonomous vehicle will arrive. Forecasts range from 2030 to 2050, but one thing is sure: autonomous development has already begun and will not stop anytime soon. The focus is now on further development and when the laws will be adapted in the world's various regions.

The question is no longer "if" but "when" the autonomous car is coming.

References

[TZ-201] https://www.tomorrowsworldtoday.com/2021/08/09/history-of-autonomous-cars/ Accessed on 19.08.2023
[TZ-202] https://www.reuters.com/graphics/AUTONOMOUS-TRUCKING/TEXAS/mopanr koova/index.html Accessed on 19.09.2023
[TZ-203] https://bmdv.bund.de/SharedDocs/DE/Artikel/DG/gesetz-zum-autonomen-fahren.html Accessed on 01.07.2023
[TZ-204] https://group.mercedes-benz.com/innovation/produktinnovation/autonomes-fahren/ drive-pilot-nevada.html Accessed on 02.07.2023
[TZ-205] https://www.golem.de/news/mehr-geschwindigkeit-autonomes-fahren-in-deutschland-mit-130-km-h-erlaubt-2301-170873.html Accessed on 02.07.2023
[TZ-206] https://traton.com/de/innovation-hub/autonomes-fahren-in-usa-und-asien.html Accessed on 19.09.2023
[TZ-207] Ivanov, Shadrin, "System of Requirements and Testing Procedures for Autonomous Driving Technologies", 2019
[TZ-208] https://www.iks.fraunhofer.de/de/projekte/ki-absicherung-vda.html   Accessed   on 03.07.2023
[TZ-209] https://bmdv.bund.de/SharedDocs/DE/Publikationen/DG/bericht-der-ethik-kommis sion.pdf?__blob=publicationFile Accessed on 02.07.203
[TZ-210] https://www.nzz.ch/mobilitaet/umfrage-zum-autonomem-fahren-in-deutschland-gro sse-skepsis-ld.1696650 Accessed on 02.07.2023

[TZ-211] https://cryptomonday.de/jede-interessante-kunstliche-intelligenz-statistik-ki-trends/ Accessed on 08.08.2023

[TZ-212] http://jmc.stanford.edu/articles/dartmouth/dartmouth.pdf Accessed on 24.09.2023

[TZ-213] https://www.welt.de/wissenschaft/article169604489/Das-erste-autonome-Auto-kostete-200-000-D-Mark.html Accessed on 09.08.2023

[TZ-214] https://www.hco.de/blog/was-ist-starke-und-schwache-kunstliche-intelligenz-ki Accessed on 05.11.2023

[TZ-215] https://writings.stephenwolfram.com/2023/02/what-is-chatgpt-ng-and-why-does-it-work/ Accessed on 05.11.2023

[TZ-216] https://arxiv.org/abs/1409.0575 Accessed on 27.07.2023

[TZ-217] https://www.continental.com/de/produkte-und-innovationen/innovationen/infotainment-im-auto/innenraumdesign/ Accessed on 28.07.2023

[TZ-218] https://industrie.de/mobilitaet/autonomous-valet-parking-kopernikus-automotive-sichert-sich-weiteres-investment/ Accessed on 12.08.2023

[TZ-219] https://group.mercedes-benz.com/innovation/produktinnovation/autonomes-fahren/intelligent-park-pilot.html Accessed on 12.08.2023

[TZ-220] https://www.cosy.sbg.ac.at/~uhl/imgProcess.pdf Accessed on 29.07.2023

[TZ-221] https://www.tensorflow.org/ Accessed on 29.07.2023

[TZ-222] https://www.bigdata-insider.de/was-ist-der-generative-pretrained-transformer-3-gpt-3-a-1011085/ Accessed on 27.07.2023

[TZ-223] https://www.stern.de/digital/-beaengstigend---chatgpt-besteht-im-zweiten-anlauf-bayerisches-abitur-33499852.html Accessed on 27.07.2023

[TZ-224] https://winfuture.de/news,135057.html Accessed on 11.08.2023

[TZ-225] https://www.heise.de/news/ChatGPT-erfindet-Gerichtsurteile-US-Anwalt-faellt-darauf-herein-9068180.html Accessed on 11.08.2023

[TZ-226] https://www.tomshardware.com/news/dark-web-chatgpt-unleashed-meet-darkbert Accessed on 05.11.2023

[TZ-227] https://www.rwkv.com/ Accessed on 28.07.2023

[TZ-228] https://www.nvidia.com/de-de/about-nvidia/ai-computing/ Accessed on 22.09.2023

[TZ-229] https://www.capital.de/wirtschaft-politik/elon-musk-und-ueber-1000-techriesen-warnen-in-offenem-brief-vor-ki-33331838.html Accessed on 11.08.2023

[TZ-230] https://www.europarl.europa.eu/news/de/headlines/society/20230601STO93804/ki-gesetz-erste-regulierung-der-kunstlichen-intelligenz?at_campaign=20226-Digital&at_medium=Google_Ads&at_platform=Search&at_creation=RSA&at_goal=TR_G&at_advertiser=Webcomm&at_audience=ki%20eu&at_topic=Artificial_intelligence_Act&at_location=DE Accessed on 11.08.2023

[TZ-231] https://www.zeit.de/arbeit/2018-10/bewerbungsroboter-kuenstliche-intelligenz-amazon-frauen-diskriminierung Accessed on 11.08.2023

[TZ-232] https://www.bosch.com/de/stories/veraendert-autonomes-fahren-die-gesellschaft/ Accessed on 30.07.2023

[TZ-233] https://umwelt-fragen.de/durchschnittliche-autostrecke-in-deutschland/ Accessed on 30.07.2023

[TZ-234] https://de.statista.com/statistik/daten/studie/36408/umfrage/durchschnittliche-neuwagenpreise-in-deutschland/ Accessed on 30.07.2023

[TZ-235]]https://www.bussgeldkatalog.org/autonomes-fahren/ Accessed on 30.07.2023

[TZ-236] https://www.springerprofessional.de/automobil---motoren/daimler-innenraumkonzept-eines-autonomen-fahrzeugs/6561722 Accessed on 30.07.2023

[TZ-237] https://prinz.de/berlin/artikel/117230-berliner-jugend-der-fuehrerschein-ist-out/ Accessed on 30.07.2023

[TZ-238] https://www.tagesschau.de/wirtschaft/unternehmen/speditionen-lkw-fachkraeftemangel-100.html Accessed on 30.07.2023

[TZ-239] https://www.automobil-industrie.vogel.de/pkw-rueckrufe-nehmen-weiter-zu-a-1068149/ Accessed on 30.07.2023

[TZ-240] https://futurezone.at/digital-life/sicherheitsforscher-hacken-tesla-mit-drohne/401
          369612 Accessed on 05.11.2023
[TZ-241] https://www.handelsblatt.com/technik/it-internet/autonomes-fahren-fiasko-bei-cruise-
          nach-robotaxi-freigabe-in-san-francisco/29330254.html Accessed on 19.09.2023
[TZ-242] https://www.nio.com/de_DE/nad Accessed on 24.09.2024
[TZ-243] https://newsroom.porsche.com/de/2023/unternehmen/porsche-mobileye-kooperation-
          teilautomatisiertes-fahren-assistenzsysteme-32249.html Accessed on 11.08.2023
[TZ-244] https://www.archivverlag.de/wissenswelt/deutsche-geschichte-ab-1945/wirtschaftsw
          under-deutschland Accessed on 30.07.2023
[TZ-245] https://www.etas.com/de/unternehmen/news-unece-regulations-for-automotive-cybers
          ecurity-adopted.php Accessed on 24.11.2023
[TZ-246] https://trends.google.de/trends/explore?date=all&geo=DE&q=ki&hl=de   search
          executed on 24.11.2023
[TZ-247] https://www.bosch-mobility.com/de/loesungen/kamera/multifunktionsk
          amera/ Accessed on 27.6.2023
[TZ-248] https://www.heise.de/news/Robotaxis-von-Cruise-Hilfe-per-Fernzugriff-alle-paar-Kil
          ometer-noetig-9353903.html Accessed on 25.11.2023
[TZ-249] https://newsroom.porsche.com/de/2023/unternehmen/porsche-mobileye-kooperation-
          teilautomatisiertes-fahren-assistenzsysteme-32249.html Accessed on 25.112023
[TZ-250] https://www.bimmertoday.de/2022/05/09/bmw-qualcomm-partner-statt-zulieferer-fur-
          autonomes-fahren/ Accessed on 25.11.2023

Digital Drive: The New Era of Connectivity

3

Abstract

This chapter examines the growing importance of vehicle connectivity in modern automobiles. It explores various aspects of connected car technology, from basic telematics to sophisticated V2X (Vehicle-to-Everything) communication systems. The chapter discusses key challenges in implementing connected vehicle systems, including data security, privacy concerns, and infrastructure requirements. It also covers emerging business models in connected mobility, the evolution of over-the-air updates, and the integration of vehicles into smart city ecosystems. The discussion concludes with an analysis of future trends and potential developments in in-vehicle connectivity.

An exciting and fast-growing trend we're exploring in this chapter is vehicle connectivity. Alongside autonomous driving and electromobility, this aspect represents the third current megatrend in the automotive industry. The topics here are various, significantly increasing the development complexity and associated security. Numerous new functions and possibilities—for example, features-on-demand or optimizing of existing functions like voice control—mean the interaction between vehicle, communication technology, backend, and data from third-party providers plays a role that makes it essential to focus on the tasks of the holistic functional chain. Not only is the growing number of stakeholders and connected functions challenging, but technical aspects such as data security and data protection are also increasing. Due to the large number of individual inputs, there is still a long way to go to achieve a uniform solution across the various car manufacturers. However, this topic must allow OEMs to expand vehicle functionality with exciting and valuable applications from manufacturers and third-party providers.

© The Author(s), under exclusive license to Springer Fachmedien Wiesbaden GmbH, part of Springer Nature 2025
R. Mildner et al., *Car IT Reloaded*, https://doi.org/10.1007/978-3-658-47691-5_3

3.1   Introduction

One of the biggest game-changers in the automotive world has been connecting vehicles to the Internet and cloud services. The use of an Internet of Things (IoT) hub or the outsourcing of required computing power to the cloud are just two examples that describe the possibilities of this powerful connectivity option. Connection to the outside world is established via the vehicle's internal Wi-Fi or the mobile network.

The connected car concept originated in Formula 1 in 1980 when the BMW team was the first to install an onboard computer with telematics [TZ-301]. Since then, the possibilities have been expanded, which means the vehicle's connectivity can generally be divided into three categories: telematics, vehicle-to-everything, and infotainment [TZ-302]. The following list shows a selection of applications that a connected vehicle makes possible:

– Real-time traffic information: Traffic information, including traffic jam warnings, road works, and road closures, can be received live. This enables the driver to take alternative routes to minimize delays. Due to the focus on navigation, this topic is also dubbed "intelligent navigation."
– Advanced driver assistance systems (ADAS): Connected vehicles use data from their sensors and other vehicles to improve driver assistance systems. Thanks to the broader database from various sources, these systems can warn of potential dangers, brake automatically, keep in lane independently, or assist with parking.
– Vehicle-to-vehicle communication (V2V): Vehicles can exchange information about their position, speed, and direction. This enables cooperative driving functions that can prevent collisions or otherwise minimize hazards.
– Vehicle-to-infrastructure communication (V2I): Information about traffic lights, traffic signs, and road conditions can be communicated with the road infrastructure to optimize traffic flow and thus increase safety.
– Automatic emergency call systems: Networked vehicles are equipped with emergency call systems that can automatically notify emergency services in the event of an accident, transmit the exact location of the accident site, and provide further essential information about the number of occupants in the car or the severity of the accident.
– Vehicle diagnostics and maintenance: Car connectivity enables remote monitoring of vehicle data (including the condition of mechanical and electronic components), allowing potential problems to be detected at an early stage and optimizing maintenance (predictive maintenance) [TZ-331].
– Infotainment and connectivity: Connected vehicles offer streamlined entertainment and information services such as music streaming, internet access, voice assistants, and smartphone integration. The latest generation of in-car infotainment allows passengers to use the Internet or streaming services while driving.
– Autonomous driving: Networking also plays a crucial role in autonomous driving. For this function, data from the vehicles must be exchanged with each other

and via the cloud to ensure interaction between cars. Additionally, the car must have real-time information about weather, traffic, and upcoming traffic via the Internet.

When the book Car IT Compact [RM-042] was written in 2015, many topics were still dreams of the future or mere visions. The above examples show what connected vehicles are already capable of today. Additional functions or expanding existing possibilities will result in even more uses that will increase safety and comfort and further advance the development of autonomous driving. According to current estimates, the number of vehicles connected to the Internet will amount to around 367 million by 2027 [TZ-303], and the trend is rising.

3.2   The Challenges of the Connected Vehicle

The possibilities for applications and increased comfort are wider than car connectivity. The significantly growing complexity is also changing the landscape. The focus here is no longer just on the mechanics' reliable interaction but expands into various areas. We'll cover the most critical aspects of this next.

Data security and data protection
Modern cars have many built-in sensors to collect and store data. In addition, for example, Volkswagen offers the function of logging in with your name and email address, both in an app provided by the vehicle manufacturer and in the car itself. This means that data collected by the vehicle can be directly assigned to a person. This mass of information must be protected against unauthorized access; data protection and security measures must be in place to prevent third parties from accessing user data. To this end, car manufacturers are developing their data protection guidelines, with data privacy protection being key. The question of who actually "owns" the data has not yet been conclusively determined. In the USA, Nissan, and Kia, for instance, collect the following information in their vehicles: age, race, color, medical condition, physical or mental disability, sex, and gender information. The vehicle owner agrees to the manufacturer's privacy policy [TZ-336].

Cybersecurity
As in the smartphone environment, connected driving generates a lot of data collected and transmitted to various locations. However, this flood of information is also of interest to criminals, so vehicle and personal data are frequently exposed to cyber-attacks. There is also a focus on preventing third parties from taking control of the car. The connection to the Internet opens the door to such attacks, which makes robust security mechanisms—not only in the vehicle but also in the infrastructure—indispensable. Cybersecurity has been emphasized, and the ISO/SAE 21434 standard "Road Vehicles—Cybersecurity Engineering" has been published. The subject of cybersecurity is discussed in more detail in Chap. 4.

Interoperability

At the current stage of development, many car manufacturers use different transmission technologies and their own protocols for networking. A harmonized technology is crucial to ensure seamless interaction between manufacturers. To this end, European and global standards are being developed to ensure interoperability in the future.

Vehicle-Infrastructure communication

To offer further new and advanced functions in connected vehicles, the vehicles must receive the data required for the functions from outside. The primary focus is communication with the infrastructure, such as pedestrians or cyclists. However, this technology is still in its infancy; a nationwide roll-out is still a long way off and requires strong cooperation between authorities and vehicle manufacturers. In the meantime, several pilot projects have been established in which different technologies and new functions are being tested, such as DTA (Digitales Testfeld Autobahn) [TZ-332]. An overview of test fields can be found on the Test Field Monitor [TZ-304]. One of the first such projects was the Adaptive VE Project [TZ-305]. Infrastructure, telematics, and driver assistance systems were examined more deeply here. A further overview of early research projects in vehicle-infrastructure communication can also be found in the study "Highly automated driving on freeways—industrial policy conclusions" by the Fraunhofer Institute [TZ-306].

Software updates and lifecycle management

The legal situation is clear: a vehicle must be developed according to the latest technology, and once it has been approved, it is allowed to be sold with precisely this state-of-the-art technology until the end of its lifecycle. However, the systems' complexity and interaction continues to increase: fast-moving technologies have been introduced, for example, smartphone connections with different operating systems such as Android from Alphabet or iOS from Apple, to which the vehicle manufacturers have to react quickly to keep the functionalities up to date. Regular software updates have become necessary for this in more and more systems, which should preferably be carried out via OTA (over-the-air) transmission. In this way, it can be guaranteed that new functions or enhancements/security updates are distributed without further action by the driver or a visit to a workshop. In 2022, the UNECE R 156 regulatory document with precise rules for OEMS dealing with OTA was published. These rules are compulsory in the EU.

Connectivity

As the definition of systems itself suggests, the range of functions available depends mainly on connectivity. Reliability of transmission technologies often lags, for example, in the event of a mobile network failure. Although the latest 5G technology is now available in 89% of regions in the European Union (as Mid of 2023) [TZ-307], there are still "dead spots" that make connectivity correspondingly tricky. 5G technology was launched in the USA in 2019. In Europe, the technology is not yet available nationwide. Furthermore, to fully exploit the speed of the 5G network, base stations must be connected to the network of the respective telecommunications provider via fiber optics. There is still room for improvement here.

Legal aspects
Legal aspects play an essential role, particularly in connected vehicles. The main focus is on secure data processing processes and the associated data protection regulations. In 2021, the European Data Protection Board published version 2.0 of the guideline "01/2020 on processing personal data in the context of connected vehicles and mobility-related applications" [TZ-308]. Another critical point is the topic of cybersecurity, which is defined in the UNECE WP.29 specifications. The topic of OTA updates was regulated in UNECE R 156 [TZ-309]. Type-approval relevant (definition: "With the type-approval, the approval authority confirms that a large number of similar vehicles or vehicle parts meet the legal minimum standards for safety and environmental behavior") [TZ-310] software updates must be possible in a vehicle.

Costs
Costs are incurred at various points for the networked functions. Among other things, suitable infrastructure and harmonizing cloud servers must be set up. The control units compatible with such an infrastructure are more expensive than those currently installed, and protection for different configurations must be set up. Collecting the associated costs is difficult, as they must be passed on to vehicle owners, drivers, dealers, and infrastructure providers. Some products in this area are already on the market, but research projects are underway to analyze the topic further. One example is regular map updates for navigation systems.

3.3   Connected System Architecture

Several components are required to offer networked functions. The following Fig. 3.1 shows the superordinate associated puzzle pieces for such a system [RM-042].

Smooth functionality requires many interfaces, which must also be defined beyond the boundaries of a single manufacturer. Today, many functions are developed directly as smartphone apps or installed via apps in the vehicle's infotainment system. This is where numerous third-party providers come into play, using the data the car or driver provides for new functions or services. The ecosystem is always the responsibility of a car manufacturer, and the apps are not transferable. This is comparable to the Android smartphone world or the Apple ecosystem.

The figure above also shows the interaction between the OEM cloud, the vehicle, the provider, and the user's touchpoint. The interfaces to the car are always accompanied by an OTA ECU, which contains the interface, including the firewall for the vehicle. This is also known as the OBD unit (**o**nboard **d**iagnostics **u**nit).

Fig. 3.1 How the cloud works in the connected vehicle

Interaction between the components allows vehicle manufacturers to develop the car platform into a device on the Internet or Internet of Things (IoT). As a result, the battle between Apple and Google over the sovereignty of vehicle data has already broken out. This is why German vehicle manufacturers develop operating systems for the "new world" *in* the vehicle to ensure that control over the car or the "user experience" remains with the OEMs and is not left to new players.

3.4   Hybrid

The industry quickly recognized that the current computing power of vehicle control units was insufficient for providing a voice control experience users are accustomed to (services like Siri or Alexa). One way to achieve a similar quality in modern cars was to link it directly to a cloud (backend). Volkswagen launched voice control in the hybrid version 2021 by introducing the Modular Infotainment Matrix 3 (MIB3) [TZ-311]. Since then, the system has received the required responses from the vehicle, i.e., offline, via the Volkswagen cloud, i.e., online. This connection was then greatly optimized regarding response times to achieve seamless operation between the individual components. There are different approaches from the individual car manufacturers. The first Android Auto applications from Google were introduced in an Audi in 2017 [TZ-327]. In addition to hybrid (cooperation of offline and online) approaches in voice control, others have also been developed in this direction, now representing the current "state of the art."

3.5   Frontend/Backend

In the interconnected world, systems are divided into two areas: the front end and the back end. The terms are explained in the next section. These two areas are crucial in networked systems, but the requirements and skills for developing each are very different. In a connected car, the functions are distributed between the backend and the embedded system in the car. This enables new functions or expands existing ones through networking with the backend, which increases quality and service availability.

Frontend is defined as direct communication with the user or, in the case of user interface (UI) and user experience (UX), as a touchpoint. Touchpoint is the place or moment a user comes into contact with a company's product. Examples include the navigation device in the car, the manufacturer's smartphone app, the homepage, etc. For the user, this type of networking provides several options for operating the car's IT functions. For example, the auxiliary heating can be operated from within the car using the key fob or via the vehicle app. The different touchpoints are very important for the user's experience and, therefore, for the individual manufacturers, as they can be used to maintain newly-won customer loyalty. With every contact the consumer has with the company, the brand can position itself and influence the buyer.

In the **backend**, processes that run in the background and do not require direct interaction with the user are implemented. For example, data is collected, processed, and made available to the frontend. The required data is collected from various sources, such as vehicle data, the Internet, or the user's smartphone. A significant advantage of these backend systems is their scalability, which is impossible in a vehicle without any connection. It enables applications to be expanded with new functions via the backend. Manufacturers or third-party providers can create and offer new functions or services anytime. Audi introduced its own Audi app store in the Modular Infotainment Matrix 3 from the VW Group in 2023, which allows apps to be installed directly on the infotainment system. The aim was to integrate the installation into the process without an existing smartphone [TZ-312].

Overall, the backend must perform many functions (list not exhaustive):

– Provision of data
– User administration
– Creation of back-ups
– Import and export of data
– User authentication
– Access authorization and access control
– Non-repudiation/commitment—no inadmissible denial of action.

Intelligent integration of frontend and backend systems and the advancement of communication technologies enable the carmaker to unlock the full potential of connected cars.

3.6   Communication Technology

Choosing the right communication channel is critical to creating a well-designed connected car. Thanks to effective and efficient communication, functions and traffic can be implemented more successfully with V2X (such as car-to-car or car-to-infrastructure). To this end, various Cooperative Intelligent Transport Systems (C-ITS) are being developed worldwide, which benefit the automotive industry and are also used in internal logistics.

Dedicated Short Range Communication (DSRC), developed to enable effective vehicle-to-vehicle (V2V) and vehicle-to-infrastructure (V2I) communication for improved traffic safety and management, was the first communication protocol published as a standard. DSRC has three versions available for the Japanese, American, and European markets. A standardization project of the International Telecommunication Union is harmonizing regional differences. The DRSC frequency range is divided into seven 10 MHz channels. The communication is based on the IEEE 802.11p WLAN standard, whereby the channels have been optimized for different applications, such as security services and control functions. The data rate can be expanded up to 27 Mbit/s, and distances of up to 300 m can be bridged. DRSC is also used in consumer electronics and billing systems for road tolls. DRSC went into series production in Japan in 2015. In Europe, Volkswagen launched DRSC on the mass market with the Golf 8 in 2020.

With the Cellular Vehicle-to-Everything (C-V2X) standard, car manufacturers aim to directly develop a global, alternative standard communication technology to I-CTS. The use case for this has stayed the same. The standard is defined in the 3GPP initiative for worldwide use, and the initiative is described on the 3GPP homepage [TZ-331]: the 3GPP specifications cover cellular telecommunications technologies, including radio access, core network, and service functions, which provide a complete system description for mobile telecommunications. The standard depends on the preparation and expansion of mobile communications technology. Four communication services will be published:

– V2V = Vehicle-to-Vehicle
– V2P = Vehicle-to-Pedestrian
– V2I = Vehicle-to-Infrastructure
– V2N = Vehicle-to-Application server.

The standard has now been extended in Release 14 to include ad-hoc networks to transmit data independently of the mobile network. This step aimed to minimize the following weaknesses in the technology:

– High latency during data transmission via a mobile network
– Increased availability with limited or non-existent mobile phone coverage
– Transmission of security-relevant data not via specific mobile network operators.

As of 3GPP standard release 15, the fifth-generation mobile network (5G network) is also supported. One of the main advantages of the C-V2X is its ability to connect to the mobile network. Furthermore, the chipsets are cheaper than the WLAN segment, positively impacting the purchase price. The first manufacturers are already offering hybrid solutions here. Cybersecurity is the focus of further development in the latest project, AutoCrypt V2X.

3.7   Edge Computing

As expectations for interconnected technology are constantly growing (for example, increasing bandwidth while minimizing latency times), vehicle manufacturers have given extensive consideration to new architectures in the area of infrastructure. The most important finding is that not all data can be transmitted via a centralized cloud, and the calculations can be done in the cloud. With new functions and greater use of technology, data volumes are becoming very high, and further infrastructure expansion is needed. To remedy this situation, Multi-Access Edge Computing (MEC) computer architecture was developed and used in research projects [TZ-313]. The aim is to provide the required computing power, resources, and services at the "network's edge" to build the infrastructure cost-effectively and efficiently. The edge technology was integrated into the 5G mobile communications standard from the outset. The realization of autonomous driving as a real-time application is only possible with this architecture. The technology was successfully tested in the "Car2Mec" research project, and all applications' latency times were significantly reduced. To realize the intelligent connected road, Intel and Capgemini have developed a solution that supports edge applications and, for example, enables a framework for the development of onboarding virtualized applications [TZ-332]; it is based on Capgemini's NSCONCE MEC platform [TZ-314].

3.8   Connected Car Implementation Examples

Telematics
The term "telematics" combines the words telecommunication and informatics. Telematics is one of the first applications in networked vehicles, even if it was not referred to as such at the time. It is mainly used in fleet management in the logistics industry, although Formula 1 is also a well-known application.

Telematics solutions consist of four areas:

- Data acquisition: for example, GPS data, speed, fuel level, etc.
- Data transmission: mobile network, real-time data, remote transmission.
- Visualization—analysis: data evaluation, definition of measures.
- Archiving: data backed up on servers.

Optimized autonomous driving
Networked technology is needed in autonomous driving to generate highly accurate digital maps and make them available to modern cars [TZ-315]. In 2015, Audi, BMW, and Mercedes decided to purchase the HERE digital map service provider to collect as much of this data as possible quickly. HERE has made the data available to all customers independently of vehicle manufacturers with a comprehensive ecosystem [TZ-316]. The resulting maps make traffic sign recognition, for example, much more accurate, as the data comes from the vehicle sensors and is transmitted from other vehicles via the backend.

OTA update
Over-the-air (OTA) update is a technology introduced to keep vehicle software up to date at all times. Previously, updates could only be carried out in workshops using the diagnostic tool on-site in a service shop. The eSync Alliance [TZ-317] was founded by suppliers and service providers to promote standardization. The eSync Alliance describes itself on its homepage: "The eSync Alliance creates a trusted, multi-vendor pathway to establish secure end-to-end OTA and data services for the connected car through a global network of cooperating providers." Functions from participating companies were also presented here to make software/functions available to the driver for a limited period. This means drivers can

only buy the functions when they need them. For example, seat heating is already available as a subscription model in America. In turn, the manufacturer can receive monetization even after selling the vehicle.

Tesla was the first manufacturer to implement the OTA technology worldwide systematically. Since then, most other OEMs, such as Audi, Jaguar Land Rover, Mercedes, and BMW, have followed suit. A small selection can be found in the list below:

– 2008: Tesla Roadster (Tesla S: 2012)
– 2013: BMW i3
– 2014: Volkswagen E-Golf
– 2014: Mercedes B-Class.

One disadvantage for manufacturers is the data transfer cost, as the number of updates and data per vehicle increases. For this reason, further solutions/improvements are currently being worked on. Aurora Labs and Infineon jointly presented AI-controlled, secure, over-the-air updates at CES in Las Vegas in 2023 [TZ-318]. Elektrobit, the Continental subsidiary, also works with Airbiquity on a "pre-integrated over-the-air solution" [TZ-319].

Travel Warning
Many German manufacturers have implemented hazard warnings via V2X in Germany [TZ-320]. Vehicle manufacturers warn of the following hazards: break-down vehicles, accidents, end-of-traffic jams, slippery roads, and fog. Here, there is a dependency on an existing mobile network. German manufacturers have different scenarios in which hazards are displayed. These are always local warnings. One manufacturer, for example, only warns of dangerous road conditions.

Audi has implemented pilot projects in this area in the United States [TZ-321]. In Alpharetta, Georgia, the aim is to protect schoolchildren by networking vehicles, school buses, and American warning beacons. To this end, information on lesson times and the distance to the school building is displayed on the warning beacons so that drivers have the necessary information and can react. In Virginia, a safety vest for construction workers is being tested to warn drivers of the construction site when approaching the danger zone.

eCall
eCall was made mandatory in Germany for vehicle models with type approval from April 1, 2018 [TZ-322]. This technology allows an automatic or manual emergency call from the vehicle. The manual emergency call is triggered by pressing the provided button and connecting the vehicle's mobile phone to the manufacturer's call center. For example, drivers can contact the call center to report an accident or a breakdown. The automatic emergency call is activated using an algorithm when the vehicle sensors detect an accident. After an emergency call is made, the following data is transmitted, for example:

- Time of the accident
- 17-digit vehicle identification number (VIN)
- Vehicle position (GPS coordinates)
- The direction of travel of the car
- Number of occupants (if seat belts were fastened).

Emergency calls are received by the integrated control centers (112 rescue centers) if they can process the eCall minimum data set (MSD). The MSD contains data such as the vehicle's position and other vital data about an accident. It is a small data package that can be transmitted quickly and inexpensively. The eCall function is installed in the vehicle as a stand-alone system. It often has a dedicated antenna and power supply, making it impossible for the driver to stop the emergency call. It also remains active even if the vehicle battery malfunctions due to a crash.

3.9   V2V Charging

In Germany, as everywhere else, electromobility is trending. Therefore, the public charging infrastructure is being expanded accordingly, supported by various funding pools. Vehicles can currently only be charged in this infrastructure. KfW "program 442" is the first funding initiative in which bilateral wall boxes are funded in Germany [TZ-334]. The hardware is not yet widespread, and only four types of wall boxes are being promoted in the initial approach.

Bilateral charging (V2V—vehicle-to-vehicle) is not just an issue in the automotive industry. In the smartphone world, Apple recently introduced bilateral charging with the iPhone 15 [TZ-323]. The vehicle stores the electricity as direct current (DC), but alternating current (AC) is used in our electricity grid. The battery can be charged bilaterally and feed power into the connected grid. For example, a separate battery storage unit must no longer be installed. One use for this could be that the vehicle, which was charged with solar power during the day, supplies the house lighting and other electrical devices with electricity in the evening. However, today's vehicle chargers and wall boxes are not yet designed for this, as the direct current in the car would have to be converted back to alternating current. One of the first pilot projects was bidirectional charging management (BDL) [TZ-325]. Another pilot project was carried out by Porsche together with TransnetBW and Intelligent Energy System Services from 2019 to 2021 as part of a research project [TZ-324]. Both pilot projects showed that this technology has enormous potential and is theoretically feasible. The concepts of "electricity from the vehicle to your own home or the general power grid" are called vehicle-to-home (V2H) and vehicle-to-grid (V2G). The IEC 61851-1 standard defines the communication of bilateral charging and makes it possible to feed electricity from the vehicle battery back into the domestic grid [TZ-233]. Several bills on bidirectional charging have been introduced in the US. At the federal level, bill HB 6178 will make the function mandatory in electric vehicles starting in 2027. 2024 California will enter the race with the bill SB 333 [TZ-337]. However, there is currently no valid bill on V2V charging in America.

3.10 Smart City

If we take the idea of networking further, we inevitably end up with the networked city, also known as the smart city. The aim is to achieve an "intelligent city" or an entirely digitized living environment. Urban planners desire to achieve a more sustainable and efficient city through digitalization. By being sustainable and efficient, we can use our resources more sparingly. An intelligent city pursues the following goals [TZ-326]:

– Low consumption of resources
– Climate protection
– Ensuring mobility
– Improved quality of life through networked technology
– Sustainable urban development
– Municipal infrastructures, such as energy, buildings, transport, water, and wastewater, should be better networked with less administrative effort
– Solve problems using information and communication technology (for example, by responding more quickly to emergencies)
– Transparency of administration, open data, digital citizens' office.

The primary goal of this innovative city sector is to improve residents' quality of life. Data protection is, therefore, an indispensable element. Human dignity remains inviolable, even in the digital space. Smart mobility also goes hand in hand with a smart city. The aim is to use all technical possibilities, such as networking, to improve citizens' mobility. Specifically, it is about improving public transport or optimizing traffic flows to reduce city congestion [TZ-235].

3.11 Business Models and Monetization

Autonomous driving and the functions of connected driving require a great deal of investment across the value chain. In the market economy, products must be developed to operate sustainably through monetization. The vehicle industry is, therefore, in a transformation phase. It aims to develop the industry away from the traditional value chain of the product and towards a data-driven business model. Recently, it has become clear that the most successful companies worldwide, such as Google and Facebook, owe their success to data-based business models. On the other hand, established car manufacturers earn their money mainly from selling vehicles and their banks. There are three levels of connected technology or telematics [TZ-328].

Telematics 1.0

These are classic subscription models, such as Opel's "OnStar," which has been discontinued, or Volkswagen's "We Connect."

Telematics 2.0

At this stage, the customer is offered remote maintenance or updates. Sometimes, the vehicle manufacturer covers the communication costs; otherwise, the customer must pay the fee.

Telematics 3.0

The backend is integrated for the associated functions. This connects the vehicle and its systems to the Internet and the world. This way, the above examples can be implemented and offered to the customer.

Telematics 4.0

The next stage is characterized by significantly larger data volumes and increased processing in the vehicle-integrated control units and the backend.

Companies constantly look for further business models to consolidate their dominance at each stage. Consequently, they aim to find additional potential in the after-market or connected cars to integrate customers into new business models. The consulting firm McKinsey estimates that a turnover of around 310 US dollars per vehicle can be generated in this area, and savings of 180 US dollars can be achieved [TZ-329]. The after-market ecosystem comprises car manufacturers, suppliers, service providers, and tech companies. Manufacturers can tap into additional business areas by unlocking existing customer features or using data to offer predictive maintenance for vehicle components.

A vehicle can also be used as a digital wallet, which enables further income sources. Mercedes, for example, has already installed a fingerprint sensor and a credit card in several model series [TZ-330]. This allows the vehicle to be further personalized after purchase or used as a payment method for different services, such as parking fees or booking a hotel room.

3.12   The Future Potential of Connected Vehicles

The journey of connected vehicles is just beginning, and their potential is immense. As this technology develops, connected cars will improve how we travel, interact with our environment, and manage our mobility. One of the most exciting benefits of connected vehicles lies in enhancing safety. With real-time data sharing and advanced communication systems, connected cars can alert drivers to potential hazards, provide collision warnings, and proactively help avoid accidents.

Car connectivity also improves efficiency and convenience. By integrating with smart city infrastructure, connected vehicles can optimize routes, reduce traffic congestion, and reduce the overall energy consumption caused by traffic. Shorter travel times, lower emissions, and a more sustainable environment are expected.

Besides, connected car technology offers unparalleled levels of personalization. Connected cars can learn about drivers' habits and preferences and provide a personalized, tailored driving experience. The possibilities seem endless, from adjusting seat positions and climate controls to suggesting preferred routes and entertainment options.

Connected cars are also likely to be essential to the future of autonomous cars. As part of the complexity of traffic with hundreds of participants on every road, proactively reacting to the dynamic traffic environment appears indispensable. Automatically and permanently observing the behavior of other traffic participants, ensuring compliance (such as avoiding traffic tickets), and predictive reactions to complicated or even dangerous road situations and road conditions—such as a sudden onset of heavy rain or hail—can prevent incidents or even save lives. An autonomous car embedded in its environment and connected with other cars and the infrastructure of the cloud and other vehicles will be an integral part of the autonomous car safety concept.

Integrating connected vehicles with the Internet of Things (IoT) opens up nearly boundless possibilities. Connected cars can communicate with other devices like home automation systems to create a seamless life experience. For example, a car could communicate with the driver's home to adjust the thermostat as they approach or send notifications about maintenance needs directly to the driver's smartphone.

The potential of connected vehicles is transformative. In the future, connected and autonomous cars will become ubiquitous. The car will be electric, interconnected, comfortable, safe, and individualized.

References

[TZ-301] https://www.geotab.com/de/blog/was-sind-vernetzte-fahrzeuge/ Accessed on 07.10.2023

[TZ-302] https://www.geotab.com/de/blog/was-sind-vernetzte-fahrzeuge/ Accessed on 31.07.2023

[TZ-303] https://www.elektronikpraxis.de/367-mio-autos-sind-bis-2027-mit-dem-internet-ver bunden-a-c3b79b0af4c15fdafe338e744e60ad5e/ Accessed on 27.10.2023

[TZ-304] https://www.testfeldmonitor.de/Testfeldmonitoring/DE/testfeld/testfeld_node.html Accessed on 08.10.2023

[TZ-305] https://www.adaptive-ip.eu Accessed on 23.10.2023

[TZ-306] https://www.bmwk.de/Redaktion/DE/Downloads/H/hochautomatisiertes-fahren-auf-autobahnen.pdf%3F__blob%3DpublicationFile%26v%3D1 Accessed on 23.10.2023

[TZ-307] https://www.deutschland-spricht-ueber-5g.de/informieren/netzausbau/europa-macht-mobil-der-5g-aktionsplan-der-eu/ Accessed on 29.10.2024

[TZ-308] https://edpb.europa.eu/sites/default/files/consultation/edpb_guidelines_202001_con nectedvehicles.pdf Accessed on 07.10.2023

[TZ-309] https://www.automotiveit.eu/technology/diese-pflichten-kommen-auf-autohersteller-zu-879.html Accessed on 08.10.2023

[TZ-310] https://www.kba.de/DE/Themen/Typgenehmigung/Informationen_TGV/Informati onen_TGV_node.html;jsessionid=E1EE9FF632D4D4B68D1072C49E062E93.liv e21304 Accessed on 23.10.2023

[TZ-311] https://www.volkswagen-newsroom.com/de/pressemitteilungen/leistungsfaehiger-und-schneller-volkswagen-hebt-die-sprachbedienung-im-golf-auf-neues-niveau-7698 Accessed on 31.07.2023

[TZ-312] https://www.audi-mediacenter.com/de/pressemitteilungen/audi-integriert-store-fuer-apps-in-eine-vielzahl-seiner-modelle-15204 Accessed on 07.10.2023

[TZ-313] https://onestore.nokia.com/asset/206173 Accessed on 27.10.2023

[TZ-314] https://www.intel.de/content/www/de/de/transportation/smart-road-infrastructure.html Accessed on 27.10.2023

[TZ-315] https://intelligente-welt.de/forschung-fuers-autonome-fahren-auto-daten-in-der-cloud/ Accessed on 01.08.2023

[TZ-316] https://www.here.com/platform Accessed on 01.08.2023

[TZ-317] https://esyncalliance.org/ Accessed on 07.10.2023

[TZ-318] https://www.pressebox.de/inaktiv/auroralabs-gmbh/Aurora-Labs-und-Infineon-Technologies-praesentieren-gemeinsame-Loesung-fuer-KI-gesteuerte-sichere-und-vollstaendig-redundante-Over-The-Air-Software-Updates-ohne-Ausfallzeiten/boxid/1140953 Accessed on 07.10.2023

[TZ-319] https://www.verbraucherzentrale.de/wissen/reise-mobilitaet/unterwegs-sein/ecall-so-funktioniert-das-automatische-notrufsystem-im-auto-32100 Accessed on 08.10.2023

[TZ-320] https://assets.adac.de/image/upload/v1595919606/ADAC-eV/KOR/Text/PDF/Umfrage_Hersteller_Car2X_dl45xm.pdf Accessed on 08.10.2023

[TZ-321] https://www.audi.com/de/innovation/autonomous-driving/car-to-x.html Accessed on 08.10.2023

[TZ-322] https://www.elektrobit.com/newsroom/elektrobit-and-airbiquity-partner-for-next-gen-over-the-air-services-in-automotive-mobility/ Accessed on 07.10.2023

[TZ-323] https://support.apple.com/de-de/HT213839 Accessed on 08.10.2023

[TZ-324] https://www.press.bmwgroup.com/deutschland/article/detail/T0338036DE/bidirektionales-lademanagement-%E2%80%93-bdl-%E2%80%93-pilotprojekt-startet-die-wichtigste-phase:-kunden-testen-erste-elektrofahrzeuge-die-ihren-strom-zurueckspeisen-koennen?language=de

[TZ-325] https://newsroom.porsche.com/de/2022/produkte/porsche-taycan-pufferspeicher-stromnetz-vehicle-to-grid-anwendungen-27527.html Accessed on 08.10.2023

[TZ-326] https://www.lpb-bw.de/smart-city#c56712 Accessed on 08.10.2023

[TP-327] https://www.autobild.de/artikel/google-audi-und-volvo-mit-android-im-cockpit-11637331.html Accessed on 23.10.2023

[TZ-328] https://intelligente-welt.de/geschaeftsmodelle-rund-ums-vernetzte-fahrzeug-und-5g/ Accessed on 28.10.2023

[TZ-329] https://www.cloudcomputing-insider.de/datenzentrierte-geschaeftsmodelle-im-connected-car-a-7a9131dab76cec9c72e6701beee738e6/ Accessed on 08.10.2023

[TZ-330] https://www.autobild.de/artikel/neuwagen-in-car-payment-bezahlvorgang-option-digital-bargeldlos-22598379.html Accessed on 08.10.2023

[TZ-331] https://www.3gpp.org/ Accessed on 10.12.2023

[TZ-332] https://www.capgemini.com/de-de/solutions/software-framework-solutions-for-automotive-connectivity/ Accessed on 16.12.2023

[TZ-333] https://www.energieloesung.de/magazin/was-ist-bidirektionales-laden-und-wie-funktioniert-das-genau/ Accessed on 16.12.2023

[TZ-334] https://bidirektionale-wallboxen.de/kfw-442-liste-foerderfaehige-wallboxen-v2h-v2g/ Accessed on 16.12.2023

[TZ-335] https://www.springerprofessional.de/smart-cities/mobilitaetskonzepte/das-muessen-sie-zu-smart-cities-wissen/23477242 Accessed on 16.12.2023

[TZ-336] https://owners.kia.com/us/en/privacy-policy.html#two Accessed on 15.07.2024

[TZ-337] https://www.microgridknowledge.com/electric-vehicles/article/33015536/bidirectional-charging-gains-ground-with-toyota-project-and-federal-california-bills Accessed on 15.07.2024

[RM-042] Car IT compakt, Volker Johanning/Roman Mildner, ISBN: 978-3-658-09967-1

Quality in the Automotive Industry—From Product to Process

4

Abstract

This chapter comprehensively overviews the automotive industry's quality standards and safety requirements in the context of automotive systems development. It traces the evolution of process improvement methodologies from CMM to CMMI and Automotive SPICE, examining their impact on development processes. The chapter focuses on critical safety and security standards, including ISO 26262 for functional safety, ISO/SAE 21434 for cybersecurity, and ISO 21448 for Safety of the Intended Functionality (SOTIF). It concludes by discussing the integration challenges of these various standards and their critical role in ensuring vehicle safety and reliability in an increasingly complex automotive landscape.

In the early days of the automotive industry, the focus was more on producing the required quantities. The introduction of assembly line work led to the first large-scale productions, such as the Model T at Ford. Over time, individualists had to make way for mass production, in which creativity was less critical than conformity. Rules, norms, and industry standards were the logical consequences. This chapter will examine essential standards pertinent to systems development and the associated improvement and assessment of the development process in the automotive industry.

4.1 Norms and Standards

Because of its extensive history, the enormous variety of standards and norms in the automotive industry is legendary. However, the development of standards has been gradual. The first industry standards were published as early as 1905 in the USA by the Society of Automotive Engineers (SAE). Ford drew on the

© The Author(s), under exclusive license to Springer Fachmedien Wiesbaden GmbH, part of Springer Nature 2025
R. Mildner et al., *Car IT Reloaded*, https://doi.org/10.1007/978-3-658-47691-5_4

first standards and then significantly expanded them. The first DIN standard in Germany, concerning so-called "taper pins", was published in 1918.

Over time, international standards have become increasingly important, with standards of all stripes and nationalities published regularly. The number of standards is, therefore, constantly increasing. Examples of organizations that publish national and, increasingly, international standards include:

- VDA: German Association of Automotive Industry (*Verband der Automobilindustrie*), founded in 1901; the first standards, however only published in the middle of the twentieth century
- IEC: International Electrotechnical Commission, founded in 1906
- ISO: International Organization for Standardization, founded in 1947
- ECE: Economic Commission for Europe, founded in 1947
- UNECE: United Nations Economic Commission for Europe, founded in 1947
- CEN: European Committee for Standardization, founded in 1961
- IEEE: Institute of Electrical and Electronics Engineers, founded in 1967 (emerged from the American Institute of Electrical Engineers (AIEE), founded in 1884)
- AIAG: Automotive Industry Action Group, founded in 1982.

This list of standards organizations is incomplete. It could also include the International Automotive Task Force (IATF), American National Standards Institute (ANSI), International Electrotechnical Commission (IEC), etc. Each of these bodies publishes many standards, the exact number of which is unknown. A rough estimate suggests that the number of standards relevant to car manufacturing in Germany, the EU, and the USA alone most likely exceeds 100,000.

Our book focuses on a small selection of norms and standards that primarily relate to systems development, project management, cybersecurity, and functional safety.

4.2 Process Improvement

Process improvement in system development aims to achieve the required product quality through higher-quality processes. This chapter examines the importance of quality in product development for automotive platforms and system components. Special attention is paid to software development. Exploring the historical background of the so-called "software crisis" is essential to understanding the challenges facing the European and global automotive industry.

4.2.1 Everyone is Talking About the "Software Crisis"

The subject of software is particularly puzzling for readers who have stumbled into the industry unprepared and are now wondering why commercial software development seems so tricky. Anyone who flippantly claims that software 'can't be that difficult after all' is in for a rude awakening., this happens not only to career changers but also to computer scientists and engineers who previously worked in less software-heavy industries.

The criticality of software systems becomes apparent when legendary software disasters occur. The book *Bits and Bugs: A Scientific and Historical Review of Software Failures in Computational Science (Software, Environments, and Tools)* [RMI-011] is a fantastic resource in which the authors document a long list of catastrophic software failures. Some disasters have direct or indirect organizational backgrounds; programming errors caused others. Here are just a few examples to illustrate how critical software quality is.

Explosion of Ariane 5: The European Space Agency's (ESA) Ariane 5 rocket exploded a few seconds after take-off in 1996. Reason: an incorrect conversion of a floating-point number. Result: destroyed satellites and rockets. Damage: approximately six hundred million euros from losing the rocket and its cargo [RMI-011].

Error in the Patriot missile defense system: This system aimed to destroy cruise missiles before they could reach their targets. During the Gulf War in 1991, the system failed sporadically. The root cause was only discovered after some time: a software error in calculating the internal system clock time. Twenty-eight soldiers lost their lives as a result of the error [RMI-011].

Irradiation overdose with the Therac-25 system: An error in the Therac-25 X-ray irradiation device caused a massive overdose of X-rays. Under certain circumstances, the device operated with a 100-fold irradiation intensity. One reason for the faulty status was an incorrect display on the device's surface and an overflow in an 8-bit variable. Three patients paid with their lives, and at least two others suffered serious physical injuries because of these software errors [RMI-012].

Boeing 737 MAX crashes: Two aircraft of this type crashed in 2018 and 2019. It was only after the second crash that the flight authorities prohibited operations. The MCAS system, which Boeing had developed for automatic flight stability, refused the pilot's command to lower the nose of the aircraft. The fatal consequence was a total loss of control of the aircraft. In addition to inadequate pilot training, the root cause was a faulty calculation. The error occurred when a so-called AOA (Angle-of-Attack) sensor failed. A second redundant sensor was not provided. As a result, the system calculated an incorrect flight angle. A total of 356 people died. In addition, the economic damage to Boeing amounted to tens of billions of US dollars, with several pending lawsuits [RM-011].

Problem with the 2014 electric power steering from General Motors: The US manufacturer recalled 1.3 million vehicles in 2014 because of a software problem that could cause a sudden failure of the electric power steering. While there were some accidents and injuries related to this problem, it is not clear how many of these incidents were directly attributable to it.

Given such dramatic events, partly or entirely caused by software errors, the fundamental question arises—regardless of the industry—of why software seems such a tough nut to crack. There are an estimated 30 million software developers worldwide [RM-144], almost one million in Germany alone, yet there never seems enough. The hunger for more software seems insatiable, yet again and again, companies complain about a shortage of skilled workers in IT professions [RM-145]. So, on the one hand, software is a frequent source of sometimes serious accidents, and on the other, the call for more software continues unabated.

The following sections attempt to demystify the enigmatic "software crisis" and how the global industry has attempted to solve this "software challenge."

4.2.2   Early Attempts to Relieve the "Software Crisis"

The "software crisis" was first proclaimed at a NATO conference in 1968. At this conference, the participants recognized that software was increasingly becoming a critical discipline that required greater attention due to its growing complexity. The following insights were published:

- Lack of reliability in data systems.
- Difficulties in adhering to schedules and specifications for extensive software projects.
- Inadequate training of software engineers.
- High systems prices associated with software and hardware.
- Software development is a young discipline that requires further, more intensive research.
- Software production needs to catch up with hardware.
- The management of software projects will only improve once a deeper understanding of the software discipline has been achieved.
- Software development needs more risk management, especially when introducing new concepts and methods.
- Software complexity must be recognized and acknowledged.
- Lessons must be learned from past projects to improve software systems.

Although more than half a century old, this summary of software-related challenges still seems surprisingly relevant. Today, the call for quality in software projects is greater than ever. One might think nothing has been done about this "software crisis," but this impression is deceptive. Scientists, engineers, and practitioners have implemented countless improvements and developed innovative concepts, some of which have already lost their relevance. A few examples of past software "hypes" prove this:

- Object-oriented programming/design/analysis
- New database concepts, such as RDBMS (SQL), object-oriented databases (such as ObjectDB or older ObjectStore or POET)
- Documentation concepts such as flow charts, structure diagrams, UML/SysML

- Hundreds of now exotic or no longer used programming languages like Fortran, Basic, Algol, Prolog, Miranda (functional programming languages), Pascal, Modula/Modula 2, Lisp, Smalltalk, and many more
- New mainstream, procedural programming languages like C, C++, C#
- Various runtime paradigms, such as compiled languages (C, C++) versus interpretive programming languages (Java, C#) or scripting languages (JavaScript, Python, PHP)
- RAD (Rapid Application Development) development environments like SQL-Base (Gupta SQLBase), Optima++ (Power++), and *Rapid Prototyping*
- Diverse process models and methodologies: waterfall, iterative waterfall, V-model, and lightweight methods such as agile development
- Design patterns for software architecture
- Numerous innovative IDEs (Integrated Development Environments) for faster programming, such as Visual Studio, Eclipse, NetBeans, etc.
- DevOps and continuous integration and collaborative cloud solutions, such as GitLap and GitHub
- Test concepts such as unit tests (e.g., JUnit) and brute force test systems such as Polyspace
- Tools like HIL and SIL
- Specialized analysis tools like ETAS INCA
- Code analyzer and reverse engineering tools, such as Understand from Scientific Toolworks, Inc.
- Model-based development tools such as Simulink
- Neural networks (software heuristics)
- RESTful API concepts for web integrations
- Architecture concepts such as AUTOSAR and SOA (Service-Oriented Architecture)
- Distributed object-oriented architectures such as CORBA
- Massive-parallel algorithms, threading, multi-processing
- Various debugging tools and profilers
- Low-code and no-code concepts, making it easier for non-experts to develop software (a reincarnation of the old RAD concept).

Last but not least, artificial intelligence (AI) is also software-based. Since the success of OpenAI's ChatGPT, numerous development tools have been released to accelerate software development further. Software has grown into an enormously diverse field of knowledge.

Older students still remember the punch strips and punches needed to punch cards. You then had to take them to the computer lab so a technician in a white coat could execute them. You received the result the following day.

Given the rapid development in software and computer technology, astonishment about this seems quite justified, raising the question: why can't we still get it right with software?

A comparison can illustrate the answer. If mechanical engineering had developed at the same speed as software engineering, we could buy cars today that travel near the speed of light, cost just a few euros, and fit on a matchstick.

Software systems are developing rapidly, and the pace of development continues to accelerate. Artificial intelligence is software made possible by hardware breakthroughs. When quantum computers become commercially available, the general AI described in Chap. 2 will only be a matter of time.

Therefore, it is hardly surprising that academics and engineers sometimes struggle to keep up with this breathtaking pace. This also applies to indirectly related disciplines, such as project management and process improvement.

If the saying that nothing is more constant than change needs proof, this applies to software.

We don't have a "software crisis"; we have a knowledge crisis that dwarfs everything else because of the rapid development of software and computer technology. This crisis can only be solved if we can endure the software change out of necessity and actively drive it forward. However, it should be noted that technological development only progresses with risks. In particular, the processes in the development and production of new technologies must be raised to a systematic level. Progress must be made carefully, and improving process quality is a popular measure to achieve this goal.

4.2.3 Alan Deming, *Total Quality Management,* and Process Quality

Many decades ago, approaches were already being taken to anchor quality as an organization's long-term goal. Alan Deming was a forerunner of this principle. His basic idea was that quality can only be guaranteed long-term if it is anchored and managed as a systematic goal in the organization. Another insight was that everyone in the organization contributes to this, including suppliers. Customers must be engaged, too, to ensure long-term market alignment of products and services.

Deming has described this organizational change based on 14 points:

1. Create constancy of purpose for improving products and services—to remain competitive and create jobs.
2. Implementation of this new philosophy in the organization.
3. Do not try to achieve quality only through mass inspections. Instead, promote the quality of your product throughout the entire process.
4. Do not base your contracts on the lowest price. Instead, minimize overall costs by working with carefully selected suppliers over the long term.
5. Improve planning, production, and service processes to increase quality and reduce costs.
6. Invest in your employees and offer them training opportunities that will help them succeed.

7. Create a management system that supports people and machines in doing their best work.
8. Take away employees' fears so that everyone can concentrate on contributing to the company's success.
9. Break down barriers between different departments. Team thinking is necessary to anticipate problems.
10. Avoid performance targets based on not being allowed to make mistakes. The system makes mistakes that are rarely within the employees' control.
11. Abolish numerical targets for staff and management.
12. Remove barriers that diminish people's pride in their work. Line managers should base performance appraisals on the quality of each individual's contribution.
13. Develop an intensive training program for all employees that motivates self-improvement.
14. Involve the entire workforce to enable change.

Deming succeeded in Japan in the 1950s with this theory and philosophy. Some American companies only adopted these principles in the 1980s after seeing the success of Japanese automobile companies in their own country.

Looking at this philosophy from today's perspective, many of these 14 points are sadly still not anchored in the systems of car manufacturers and suppliers. Often, they only want to get crucial projects to the finish line and forget about the overarching strategy and employee motivation (see also examples in the chapter on VDA Automotive SPICE®).

Audits and maturity levels are often conducted purely to create a paper trail. One of our favorite anecdotes in the assessment guild was a statement from a quality representative of one of our authors' superiors after the organization had reached "Maturity Level 3" (CMMI): 'From now on, quality is no longer the focus!'.

Looking at point 3, malfunctions are still usually only noticed during testing. If we were to take care of quality in the development process (for example, through better employee training, greater sensitization, deeper motivation, and, above all, a continuous and reliable strategy), we would save a lot of senseless and frustrating work, resources, time, and costs. One might think that learning from our past mistakes would be an excellent idea, but in reality, such crucial feedback loops, although essential for improvement, are usually ignored.

Regarding suppliers and point 4, the buyers almost always decide when contracts are awarded; the focus is clearly on the price (the so-called "best price") and not primarily on the supplier's expertise or business value.

Point 9 faces a similar fate: responsibility is spread across numerous departments or groups when tackling a large project today. Here, the project manager constantly battles with line managers to complete the project. Responsibilities are diluted once everyone seems to be responsible for something.

And so, while Deming argued for long-term quality and sustainable organizational success, our industry *still* often prioritizes short-term gain over sustainable practice. This is a method-agnostic, universal observation. It does not matter if we look at recent ideas, such as agility, or the traditional process improvement principles. The point few seem to be willing to internalize is the fundamental insight that quality is not achieved through superficial audits, more complex and cryptic standards, or cost-cutting measures but by nurturing and embedding a culture of continuous improvement, responsibility, and collaboration at every level. These are values that, decades later, remain as relevant—if not more so—than ever.

4.3   Process Models and Maturity Levels

Process models and ideas are tools for achieving quality improvement by reviewing and auditing them to define and implement rational process improvement measures. This chapter will examine process models and methods for determining process maturity levels. We will examine earlier systematics, such as CMM, to provide a broader context for today's assessment models.

4.3.1   The Challenge of Process Quality: From CMM®
to Automotive SPICE®

The development of complex systems is not only a technical and organizational challenge but also entails a significant risk in terms of the so-called "project management triangle" (also known as the "magic triangle" or, jokingly, the "Bermuda Triangle") (Fig. 4.1).

These aspects are a matter of negotiation at the beginning and during project implementation, at least in principle. If, for example, the customer demands more or different services during the project, the costs could be structured differently, or a more extended timeline could be negotiated. In the automotive supply industry, however, this is not usually possible, as the end customer (OEM, such as Volkswagen, etc.) cannot allow any deviations for a single project, as the dependencies between hundreds of components in a vehicle platform generally do not

Fig. 4.1 The "Bermuda triangle"

allow for any variability (see also Chap. 5 on the challenge of automotive project management).

For this reason, rigorous management of project risk is critical. The aim is to reduce the project risk as much as possible. This is particularly important when a manufacturer or government agency needs to work with many suppliers to deliver a large project. Each of these components can potentially have a high level of complexity. Therefore, the project risk for each of these sub-projects must be minimized so the overall risk for the project can be minimized. However, selecting potential suppliers is a huge challenge, which becomes particularly critical if there is no uniform standard to evaluate these suppliers and then select as rationally as possible the one most suitable to deliver a given component according to specification.

As early as the 1960s, there was a trend toward installing and integrating mechanical, electrical, and increasingly complex electronic systems. One particular problem was the new discipline of software, which appeared particularly complex and could no longer be tamed using traditional project management methods. Hence, the buzzword "software crisis" began to appear in the rounds. During the Cold War, this situation became a national security problem. It was unacceptable that the Soviets could become faster with their increasingly complex systems than the Western alliance.

Software was identified as the most significant problem, and it was no longer acceptable that coding was still viewed as an "art." When a piece of software could no longer be coded by one or a few "heroes" due to its growing size, something had to be done urgently. For this reason, the US Department of Defense (DoD) decided in 1986 to commission a suitable uniform project quality standard (Fig. 4.2).

CMM—a Capability Maturity Model—was born.

4.3.2 CMM®

Development of the Capability Maturity Model (CMM) was significantly influenced by Watts S. Humphrey, who was involved in founding the Software Engineering Institute in Pittsburgh in the 1980s. This model marked a decisive step forward in solving the software crisis. At the time, CMM provided an innovative framework for process maturity and process improvement. Although it has been replaced by newer models such as Automotive SPICE (ASPICE), knowledge of CMM makes it easier to understand the background against which models such as ASPICE are still used. In the following section, we will shed light on the history of CMM.

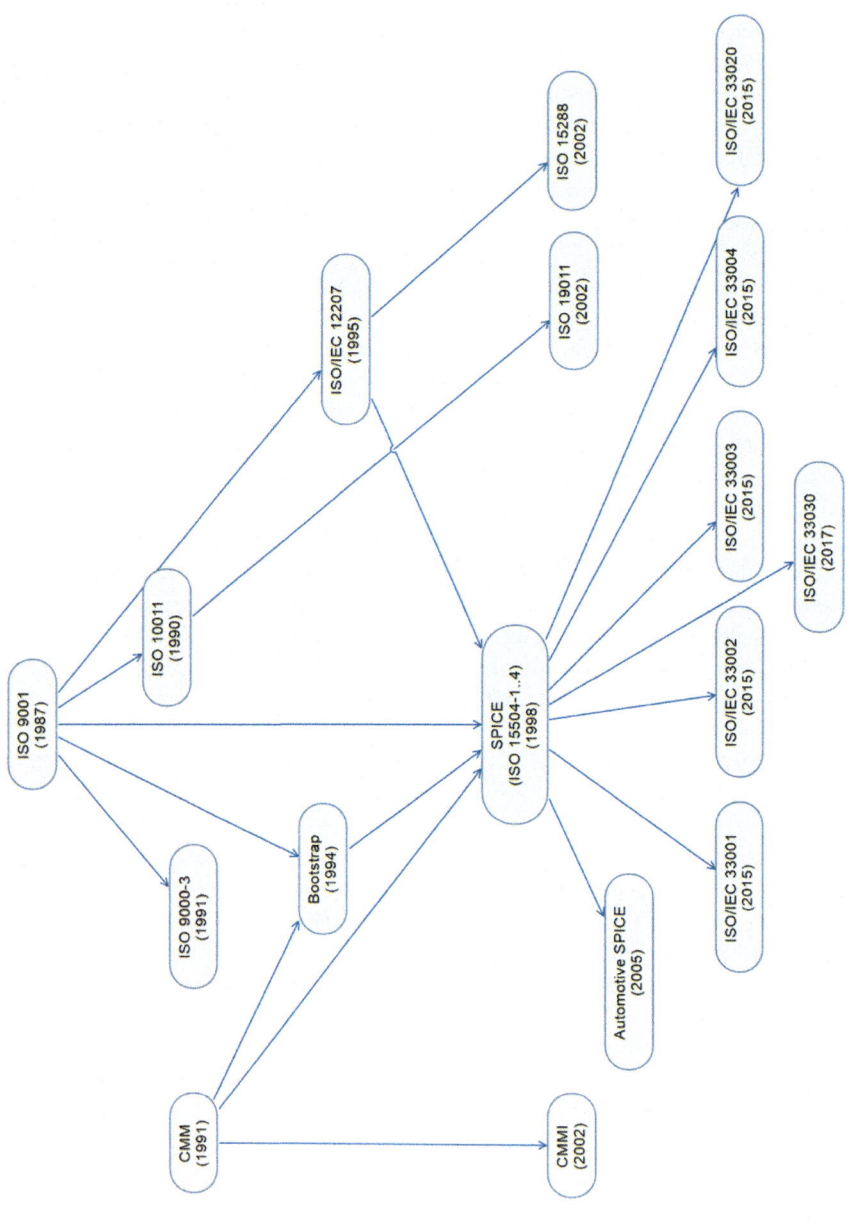

Fig. 4.2 Development and interrelationships of standards and models

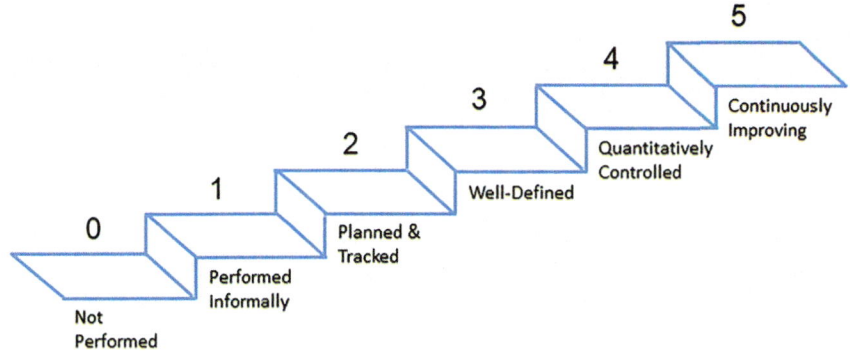

Fig. 4.3 The CMM level model

4.3.2.1 Overview of the CMM® Framework

The Capability Maturity Model (CMM 1.0) was designed by the SEI (Software Engineering Institute) at Carnegie Mellon University on behalf of the US Department of Defense. The fully defined version of CMM was developed from 1986 to 1991 and formally released under the name SE-CMM [RM-002].[1]

Even though CMM is now quite outdated and has long since been replaced by new process models of this type (such as VDA Automotive SPICE®), it is a rewarding subject to study, as it still reflects the basic ideas of this type of assessment model.

CMM, like the current Automotive SPICE assessment model, was designed as a level-based framework with maturity levels (capabilities) (Fig. 4.3).

The concept was that the first level ("zero") represents a chaotic, unpredictable project. In contrast, the fifth level represents the pinnacle of development—a near-perfect, almost utopian process where everything runs perfectly.

The stages of the CMM 1.0 model are illustrated in Table 4.1.

Level 0: Not Performed

There needs to be planning and a realistic estimate of the effort involved. Everything depends on randomly available "heroes." Deadlines are not met as they are typically not even set. It is unclear what needs to be done and in what order. Also, no one seems to be accountable for anything. When the software is delivered to the customer, it is unknown what it is and whether anyone requested it. If it does not work, the reason for the software errors is not apparent.

Conclusion: Chaos and despair reign in the project.

---

[1] A little-known predecessor of CMM was the "Stages of Growth" model developed by Richard L. Nolan in 1973. However, CMM was not comparable with this model.

Table 4.1 CMM maturity levels

	Maturity (capability)	Definition
0	Not performed	(Undefined)
1	Performed informally	– Base practices performed
2	Planned and tracked	– Base practices performed
3	Well-defined	– Committing to perform – Planning performance – Disciplined performance – Tracking performance – Verifying performance
4	Quantitatively controlled	– Established measurable quantity goals – Determined process capability to achieve goals – Objectively managing performance
5	Continuously improving	– Establishing the quantitative process – Effectiveness goals – Improving process effectiveness

Level 1: Performed Informally

It is clear what needs to be delivered and how this should be done. Necessary core plans, such as configuration management or project plans, have been created sensibly, although they have only sometimes been adequately coordinated. Software requirements are in place. A test plan is also in place, and the test cases are executed according to the previously agreed requirements. However, if the customer "just happens" to want to change the project scope (such as new requirements), these requests are sometimes implemented without consulting other project team members. You often hear: 'We'll probably never learn.' Deliveries are usually made according to the principle of hope: sometimes on time, sometimes not, and it is not entirely clear what it depends on.

Responsibilities in the project still need to be effectively defined. Required project resources, such as qualified staff, test benches, licenses, programming tools, etc., are not planned. A "heroic" work effort continues to be demanded, often in the form of overtime. Key employees are called back from vacation because only they know "how to do it." And woe betide them if a key employee suddenly falls ill or resigns. In such a project, new developers are often hastily thrown into the project on behalf of the desperate customer.

Conclusion: The team collectively knows "how to do it" but is always surprised anew.

Level 2: Planned and Tracked

In addition to the process characteristics described at the first maturity level, all project activities are known and planned. Deliveries are systematically checked in advance. Project roles are defined and distributed and are adequately practiced. Project resources are procured on time and as required, and the team is trained if necessary. A project budget is available, and adherence to it is monitored. Project defects (bugs) are systematically managed and fixed. This also applies to all other critical work products, such as requirements, test cases, technical documentation and drafts, project plans, etc. A quality representative (quality manager) independent of the project is assigned as a control body from this maturity level onwards to ensure that a minimum level of process quality can be maintained objectively. Progress is systematically reported to the defined project participants and the customer. All project risks are known, and their responsibilities are defined. Customer change requests are only implemented with a formal impact analysis. The project delivers according to plan, within budget, of appropriate quality, and within the time agreed upon with the customer. This is monitored using appropriate, predefined metrics.

However, there is no project management standard for new projects. Thus, in level 2 organizations, new projects are often organized from scratch. The entire organizational effort must be repeated. Such organizations must frequently reinvent when a new project is due.

Conclusion: The project runs like a well-oiled machine, but it is still questionable if the next one will also work.

Level 3: Well-Defined
In addition to the process characteristics described at the second maturity level, there is a well-defined process at the third maturity level. Before a project is started, its nature is determined. For example, if functional safety ("FuSa") is relevant, software/hardware/mechanical parts are required, and it is a customer project, specially prepared project templates are used. A suitable process description is available for this purpose. However, suppose the project is an internal software development project in which end customer data will be managed. In that case, only software-relevant templates without FuSa—but with cybersecurity content— are used. In other words, different process configurations are predefined, and the process of "tailoring" is carried out based on defined rules ("tailoring rules"). For each project, all procedural instructions, templates, process descriptions, checklists, and so on, appropriate to the project type, are already agreed upon and approved in advance so these "project assets" can and must be used immediately. The process definition is not developed during the project from scratch so that a new project can start without delay. Such a development process is officially known, supported, approved at all management levels, and known to all project stakeholders. The process is, therefore, "institutionalized," meaning that, in project management, these process artifacts are always used the same way for all projects within the organizational scope (e.g., department or division). All project staff are trained using the same specially created training materials. In addition, key process indicators (KPIs) are set up in all process areas, for example, by systematically counting and categorizing bugs so that project management can provide information at all times.

Conclusion: In such businesses, project organizations are standardized, and project processes are defined and practiced the same way everywhere.

Level 4: Quantitatively Controlled

In addition to the process characteristics defined at the third maturity level, process quality is quantitatively defined and systematically monitored. This makes it clear when the process is not running as expected. The reasons for this are analyzed and substantiated with objective data. If necessary, the project organizations are then improved across the board. For example, the "process assets"—checklists or work instructions—are revised and approved by senior management. Quality management ensures that the process descriptions are appropriately managed.

Conclusion: In such organizations, it is clear which processes should be improved, and which process weaknesses should be eliminated.

Level 5: Continuously Improving

In addition to the process characteristics described at the fourth maturity level, continuous improvement of all processes is ensured at this highest maturity level. It is a proactive "metaprocess" that enables ever-greater project management efficiency in all projects. This is achieved through a "causality analysis" in which process deviations are cataloged, analyzed, and systematically managed. As process errors can now be planned as tasks, a "roadmap" is drawn up in such cases, which is constantly developed and published within the organization.

Conclusion: This is the ideal state of project bliss: a "process zen." We know how good we are; we constantly improve and can even plan for when we will be even better.

The six-level model was adopted in later variants and used in the current Automotive SPICE® process model (capability level 0 to capability 5).

The aforementioned Base Practices (BP) are used as a concrete, fundamental structure from the first maturity level onwards. They are structured in process areas (PA, also called KPA—Key Process Areas) and divided into three blocks: Engineering, Project, and Organizational (see Table 4.2).

Several Base Practices (BP) are defined for each process area. We will only demonstrate one to give you a feel for the overall model. The details are in a PDF [RM-02].

Example:

PA 02: Derive and Allocate Requirements

PA 02 analyzes customer requirements to derive a more detailed and precise set of software designs. Traceability is also required.

Several base practices are defined in PA 02 (see Table 4.3).

If the above content seems incomprehensible, it is not due to the language but rather the often difficult-to-digest way such standards are formulated. Both CMM® and derived standards such as CMMI®, ISO 15504 "SPICE," and VDA Automotive SPICE® use jargon that makes the intention of such documents difficult to understand. Therefore, in the right-hand column of the table above, we have

Table 4.2 Process areas

Engineering areas	Project process areas	Organizational process areas
PA 01: Analyze candidate solutions	**PA 08**: Ensure quality	**PA 13**: Define organization's systems engineering processes
PA 02: Derive and allocate requirements	**PA 09**: Manage configurations	**PA 14**: Improve organization's systems engineering processes
PA 03: Develop physical architecture	**PA 10**: Manage risk	**PA 15**: Manage product line evolution
PA 04: Integrate disciplines	**PA 11**: Monitor and control technical effort	**PA 16**: Manage systems engineering support environment
PA 05: Integrate system	**PA 12**: Plan technical effort	**PA 17**: Provide ongoing skills and knowledge
PA 06: Understand customer needs and expectations		**PA 18**: Coordinate with suppliers
PA 07: Verify and validate system		

attempted to "humanize" the practices to reflect at least a fundamental purpose of these base practices in simple terms (see Table 4.3).

The base practices are a fundamental framework with which the first level of maturity can be achieved. It is now possible to define what needs to be done in the project, but how it can be achieved still needs to be determined. Generic practices have been described for this purpose, summarized into standard features for each maturity level (ML) (Table 4.4).

Process objectives were not well-defined in the original CMM document. Instead, the SEI published a separate document in which a process objective was defined for each process in the CMM [RM-003]. The number of objectives and their description are too extensive to quote in full here (see [RM-003], where the document can be viewed as a PDF). Here is, therefore, just an example in which the goals of requirements management are listed:

Requirements Management

Goal 1: System requirements allocated to software are controlled to establish a baseline for software engineering and management use.

Goal 2: Software plans, products, and activities are kept consistent with the system requirements allocated to software.

However, these goals are incompatible with the process area "PA 02 Derive and Allocate Requirements." Unfortunately, while further developing the CMM model, the SEI published another document called "Capability Maturity Model for Software," which describes the common features of the actual CMM model [RM-04], which appears to be partially redundant (Table 4.5).

Table 4.3 CMM *base practices*—example

BP	Simple translation
BP.02.01 Develop a detailed operational concept of the interaction of the system, the user, and the environment that satisfies the operational need	Document interfaces and their interaction with the system
BP.02.02 Identify key requirements that have a strong influence on cost, schedule, functionality, or performance	Describe customer key requirements
BP.02.03 Partition requirements into groups of requirements based on established criteria, such as similar functionality, performance, or coupling, to facilitate and focus the requirements analysis	Sort the customer requirements according to a defined pattern
BP.02.04 Derive, from the system and other (e.g., environmental) requirements, requirements that may be logically inferred and implied as essential to system effectiveness	Identify other requirements that may not have been explicitly specified
BP.02.05 Identify the requirements associated with external interfaces to the system and interfaces between functional partitions	Describe interface requirements
BP.02.06 Allocate requirements to functional partitions, system elements, people, and support elements to support synthesis of solutions	Document the resulting software requirements
BP.02.07 Analyze requirements to ensure that they are verifiable by the methods available to the development effort	Analyze the software requirements in detail
BP.02.08 Maintain requirements' traceability to ensure that lower-level (derived) requirements are necessary and sufficient to meet the objectives of higher-level requirements	Ensure the traceability of requirements (traceability)
BP.02.09 Capture system and other requirements, derived requirements, derivation rationale, allocations, traceability, and requirements status	Manage the requirements

Table 4.4 CMM *common features*

ML	Common feature	Generic practices
1	**1.1** Base practices are performed	**1.1.1** Perform the process Base practices are performed
2	**2.1** Planning performance	**2.1.1** Allocate resources Allocate adequate resources (including people) for performing the process area
		2.1.2 Assign responsibilities Assign responsibilities for developing the work products and/or providing the services of the process area
		2.1.3 Document the process Document the approach to performing the process area in standards and/or procedures
		2.1.4 Provide tools Provide appropriate tools to support performance of the process area
		2.1.5 Ensure training Ensure that the individuals performing the process are appropriately trained in how to perform the process
		2.1.6 Plan the process Plan the performance of the process area
	2.2 Disciplined performance	**2.2.1** Use plans, standards, and procedures
		2.2.2 Do configuration management
	2.3 Verifying performance	**2.3.1** Verify process compliance
		2.3.2 Audit work products
	2.4 Tracking performance	**2.4.1** Track with measurement
		2.4.2 Take corrective action
3	**3.1** Defining a Standard process	**3.1.1** Standardize the process
		3.1.2 Tailor the standard process
	3.2 Perform the defined process	**3.2.1** Use a well-defined process
		3.2.2 Perform defect reviews
		3.2.3 Use well-defined data
4	**4.1** Establishing measurable quality goals	**4.1.1** Establish quality goals
	4.2 Objectively managing performance	**4.2.1** Determine process capability
		4.2.2 Use process capability
5	**5.1** Improving organizational capability	**5.1.1** Establish process effectiveness goals
		5.1.2 Continuously improve the standard process
	5.2 Improving process effectiveness	**5.2.1** Perform causal analysis

(continued)

Table 4.4 (continued)

ML	Common feature	Generic practices
		5.2.2 Eliminate defect causes
		5.2.3 Continuously improve the defined process

Table 4.5 Generalized CMM common features

Common feature	Description
Commitment to perform	Commitment to perform describes the organization's actions to ensure that the process is established and will endure. Commitment to Perform typically involves establishing organizational policies and senior management sponsorship
Ability to perform	Ability to perform describes the preconditions that must exist in the project or organization to implement the software process competently. Ability to Perform typically involves resources, organizational structures, and training
Activities performed	Activities performed describes the roles and procedures necessary to implement a key process area. Activities Performed typically involves establishing plans and procedures, performing the work, tracking it, and taking corrective actions as necessary
Measurement and analysis	Measurement and analysis describes the need to measure the process and analyze the measurements. Measurement and Analysis typically includes examples of the measurements that could be taken to determine the status and effectiveness of the Activities Performed
Verifying implementation	Verifying implementation describes the steps to ensure that the activities are performed in compliance with the established process. Verification Implementation typically encompasses reviews and audits by management and software quality assurance

The five *common features* formulated this way represent a basic idea of how a "good development process" should be structured. In simpler terms, they could be paraphrased as follows:

1. Do we seriously want to implement this process?
2. Have we checked whether this is even possible?
3. Basically, it works. Now, we are implementing the process!
4. Is the implementation running correctly? Let's measure it objectively!
5. Let's check whether the process has been fully implemented!

If this approach is consistently followed, a project structured this way is systematically implemented in all its process areas.

It could be argued that only these five common features are needed to implement the full beauty of the project management Bermuda Triangle. We will return to this idea in the chapter on CORE SPICE.

4.3.2.2 CMM® Appraisals

Like all other standards in this chapter, CMM also defines an evaluation system for evaluating conformity with CMM. In CMM, these assessments are called "appraisals." The separate SEI document "CMM Appraisal Framework" (CAF) [RM-005] describes the procedure in detail (Fig. 4.4).

The diagram above shows the appraisal process, which is mainly self-explanatory. The numbers represent the chapters in the CAF document. Each step contains a detailed description of the appraisal process (see [RM-005] for the entire document).

In the first phase, step 3.3, the appraisal's scope is determined, the appraisal team is defined, and the schedule is set. The scope concerns the organizational focus (e.g., department, project) and the expectations regarding the process areas and maturity level. The expected artifacts are also identified and reviewed in advance of the appraisal.

The appraisal itself is carried out in step 3.4. All results are documented in writing, consolidated, and evaluated on the maturity scale from 1 to 5 regarding the achievement of goals (see [RM-003]) in the context of KPA (Key Process

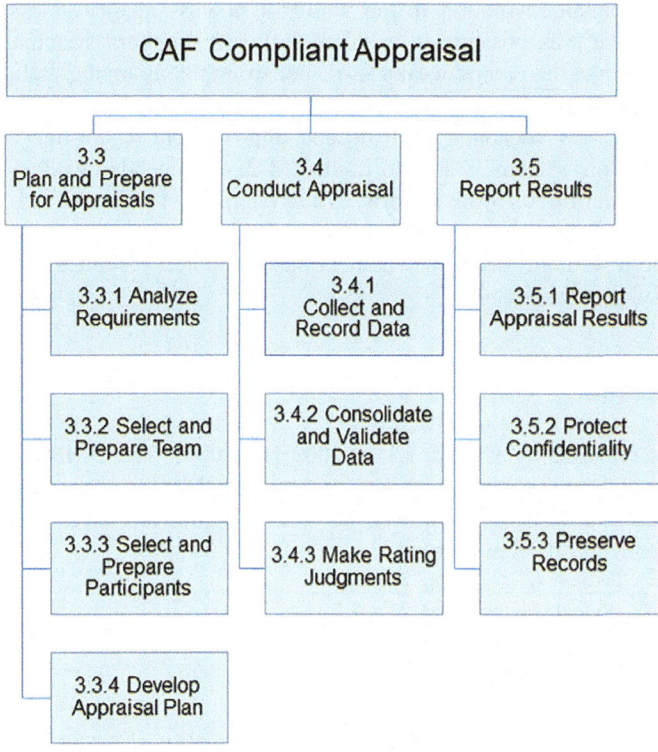

Fig. 4.4 CMM appraisal model

Areas). For this purpose, a complex mapping between process areas and process goals is defined in [RM-005], Appendix B, which is, in turn, divided into four common features.

In the final step, 3.5, the appraisal results are reported and preserved for future use.

In the CAF document [RM-005], the CMM model and the Common Features are brought together and used as the basis for appraisals. In other words, to be able to use the CMM model in a qualified manner, the three documents mentioned must be internalized:

1. The SE CMM model itself [RM-002]
2. Common Features document [RM-003]
3. CAF document [RM-005].

Together, the three documents are sufficient to understand all aspects of CMM. They are approximately 350 pages long in total, which is manageable compared to the successor models.

4.3.2.3 CMM®—Conclusion

CMM was a small revolution in the world of process quality assessment. For the first time, it was possible to evaluate software suppliers systematically. Of course, CMM also had some weaknesses. For example, assessing maturity levels in the context of project goals and common features was not easy to apply. For instance, the CMM vocabulary still needed improvement. Using the term "software" instead of "system" was also fluid. CMM was intended exclusively as an evaluation model for software maturity. The system level—for example, electronics development—was not assessment-relevant. There was no "system" in CMM in the modern sense of this term. The next major "update"—the CMMI—rectified these and other weaknesses.

4.3.3   CMMI®

CMM was replaced by CMMI 1.1 in 2002 [RM-001]. The CMMI 1.0 version was never published as a standard. The structure of CMMI was considerably more complex than that of CMM and, over the years, included more and more different model types with different focuses:

– CMMI for Development (CMMI-DEV)
– CMMI for Services (CMMI-SRV)
– CMMI for Acquisition (CMMI-ACQ).

CMMI-DEV is the core model for assessing the quality of system development processes. Unlike CMM, which focuses solely on software, CMMI theoretically covers all development disciplines. Nonetheless, in practice, it primarily concerns system and software levels.

CMMI-SRV was an attempt to measure the process quality of services (e.g., IT services). It could be viewed as a competitor to ITIL.

CMMI-ACQ is a supplier management model that can be used to measure the quality of procurement processes.

The scope of CMMI was intended to be expanded beyond the three variants. However, this scope expansion has been abandoned, and its practical focus largely remains on system and software processes.

4.3.3.1 CMMI® Structure

We will deal exclusively with the CMMI-DEV in the following, specifically with its latest version, CMMI-DEV v1.3 [RM-007].

First, let's look at the key differences between CMM and CMMI (Table 4.6).

The CMMI structure defines objectives and spans them across process areas (Fig. 4.5).

CMMI provides two dimensions in terms of maturity and capability levels:

– Capability dimension (Continuous)
– Maturity dimension (Maturity levels).

The first dimension primarily supports process improvement activities, while the latter is intended to enable process quality assessment. Unfortunately, CMMI-DEV 1.3 only defines three associated generic objectives (Table 4.7).

Table 4.6 Comparison CMM versus CMMI

CMM	CMMI
Applies only to software	Deals with all development activities, especially systems and software
18 process areas	22 process areas
Only whole maturity levels are assessed	Maturity levels or selected processes can be evaluated (for example, project management and verification, but not validation and configuration management, etc.). This process is referred to as "continuous." For "continuous", only four levels were possible instead of five
There is only one variant from CM: SE-CMM	There are three cuts: CMMI-DEV, CMMI-SRV, CMMI-ACQ
Appraisals were carried out by SCE	Appraisals were carried out in accordance with SCAMPI. Different classes of appraisals are defined: SCAMPI Class A, B, and C (described below)

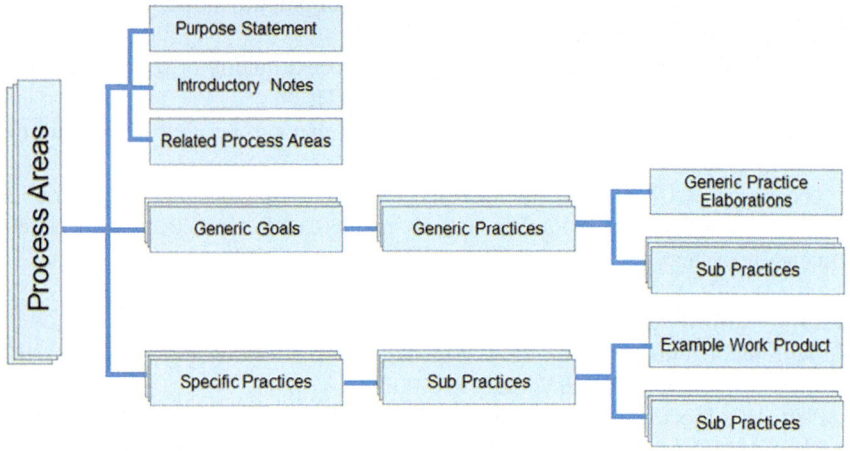

Fig. 4.5 CMMI structure

Table 4.7 CMMI: generic goals and maturity levels

Continuous	Generic goals	Level	Staged, maturity dimension
Incomplete	–	0	–
Performed	GG 1: Performed process	1	Initial
Managed	GG 2: Managed process	2	Managed
Defined	GG 3: Defined process	3	Defined
–	–	4	Quantitatively managed
–	–	5	Optimizing

The two concepts are not easily comprehensible. For this reason, it is advisable to consult other specialist literature, such as [RM-008] or [RM-009], to fully understand CMMI.

In practice, the five-level maturity scale was mainly used, as it was used for the CMMI appraisals, and only these were used as the subject matter in contractual obligations between the client and the supplier. We will, therefore, limit ourselves to the maturity level dimension.

The maturity levels are structured similarly to CMM®, and their semantics are similar (Table 4.8).

The maturity levels are distributed across 22 process areas, which are also divided into process groups:

– Process Management
– Project Management
– Engineering
– Support.

Table 4.8 CMMI maturity levels

Maturity level	Brief description
1: Initial	Chaotic and heroic, deadlines are exceeded, and sometimes successful, but the success cannot be repeated
2: Managed	In addition to the process characteristics of the first maturity level, project plans are created and tracked. Deviations from the plan are resolved, work products are systematically managed, work instructions are developed and followed, and stakeholders are systematically involved. Project successes are repeatable
3: Defined	In addition to the process characteristics of the second maturity level, processes are now well-defined, described, and understood by all involved. Process descriptions, guidelines, and other process documents are set out in writing and are systematically improved. The process definition typically applies to all relevant projects. Certain process sections are determined and used as required (tailoring rules)
4: Quantitatively managed	In addition to the process characteristics of the third maturity level, quantitative targets are defined, and their achievement is systematically measured. The measurement results are evaluated and used for process improvement. The process quality can be predicted on this basis
5: Optimizing	In addition to the process characteristics of the fourth maturity level, processes are constantly and, if necessary, creatively (innovatively) improved. This enables not only continuous but also disruptive process improvement, which can also be planned

The CMMI process areas are more extensive than in CMM. The table below lists all CMMI process areas and their assignment to the maturity levels (Table 4.9).

Therefore, the following process areas are relevant for the second maturity level: CM, MA, PMC, PP, PPQA, and SAM. Other process areas, such as TS (Technical Solution), are optional for the second maturity level. The V-model is also not a mandatory requirement in CMMI. A lifecycle must be defined as a V-model, spiral, waterfall, or something similar.

The individual process areas are not linked by any systematic structure, even if implicit dependencies exist between them. They are all regarded as indispensable, depending on their level of maturity. As CMMI defines 22 process areas, we cannot reproduce them all in this book. We chose one of the more complex process areas to illustrate the CMMI style: TS (Technical Solution).

The purpose of the TS process is defined as follows:

Purpose: Technical Solution (TS) aims to design, develop, and implement solutions to requirements. Solutions, designs, and implementations encompass products, components, and product-related lifecycle processes singly or in combination as appropriate.

The TS process area is shown in the table below (Table 4.10).

Table 4.9 CMMI PA—process areas

Abbr	Process area	Brief description	Category	Maturity level
CAR	Causal analysis and resolution	Identifying the causes of defects and problems and implementing measures to prevent problems in the future	Process management	5
CM	Configuration management	Managing the configuration of the project's work products (configuration management)	Support	2
DAR	Decision analysis and resolution	Analysis of possible decisions through a formal decision-making process	Support	3
IPM	Integrated project management	Establishment and maintenance of interest groups (stakeholder management)	Project management	3
MA	Measurement and analysis	Development and reporting of process-related measured values (KPI—key process indicators) for project, product, and process management	Support	2
OPD	Organizational process definition	Establishment and maintenance of the organization's standard processes (process management)	Process management	3
OPF	Organizational process focus	Management of organizational process improvements	Process management	3
OPM	Organizational performance management	Proactive performance management of the organization to achieve business goals	Process management	5
OPP	Organizational process performance	Building a quantitative understanding of process performance to predict future results and achieve business goals	Process management	4

(continued)

Table 4.9 (continued)

Abbr	Process area	Brief description	Category	Maturity level
OT	Organizational training	Development and support of employee skills (skills management)	Process management	3
PI	Product integration	Integration of product components, ensuring the proper functionality and quality of the integrated product and its delivery	Engineering	3
PMC	Project monitoring and control	Overview of project progress for early identification of deviations from plan and initiation of suitable corrective measures	Project management	2
PP	Project planning	Creation and maintenance of project plans	Project management	2
PPQA	Process and product quality assurance	Provide objective insight into processes and related work products for employees and management	Support	2
QPM	Quantitative project management	Quantitative management of the project to achieve the defined quality and process performance targets	Project management	4
RD	Requirements development	Creation and analysis of customer, product, and product component requirements	Engineering	3
REQM	Requirements management	Requirements management of product and project requirements and ensuring the quality of requirements	Engineering	2

(continued)

Table 4.9 (continued)

Abbr	Process area	Brief description	Category	Maturity level
RSKM	Risk management	Early detection of potential planning problems and definition of risk management measures to minimize negative effects on project objectives	Project management	3
SAM	Supplier agreement management	Supplier management for all services	Project management	2
TS	Technical solution	Selection, design, and implementation of technical solutions based on requirements	Engineering	3
VAL	Validation	Proof that a product or product component fulfills the intended purpose in the intended environment	Engineering	3
VER	Verification	Ensuring that selected work products meet the specified requirements	Engineering	3

Table 4.10 TS—technical solution

SG—Specific goals	Specific practices
SG 1: Select product component solutions	**SP 1.1:** Develop alternative solutions
	SG 1.2: Select product component solutions
SG 2: Develop the design	**SP 2.1:** Design the product or product component
	SP 2.2: Establish a technical data package
	SP 2.3: Design interfaces using criteria
	SP 2.4: Perform make, buy, or reuse analyses
SG 3: Implement the product design	**SP 3.1:** Implement the design
	SP 3.2: Develop product support documentation

TS was often a problematic appraisal topic, as this process area was very broadly defined. Therefore, the range of interpretation in CMMI appraisals was quite broad. One reason CMMI struggled to establish itself long-term was the lack of clear design area specifications.

4.3.3.2 CMMI® Appraisals

The assessment model (SCAMPI—Standard CMMI Appraisal Method for Process Improvement) follows a well-defined assessment process. CMMI offers three appraisal classes: SCAMPI A, B, and C.

SCAMPI A is the most rigorous CMMI assessment scheme. The aim is to systematically review the maturity level in the previously defined scope (e.g., organization, project, etc.). The appraisal may only be carried out by a SCAMPI Lead Appraiser certified by the SEI. The SCAMPI A assessment is usually carried out by a team (and not a single assessor) so that as many different perspectives as possible are given.

SCAMPI B is a less formal and detailed assessment scheme than SCAMPI A. It is used to identify areas for improvement and prepare the organization for a SCAMPI A assessment. SCAMPI B allows for greater flexibility and adaptation to the organization's needs. One or more qualified appraisers typically conduct it to gather initial insight into process maturity and potential areas for improvement.

SCAMPI C is the most flexible and least formal of the assessment schemes. It is used to obtain an overview of strengths and weaknesses in specific areas quickly and with little effort. SCAMPI-C assessments are often less structured and can be carried out by internal teams without formal certification.

From OEMs' perspectives, the SCAMPI A model—the most comprehensive SCAMPI—was the preferred assessment type sanctioned in development project contracts.

4.3.3.3 CMMI®: Conclusion

CMMI was a relatively short but important episode in the saga of process quality assessment methods. It was mainly popular in the US, but many offshore providers could use certified CMMI maturity as a selling point at the fifth CMMI level. The fifth maturity level suggested that an outsourcing company had excellent, constantly improving and optimizing process efficiency. However, the inflation of CMMI Level 5 certified projects suggested that CMMI certification had lost its reliability as a measure of suppliers' capabilities. Among other reasons, Automotive SPICE was supposed to be an answer to this problem.

Figure 4.6 shows the history of CMM-/CMMI-based maturity models.

4.3.4 ISO/IEC TR 15504 (SPICE)

Another capability assessment model, ISO/IEC TR 15504 (SPICE), was published by ISO [BF-005] in 1998 as a "Technical Report" (TR). SPICE stands for Software Process Improvement and Capability Evaluation. The standard was composed of several parts.

Part 2 of the standard defined the requirements for a process assessment model and a process reference model. The assessment model must be based on indicators that prove the achievement of the process attributes of the individual capability levels.

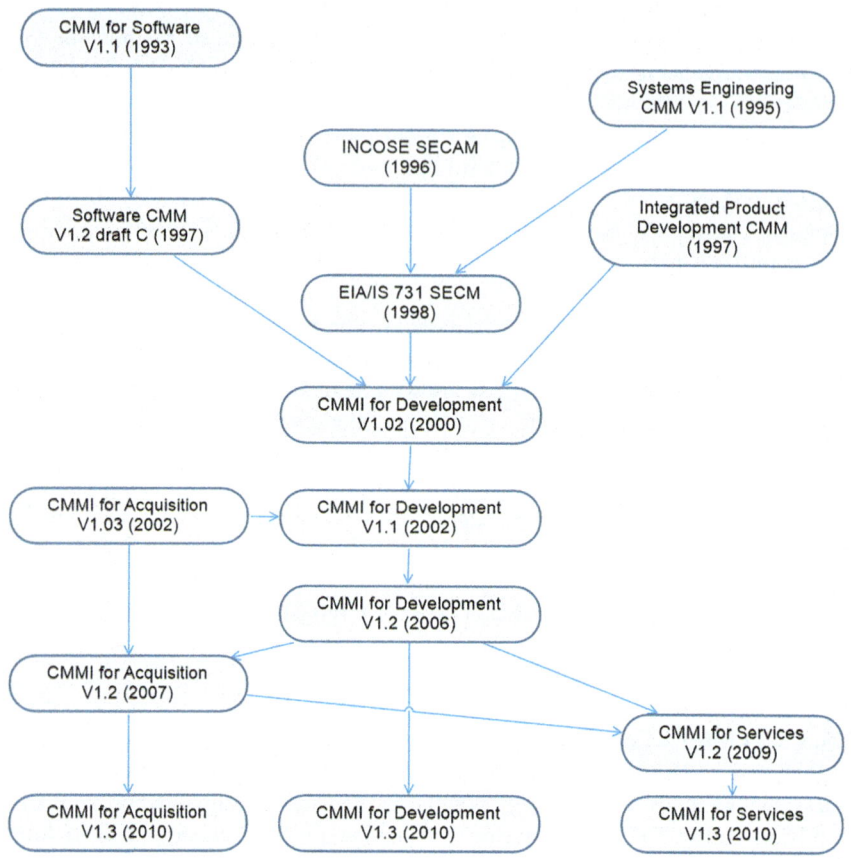

Fig. 4.6 CMMI history

Based on this framework, it was possible to develop industry-specific models. Examples of such models were BankingSPICE and SPICE4SPACE. SPICE4SPACE still exists, and the corresponding framework is described in the *ECSS-Q-HB-80–02 Part 1A Software process assessment and improvement—Part 1: Framework* manual of the ECSS (European Cooperation for Space Standardization) [BF-015]. The processes are described in the standards "ECSS-E-ST-40C—Software" and "ECSS-Q-ST-80C Rev. 1—Software product assurance."

Medical SPICE is another industry-specific model for software development in medical technology. It is known as Guideline VDI 5702 Sheet 1 [BF-013] in Germany.

In our book, we will only focus on the automotive sector.

Part 5 of the standard was first published in 1999 and contained three lifecycle processes with subordinate processes in Chap. 5, "The Process Dimension" (Fig. 4.7).

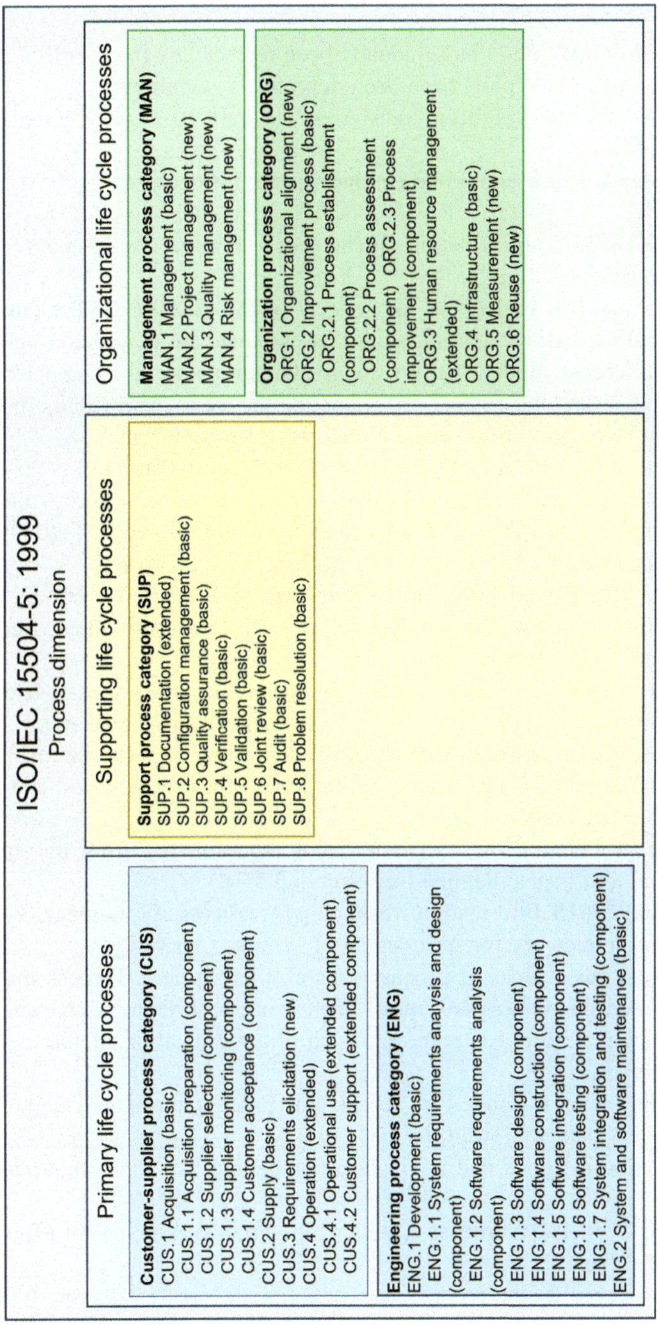

ISO/IEC 15504-5: 1999
Process dimension

Primary life cycle processes

Customer-supplier process category (CUS)
CUS.1 Acquisition (basic)
CUS.1.1 Acquisition preparation (component)
CUS.1.2 Supplier selection (component)
CUS.1.3 Supplier monitoring (component)
CUS.1.4 Customer acceptance (component)
CUS.2 Supply (basic)
CUS.3 Requirements elicitation (new)
CUS.4 Operation (extended)
CUS.4.1 Operational use (extended component)
CUS.4.2 Customer support (extended component)

Engineering process category (ENG)
ENG.1 Development (basic)
ENG.1.1 System requirements analysis and design (component)
ENG.1.2 Software requirements analysis (component)
ENG.1.3 Software design (component)
ENG.1.4 Software construction (component)
ENG.1.5 Software integration (component)
ENG.1.6 Software testing (component)
ENG.1.7 System integration and testing (component)
ENG.2 System and software maintenance (basic)

Supporting life cycle processes

Support process category (SUP)
SUP.1 Documentation (extended)
SUP.2 Configuration management (basic)
SUP.3 Quality assurance (basic)
SUP.4 Verification (basic)
SUP.5 Validation (basic)
SUP.6 Joint review (basic)
SUP.7 Audit (basic)
SUP.8 Problem resolution (basic)

Organizational life cycle processes

Management process category (MAN)
MAN.1 Management (basic)
MAN.2 Project management (new)
MAN.3 Quality management (new)
MAN.4 Risk management (new)

Organization process category (ORG)
ORG.1 Organizational alignment (new)
ORG.2 Improvement process (basic)
ORG.2.1 Process establishment (component)
ORG.2.2 Process assessment (component) ORG.2.3 Process improvement (component)
ORG.3 Human resource management (extended)
ORG.4 Infrastructure (basic)
ORG.5 Measurement (new)
ORG.6 Reuse (new)

Fig. 4.7 Lifecycle processes with subordinate processes from ISO/IEC 15504-5:1999

Ten parts of this standard have been published and updated so far. The basis for a process reference model and for assessing these processes' maturity has been defined.

Since 2015, ISO/IEC 15504 has gradually been replaced by the ISO/IEC 33000 family [BF-006], and some parts have been definitively withdrawn.

The part that has not yet been transferred is "Part 10 (Safety Extension)" [BF-012] (Fig. 4.8).

The following standards should be emphasized in this context:

- ISO/IEC 33001:2015 Information Technology—Process Assessment—Concepts and Vocabulary:
  Here, the terminology is defined, the concept of the ISO/IEC 330xx family is described, and the basic requirements for assessments are described, including the process reference model, the process assessment model, the competence of the lead assessor, and the context for evaluating the assessment results (process improvement, evaluation of process-related risks, benchmarking).
  The structure of the ISO/IEC-330xx family is as follows (Fig. 4.9).
  So far, only the process reference models for information security management (ISO/IEC 33052:2016), quality management (ISO/IEC 33053:2019), and service management (ISO/IEC 33054:2020) have been created.
  The software lifecycle processes are described in ISO/IEC 33061 (Chap. 5).
- ISO/IEC 33002:2015 Information Technology—Process Assessment: Requirements for Performing Process Assessment
  This standard defines the requirements for conducting assessments. It covers all topics from planning to creating the assessment report. Furthermore, three different classes of assessments are defined, depending on how the results are used (external or internal benchmarking, improvement opportunities, and areas with process-related risks).
  Furthermore, the degree of independence of the assessors from the audited organizational unit is also defined (see Sect. 4.3.5.3).
- ISO/IEC 33020:2015 Information Technology—Process Assessment: Process Measurement Framework for Assessment of Process Capability
  The framework for evaluating process capability is set here. Process maturity is defined based on process attributes. Each process attribute is characterized by corresponding results (outcomes), which allow evaluation of the assigned process attributes.
  Each capability level is defined by at least one process attribute. There are six maturity levels (maturity level 0 to 5).
  Three evaluation methods and aggregation procedures for the results are also provided.
- ISO/IEC TS 33061:2021 Process Assessment—Process Assessment Model for Software Life Cycle Processes
  This standard describes the process assessment model for the software lifecycle processes. Chapter 5 describes the individual process groups and processes.

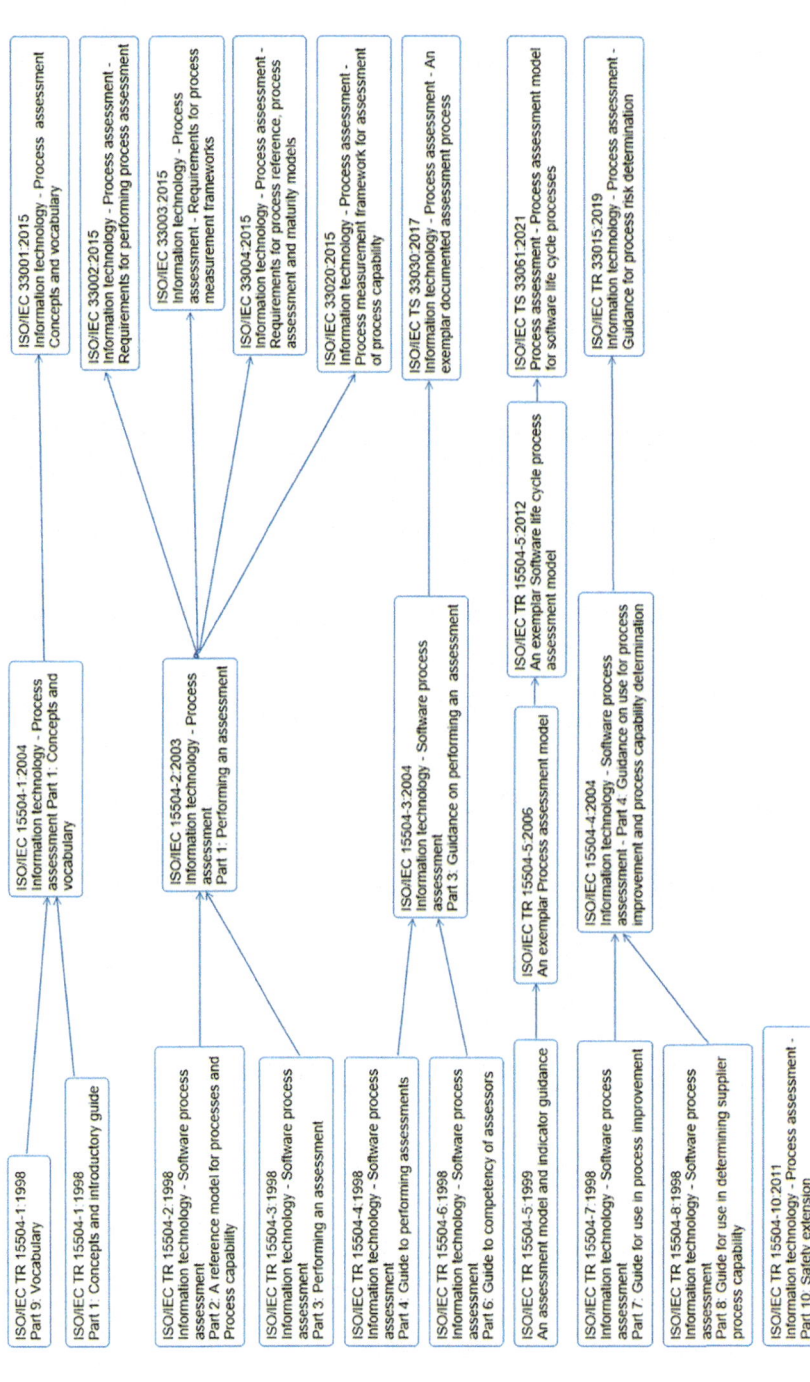

Fig. 4.8 Overview of the development of ISO/IEC TR 15504 and gradual replacement by ISO/IEC 33000 family

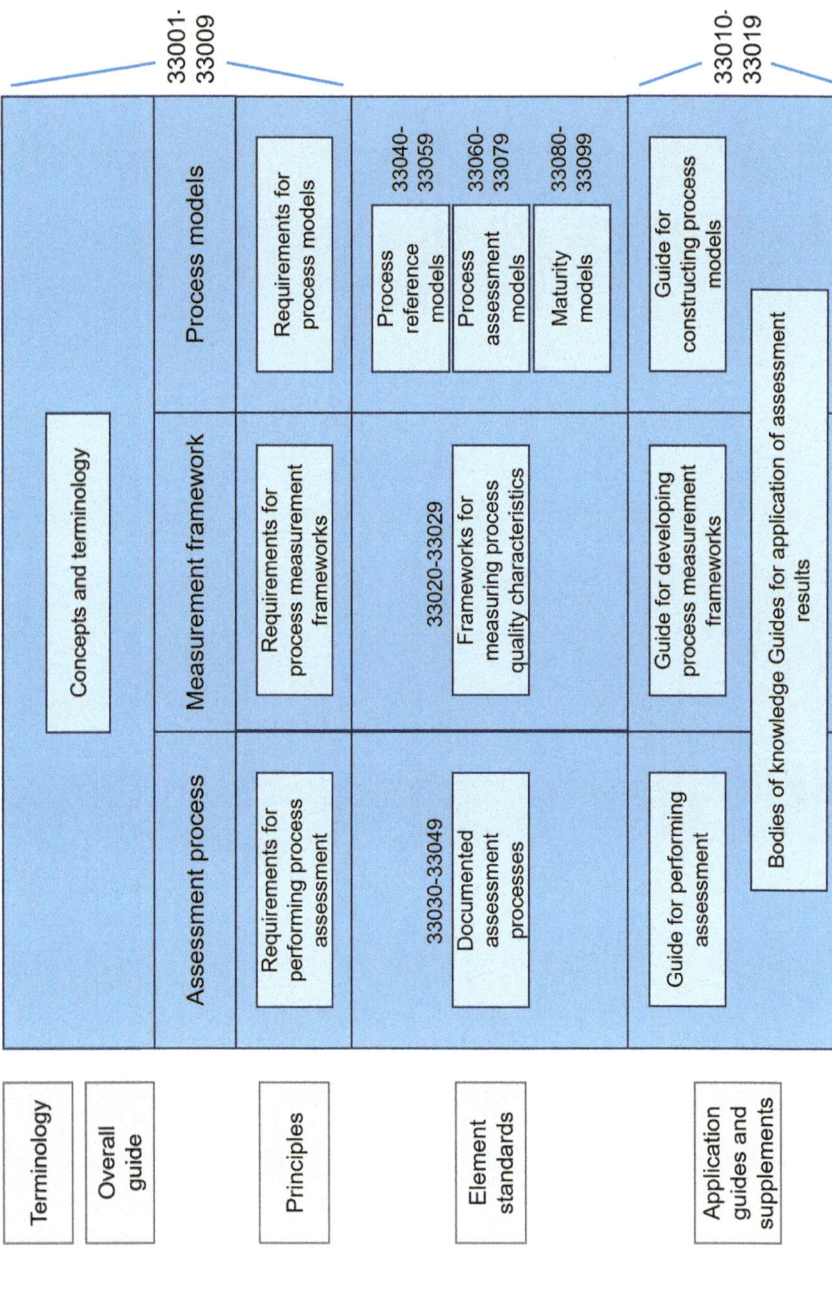

Fig. 4.9 Structure of the ISO/IEC 330xx standards for process assessments

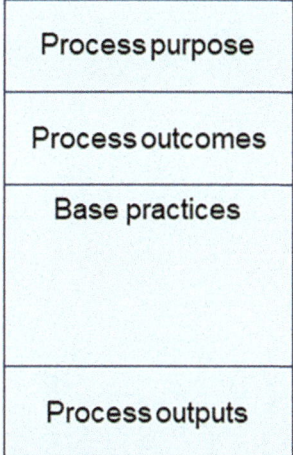

Fig. 4.10   Structure of a process in ISO/IEC TS 33061

Each process is described in terms of purpose, outcomes, base practices, and work products (Fig. 4.10).

The defined processes are shown in the following diagram (Fig. 4.11).

4.3.5   VDA Automotive SPICE®

As described in the previous chapter, Automotive SPICE® is another industry-specific model that uses the framework of ISO/IEC 15504 or, today, the ISO/IEC 33000 family.

4.3.5.1   Introduction

Like CMMI, Automotive SPICE® [BF-001] (commonly called "ASPICE") was developed as a model to evaluate software development processes, but in this case, it was developed specifically for the automotive sector. The Automotive Special Interest Group (AUTOSIG) developed the first version, which was released in 2005. Car manufacturers Audi, BMW, Daimler, Porsche, Volkswagen, Fiat, Ford, Jaguar, Land Rover, and Volvo were members of AUTOSIG. Later, in 2007, the German car manufacturers agreed upon a minimal set of processes to be assessed (formerly HIS scope, now VDA scope) and required their suppliers to be assessed according to Automotive SPICE®.

Depending on the Automotive Safety Integrity Level (ASIL, see Sect. 4.4.3), some car manufacturers require a specific capability level (e.g., level 2) if the system is safety-relevant (see explanations about the capability level in the next chapter).

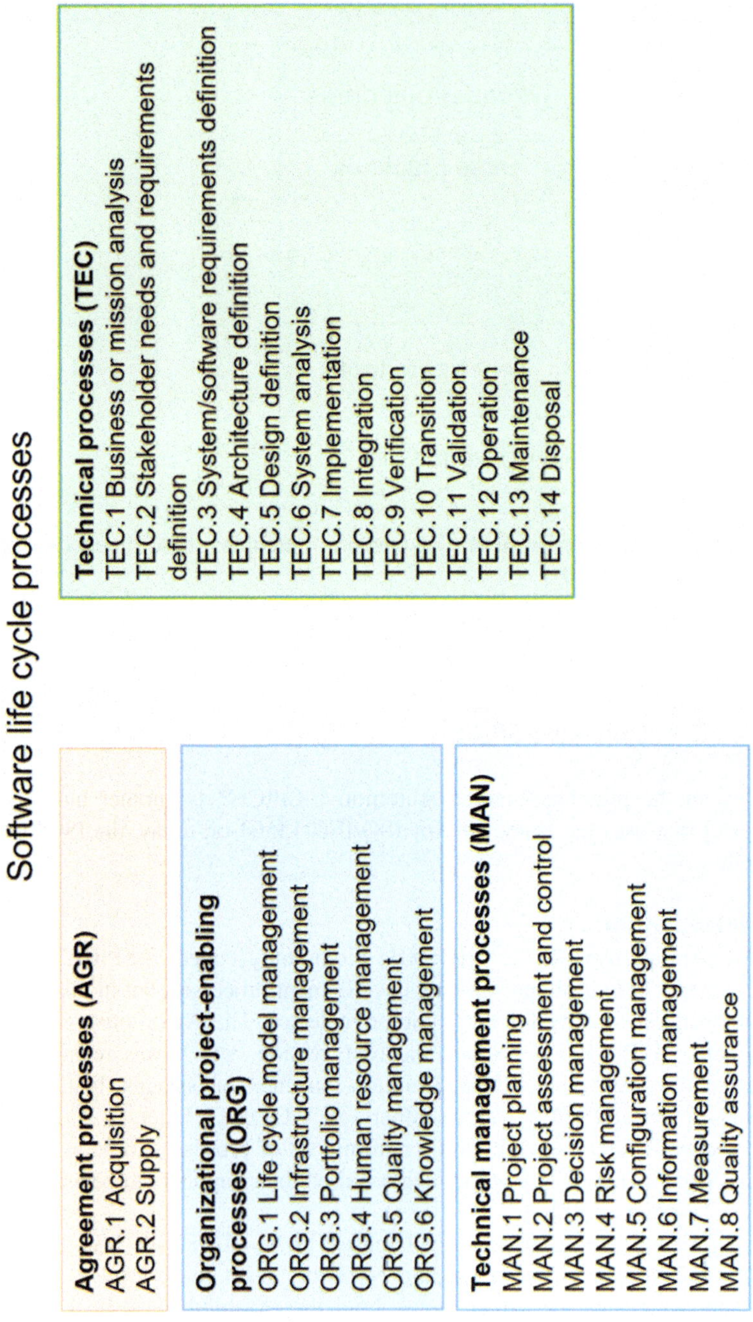

Software life cycle processes

Agreement processes (AGR)
AGR.1 Acquisition
AGR.2 Supply

Organizational project-enabling processes (ORG)
ORG.1 Life cycle model management
ORG.2 Infrastructure management
ORG.3 Portfolio management
ORG.4 Human resource management
ORG.5 Quality management
ORG.6 Knowledge management

Technical management processes (MAN)
MAN.1 Project planning
MAN.2 Project assessment and control
MAN.3 Decision management
MAN.4 Risk management
MAN.5 Configuration management
MAN.6 Information management
MAN.7 Measurement
MAN.8 Quality assurance

Technical processes (TEC)
TEC.1 Business or mission analysis
TEC.2 Stakeholder needs and requirements definition
TEC.3 System/software requirements definition
TEC.4 Architecture definition
TEC.5 Design definition
TEC.6 System analysis
TEC.7 Implementation
TEC.8 Integration
TEC.9 Verification
TEC.10 Transition
TEC.11 Validation
TEC.12 Operation
TEC.13 Maintenance
TEC.14 Disposal

Fig. 4.11 Process groups ISO/IEC TS 33061:2021

Process Identification	PRM process name	Source
ACQ.4	Supplier monitoring	ISO/IEC 12207 Amd. 1 (Annex 1; §F.1.1.3)
ENG.2	System requirements analysis	ISO/IEC 12207 Amd.1 (Annex F; §F.1.3.2)
ENG.3	System architectural design	ISO/IEC 12207 Amd.1 (Annex F; §F.1.3.3)
ENG.4	Software requirements analysis	ISO/IEC 12207 Amd.1 (Annex F; §F.1.3.4)
ENG.5	Software design	ISO/IEC 12207 Amd.1 (Annex F; §F.1.3.5)
ENG.6	Software construction	ISO/IEC 12207 Amd.1 (Annex F; §F.1.3.6)
ENG.7	Software integration test	ISO/IEC 12207 Amd.1 (Annex F; §F.1.3.7)
ENG.8	Software testing	ISO/IEC 12207 Amd.1 (Annex F; §F.1.3.8)
ENG.9	System integration test	ISO/IEC 12207 Amd.1 (Annex F; §F.1.3.9)
ENG.10	System testing	ISO/IEC 12207 Amd.1 (Annex F; §F.1.3.10)
SUP.1	Quality assurance	ISO/IEC 12207 Amd.1 (Annex F; §F.2.3)
SUP.8	Configuration management	ISO/IEC 12207 Amd.2 (§F2.2)
SUP.9	Problem resolution management	ISO/IEC 12207 Amd.2 (§F2.8)
SUP.10	Change request management	ISO/IEC 12207 Amd.2 (§F2.11)
MAN.3	Project management	ISO/IEC 12207 Amd. 1 (Annex F; §F.3.1.3)

Fig. 4.12 Processes from PRM v4.5 and their sources in ISO/IEC 12207

Process reference model	Process ID Process name Process purpose Process outcomes		The individual processes are identified with a unique process identifier and a process name. A process purpose statement is provided, and process outcomes are defined to represent the process dimension of the Automotive SPICE process reference model. The background coloring of process ID's and names are indicating the assignment to the corresponding process group.
Process performance indicators	Base practices		A set of base practices for the process providing a definition of the activities to be performed to accomplish the process purpose and fulfill the process outcomes. The base practice headers are summarized at the end of a process to demonstrate their relationship to the process outcomes.
	Output information items		The output information items that are relevant to accomplish the process purpose and fulfill the process outcomes summarized at the end of a process to demonstrate their relationship to the process outcomes. Note: Refer to Annex B for the characteristics of each information item.

Fig. 4.13 Relationship process ID and process performance indicators from automotive SPICE

In the first version of ASPICE, the processes of the process reference model were derived from ISO/IEC 12207 [BF-002] AMD ("Amendment") 1 Annex F and H and AMD 2. The processes are summarized in the same primary lifecycle processes as in ISO/IEC 12207 AMD 1:

– Primary Lifecycle Processes
– Supporting Lifecycle Processes
– Organizational Lifecycle Processes.

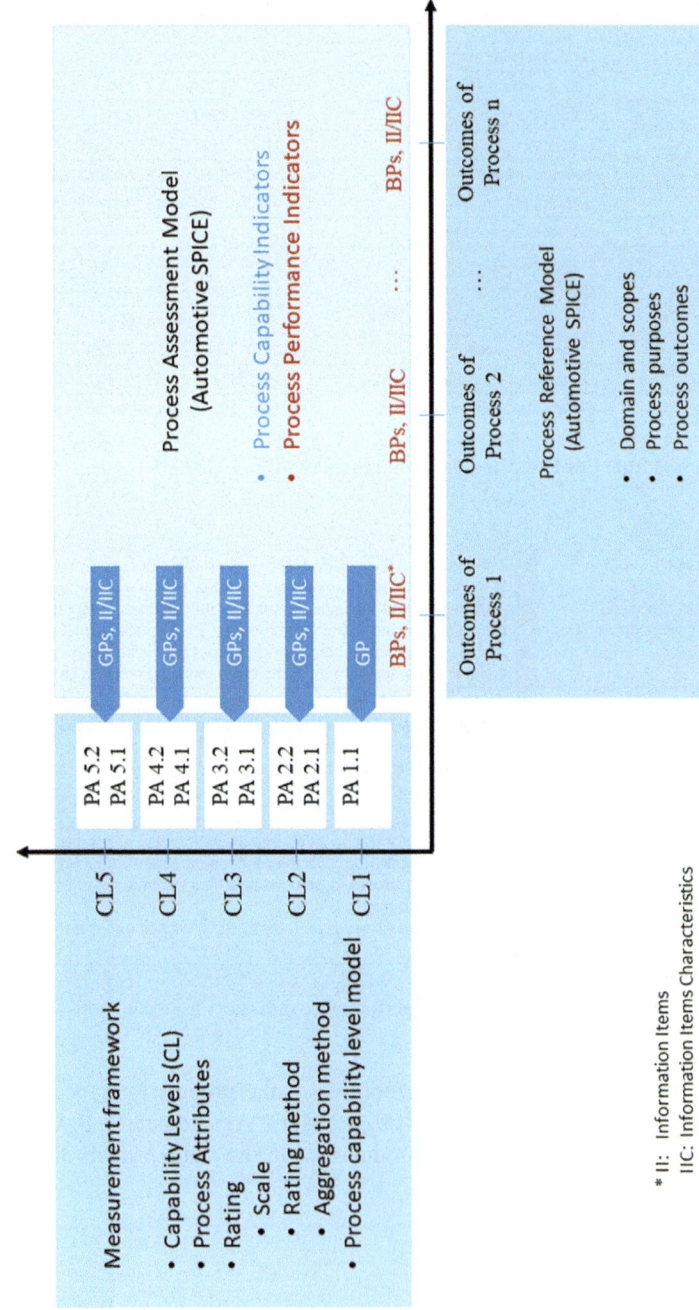

Fig. 4.14 Relationship between assessment indicators and processes from automotive SPICE

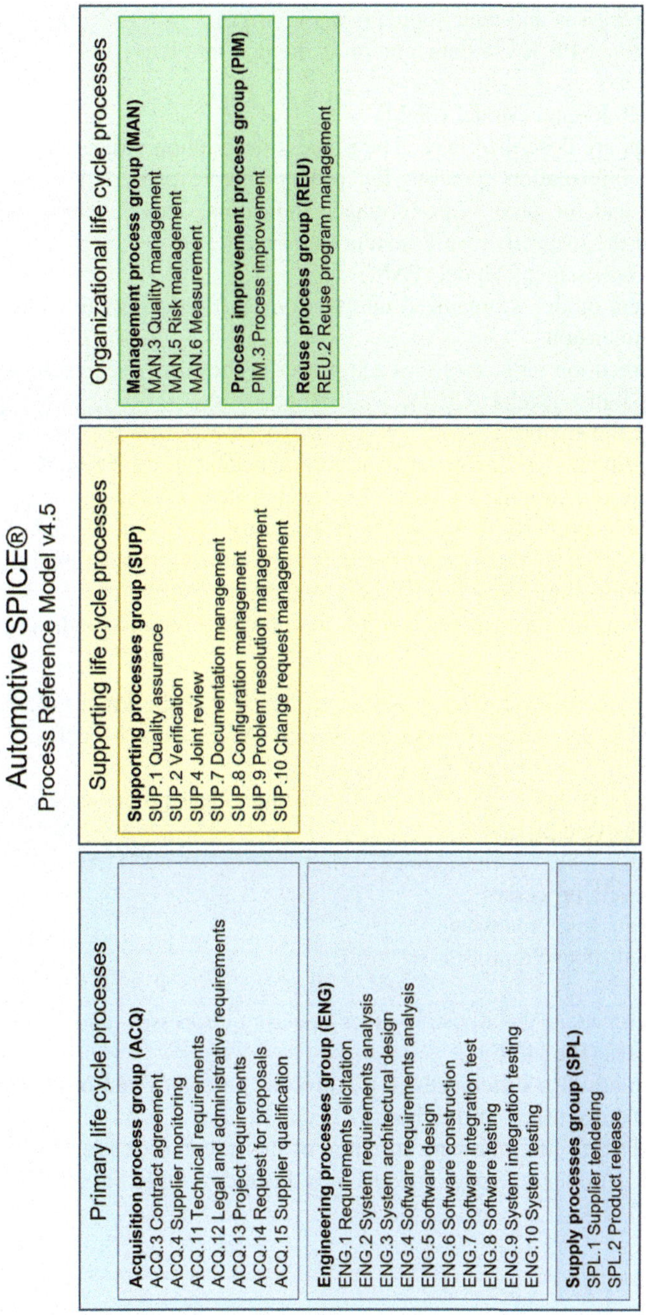

Automotive SPICE®
Process Reference Model v4.5

Primary life cycle processes

Acquisition process group (ACQ)
ACQ.3 Contract agreement
ACQ.4 Supplier monitoring
ACQ.11 Technical requirements
ACQ.12 Legal and administrative requirements
ACQ.13 Project requirements
ACQ.14 Request for proposals
ACQ.15 Supplier qualification

Engineering processes group (ENG)
ENG.1 Requirements elicitation
ENG.2 System requirements analysis
ENG.3 System architectural design
ENG.4 Software requirements analysis
ENG.5 Software design
ENG.6 Software construction
ENG.7 Software integration test
ENG.8 Software testing
ENG.9 System integration testing
ENG.10 System testing

Supply processes group (SPL)
SPL.1 Supplier tendering
SPL.2 Product release

Supporting life cycle processes

Supporting processes group (SUP)
SUP.1 Quality assurance
SUP.2 Verification
SUP.4 Joint review
SUP.7 Documentation management
SUP.8 Configuration management
SUP.9 Problem resolution management
SUP.10 Change request management

Organizational life cycle processes

Management process group (MAN)
MAN.3 Quality management
MAN.5 Risk management
MAN.6 Measurement

Process improvement process group (PIM)
PIM.3 Process improvement

Reuse process group (REU)
REU.2 Reuse program management

Fig. 4.15   Process reference model v4.5 from May 8th, 2010

The source of the processes from ISO/IEC 12207 is documented in the first versions of the process reference model in corresponding tables (PRM v4.5). Here is a selection of processes and their source (see Fig. 4.12).

The Automotive SPICE® documentation is divided into two parts:

- The Process Reference Model (PRM)
  The processes are described here. The process description structure is divided into general information (process ID, process name, process purpose, process results) and the process performance indicators. These include the base practices and the respective work results (see Fig. 4.13).
- The Process Assessment Model (PAM)
  The assessment model is organized into two types of indicators (dimensions):
  - Process dimension
    Process execution indicators for each process. These are used exclusively to assess capability level 1 (CL 1).
  - Capability dimension
    Process attributes consisting of generic practices (GP) and resources (GR) are defined to determine the capability level. These process attributes define a measurable property of the process capability (see Sect. 5.1 in ISO/IEC 33020). At least one process attribute must be defined for each maturity level. The exception is maturity level 1: only one process attribute is defined, containing a generic practice referencing the process execution indicators (see Fig. 4.14).

The following processes were described in version 4.5 of the Process Reference Model (released at the same time as the Process Assessment Model v2.5) (see Fig. 4.15).

In the newer versions (e.g., v3.1, v4.0), the individual processes are divided into the same lifecycle processes:

- Primary lifecycle processes
- Supporting lifecycle processes
- Organizational lifecycle processes.

The main difference is the division of engineering processes into system and software processes (Fig. 4.16).

Of course, there are also differences in the process outcomes and base practices, which will not be discussed further here.

These changes lead to the following structure of the reference model (Fig. 4.17).

In the meantime, further reference models have been developed for ASPICE:

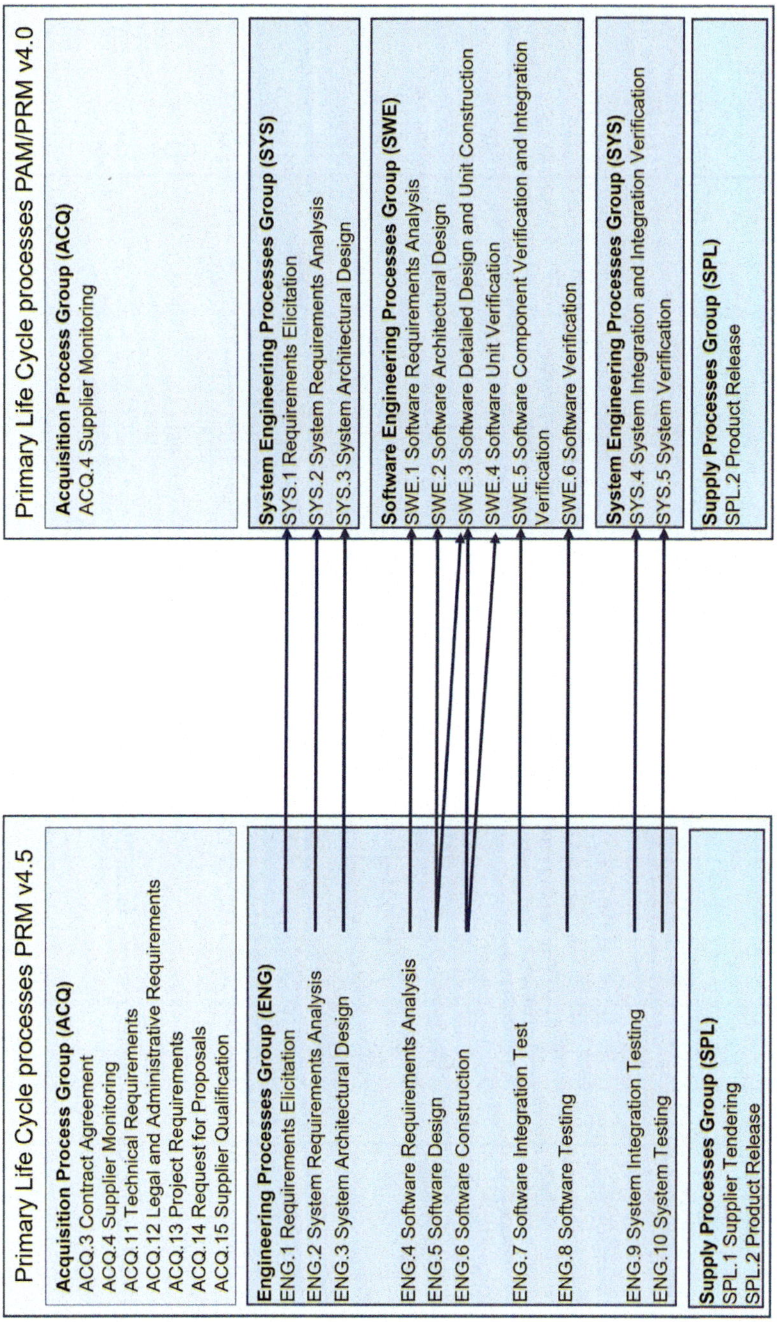

Fig. 4.16 Changes of the engineering processes from PRM v4.5 to PAM/PRM v4.0

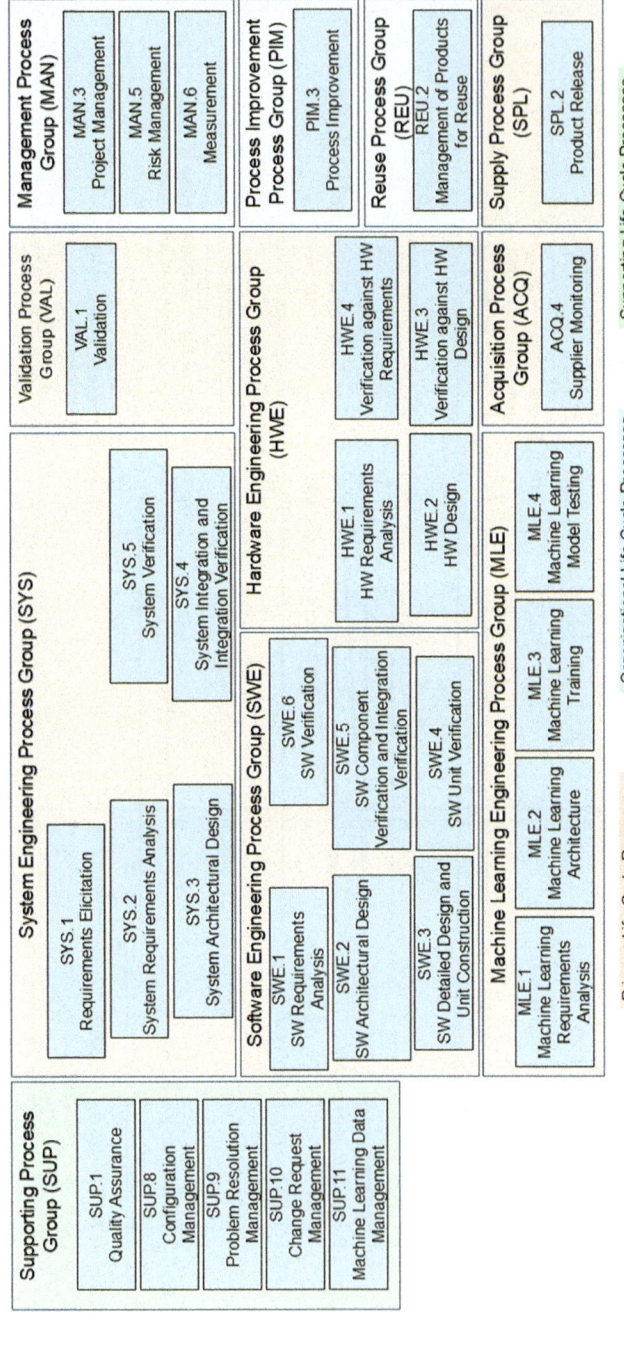

Fig. 4.17 Automotive SPICE: overview reference model (PAM/PRM v4.0)

– Hardware SPICE:

This model only contains pure hardware engineering processes and is based on the system engineering processes of the reference model. Hardware engineering processes have been integrated into PAM v4.0.
– Cybersecurity:

The model is based on the reference model and introduces a total of six new processes:
   – Supplier request and selection
   – Cybersecurity risk management
   – Four cybersecurity engineering processes (Fig. 4.18).
– Machine Learning:

Four processes cover the area of machine learning and are integrated in PAM/PRM v4.0.

As the reference model contains a considerable number of reference processes, the capability level of a project is usually determined based on an excerpt of these processes. This selection is now referred to as the VDA scope and is defined in *Automotive SPICE® Guidelines Process Assessment Using the Automotive SPICE PAM 4.0 (2nd revised edition, November 2023)* [BF-074]. There is a basis with five processes. At least one plug-in process must be added to these; further processes can be added depending on the focus of the assessment (Fig. 4.19).

4.3.5.2 Capability Level
The capability level is divided into six different stages (see Fig. 4.20).

These capability levels are based on the definition from ISO/IEC 33020 Information Technology—Process Assessment: Process Measurement Framework for Assessment of Process Capability [BF-008]. The process attributes are used to assess the maturity level.

This combination of capability level and process attributes is defined as follows (see Sect. 5.2 of ISO/IEC ISO/IEC 33020):

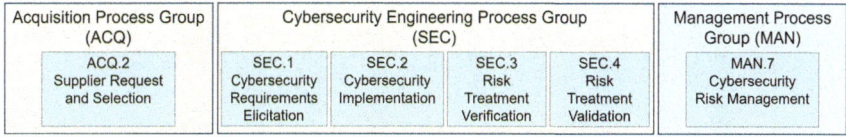

Fig. 4.18 Cybersecurity processes (v1.0)

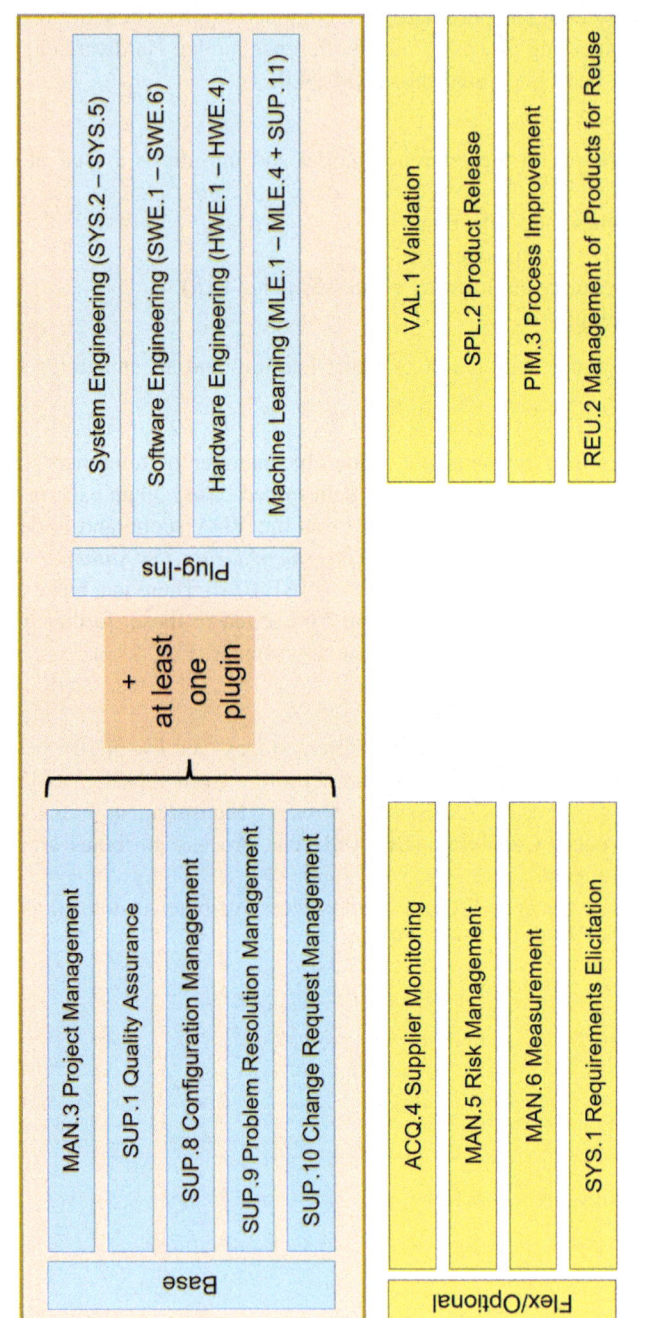

Fig. 4.19 VDA scope (PRM/PAM v4.0)

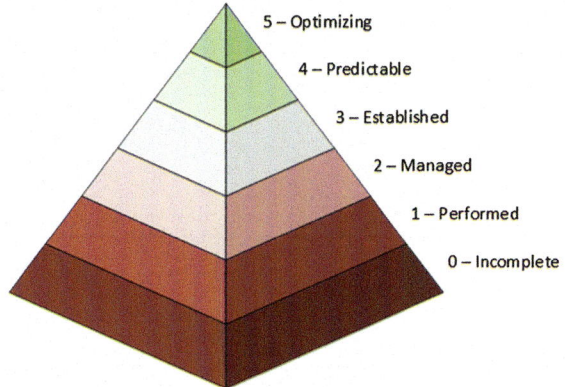

Fig. 4.20 Capability levels according to ISO/IEC 15504

5 – Optimizing
4 – Predictable
3 – Established
2 – Managed
1 – Performed
0 – Incomplete

- Capability Level 0: Incomplete Process. The development organization does not work systematically and can produce the required products "by luck."
- Capability Level 1: Performed Process. The development organization delivers, but it is still not systematically planned and is not of consistent quality. The assessment is based on the process attribute "1.1 Process Execution."
- Capability Level 2: Managed Process. The development organization delivers as planned but does not yet work consistently and has problems with particular project types and cases. The attributes "2.1 Management of Process Execution" and "2.2 Management of Work Products" are additionally assessed.
- Capability Level 3: Established Process. The development organization has a defined (well-documented) process and processes adapted or adaptable for particular tasks. Additionally, the attributes "3.1 Process Definition" and "3.2 Process Application" are assessed.
- Capability Level 4: Predictable Process. The development organization has planned and controlled the required performance of individual process steps. Additionally, the attributes "4.1 Quantitative Process Analysis" and "4.2 Quantitative Process Control" are assessed.
- Capability Level 5: Innovating Process. The development organization plans its continuous improvement and implements it according to plan. Additionally, the attributes "5.1 Process Innovation" and "5.2 Implementation of Process Innovation" are assessed.

4.3.5.3 Assessments

The capability level is checked on a project-specific basis in an assessment. This means that a project covering a specific product development is assessed, and only the affected project 'unit.' Based on ISO/IEC 33002 Information Technology—Process Assessment: Requirements for Performing Process Assessment [BF-008] the assessment process could be represented as follows, starting with planning (Fig. 4.21).

There is also the possibility of assessing the process maturity of an organizational unit. For this purpose, a specific process reference and process assessment

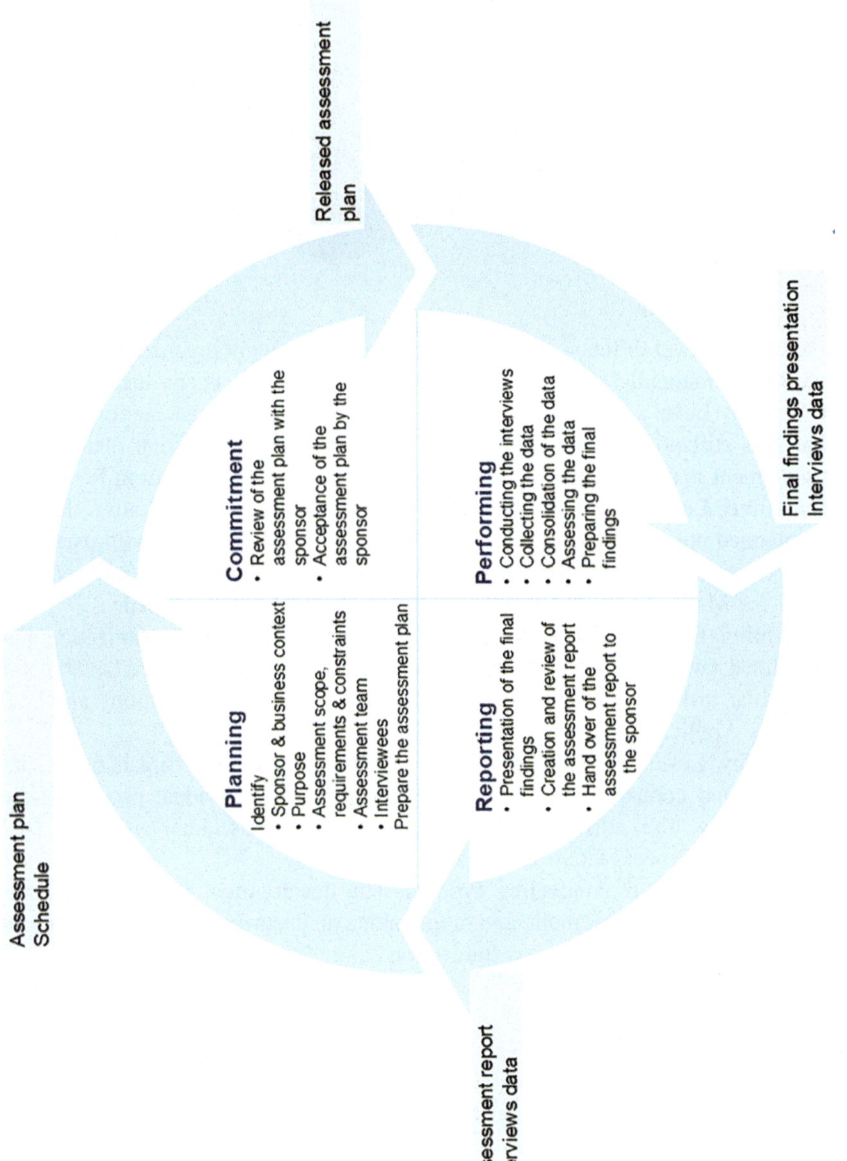

Assessment plan
Schedule

Released assessment
plan

Final findings presentation
Interviews data

Commitment
• Review of the
  assessment plan with the
  sponsor
• Acceptance of the
  assessment plan by the
  sponsor

Performing
• Conducting the interviews
• Collecting the data
• Consolidation of the data
• Assessing the data
• Preparing the final
  findings

Planning
Identify
• Sponsor & business context
• Purpose
• Assessment scope,
  requirements & constraints
• Assessment team
• Interviewees
Prepare the assessment plan

Reporting
• Presentation of the final
  findings
• Creation and review of
  the assessment report
• Hand over of the
  assessment report to
  the sponsor

Assessment report
Interviews data

Fig. 4.21 Assessment process

model has been developed (Organization SPICE PAM/PRM, currently v3.00 [BF-078]). As stated before, car manufacturers require projects to be assessed. So, apart from the fact that some of the assessed organizational maturity levels seem doubtful, they do not avoid assessments for a specific project.

In an assessment, the base practices for each process in the scope are examined by an assessment team in interviews, and the level of maturity achieved is evaluated. (see Sect. 4.3.5.5).

The assessment team consists of at least one lead assessor and, optionally, another assessor. According to ISO/IEC 33001 Information Technology—Process Assessment: Concepts and Terminology [BF-007], a lead assessor is an assessor who has demonstrated the competence to conduct an assessment and to monitor and verify the conformity of a process assessment. This competence is underpinned by appropriate training, proof of participation as an assessor in assessments for a minimum number of hours, and passing an observed assessment.

The assessment requirements are described in ISO/IEC 33002 Information Technology—Process Assessment: Requirements for Performing Process Assessment [BF-008].

A basic requirement is that the assessors are independent of the audited organization. This is defined in Annex A of ISO/IEC 33002 as follows (Fig. 4.22).

The assessment is carried out in accordance with ISO/IEC 33020 Sect. 5.3, "Process Attribute Rating Scale", based on the classifications N, P, L, and F (Fig. 4.23).

There is an even more refined classification of the rating, but it is only sometimes used.

The assessment is based on achieving the purpose of the base practices (in PA 1.1) and the generic practices (from PA 2.1 on). The information obtained in the interview is documented in a report. It can be structured as follows:

– Observations
– Weaknesses
– Recommendations (as far as possible)
– Strengths (if available).

The evidence shown is also recorded and assigned to the individual practices.

When rating, the rules from VDA QMC Band Automotive SPICE Guidelines— Process Assessment using the *Automotive SPICE PAM 4.0, 2nd revised edition, November 2023* [BF-003] are also observed. This volume contains both assessment rules and assessment recommendations. In an assessment, the assessment team applies the assessment rules. The usual assessment tools (e.g., Trace, Capability Adviser, etc.) indicate a violation of the rules when rating the individual practices. It would otherwise be difficult to keep track of the almost 300 pages, even if some rules follow a logical pattern.

	Category A	Category B	Category C	Category D
Body performing the assessment	The body performing the assessment shall be independent of the organization being assessed.		The body performing the assessment shall be part of the organization being assessed.	
Lead Assessor	The lead assessor shall be independent of the organization being assessed.		There shall be adequate separation of responsibilities of the lead assessor from personnel in the organizational unit being assessed.	The lead assessor can be part of the organizational unit being assessed.
Co-Assessors	The assessors shall be independent of the organization being assessed.	There shall be a separation of the responsibilities of the assessors from personnel in the organizational unit being assessed.	There shall be a separation of the responsibilities of the assessors from personnel in the organizational unit being assessed.	The assessors can be part of the organizational unit being assessed.

Fig. 4.22 Independence of the assessors

Code	Rating	Level of Achievement
N	Not Implemented	0% - 15% achievement
P	Partially implemented	>15% - 50% achievement
L	Largely implemented	>50% - 85% achievement
F	Fully implemented	>85% - 100% achievement

Fig. 4.23 Rating scale according to ISO/IEC 33020

4.3.5.4 Implementation

As with implementing any model or standard, organizational aspects must first be considered.

Some OEMs only commission/approve a supplier if a specified capability level in the VDA scope is achieved (e.g., classification as an A-class supplier). If a supplier fails an assessment, some OEMs suggest a form of support/supervision. This is usually carried out by improvement experts who monitor and evaluate the improvement process at the supplier with a focus on completing the following assessment (e.g., SQIL at VW: Software Quality Improvement Leader).

As a result, projects for these OEMs receive greater attention and pressure from their management. A project with limited resources must develop additional processes, tools, and templates to reach the required capability level. If the project cannot manage this independently, a task force is often created to enforce it. Eventually, once this measure has been successful the pressure can yield and, frequently, what has been learned is not transferred back into the organization. This means the next project starts at the same stage and must repeat the same learning process. Such inefficiencies are usually not emphasized because no one can adequately comprehend them. That's an unfortunate circumstance since the signs of process-related distress are often quite apparent:

- Project managers are frequently replaced.
- The project relies heavily on external experts because turnover makes internal specialists scarce.
- The purpose, content, or status of work products are unknown to anyone in the project.
- Numerous outdated documents reference persons who no longer belong to the project or left the company long ago.
- It is usually apparent that there has been no proper handover from the predecessor.

Unfortunately, organizations that suffer from such symptoms are the rule. Reports like the Gallup *State of the Global Workplace: 2023 Report* [BF-004] provide insights into the state of mind of employees that correlates with the above observations. The report includes data on employee engagement in Europe (Fig. 4.24, reproduced with permission from Gallup, Inc.).

According to this study, employees' mindset can be categorized as follows:

- **Thriving at work**:
These employees find their work meaningful and feel connected to the team and their organization. They feel proud of their work and take ownership of their performance, going the extra mile for teammates and customers.
- **Quiet Quitting**:

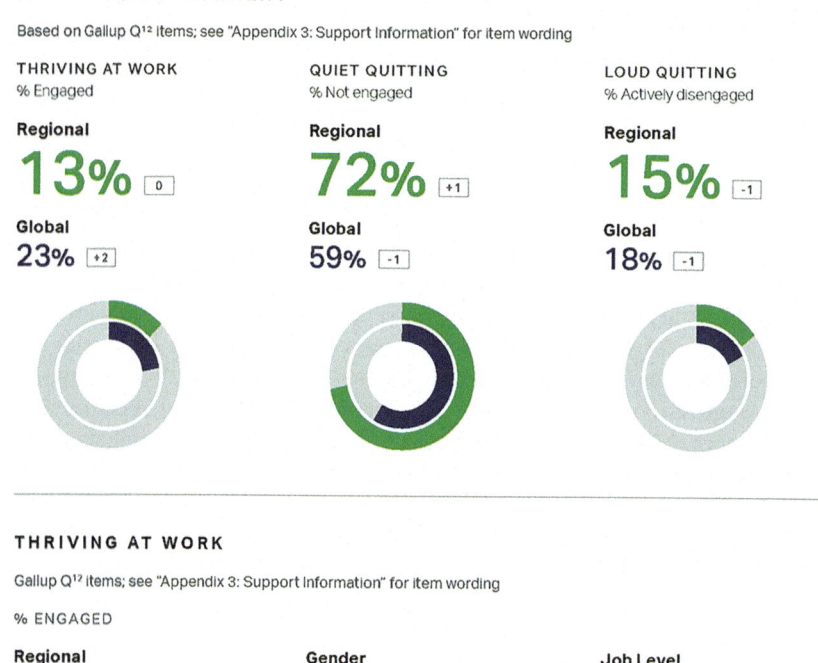

Fig. 4.24 Employee engagement

These employees are filling a seat and watching the clock. They put in the minimum effort required and are psychologically disconnected from their employer. Although minimally productive, they are more likely to be stressed and burnt out than engaged workers because they feel lost and disconnected from their workplace.

– **Loud Quitting**:
These employees take ations that directly harm the organization, undercutting its goals and opposing its leaders. At some point, the trust between employee and employer is severely broken, or the employee is woefully mismatched to a role, causing constant crises.
(Source: Gallup *State of the Global Workplace: 2023 Report*).

According to this report, only 13 percent of European interviewees are satisfied with their job.

From our perspective, a more systematic approach would be far more effective. Even if the initial response from managers is frustration over resource shortages, have they ever analyzed how many resources are wasted on reworking and reducing "technical debt"?

A more solid approach would be to establish a plan, as outlined in Fig. 4.25, to implement a solution across the entire organization or within a specific product area.

The improvement process is explained as follows.

Analysis of the current status
The first step is to analyze the current processes based on the model. This can be achieved by having experts analyze current processes and identify existing gaps and possible improvements (gap analysis).

For this purpose, a pilot project can be selected to perform such gap analysis by conducting a similar survey to an assessment and documenting the identified gaps. All kinds of issues can be found here, such as gaps in the process, gaps in tools and templates, gaps in the application of the processes in the project, or inconsistencies in the work products.

These gaps form the basis for further planning (Fig. 4.26).

Realistic effort estimation
Once the analysis has been performed and the gaps and possible improvements have been documented, the activities needed to implement them must be estimated realistically. These activities should be described in plain English, which is easily understandable for everyone. Success and acceptance criteria should also be defined.

The improvements should also include the definition of relevant tools. In ASPICE, traceability is a significant challenge, and an inconsistent tool landscape is a frequently observed impediment to this requirement.

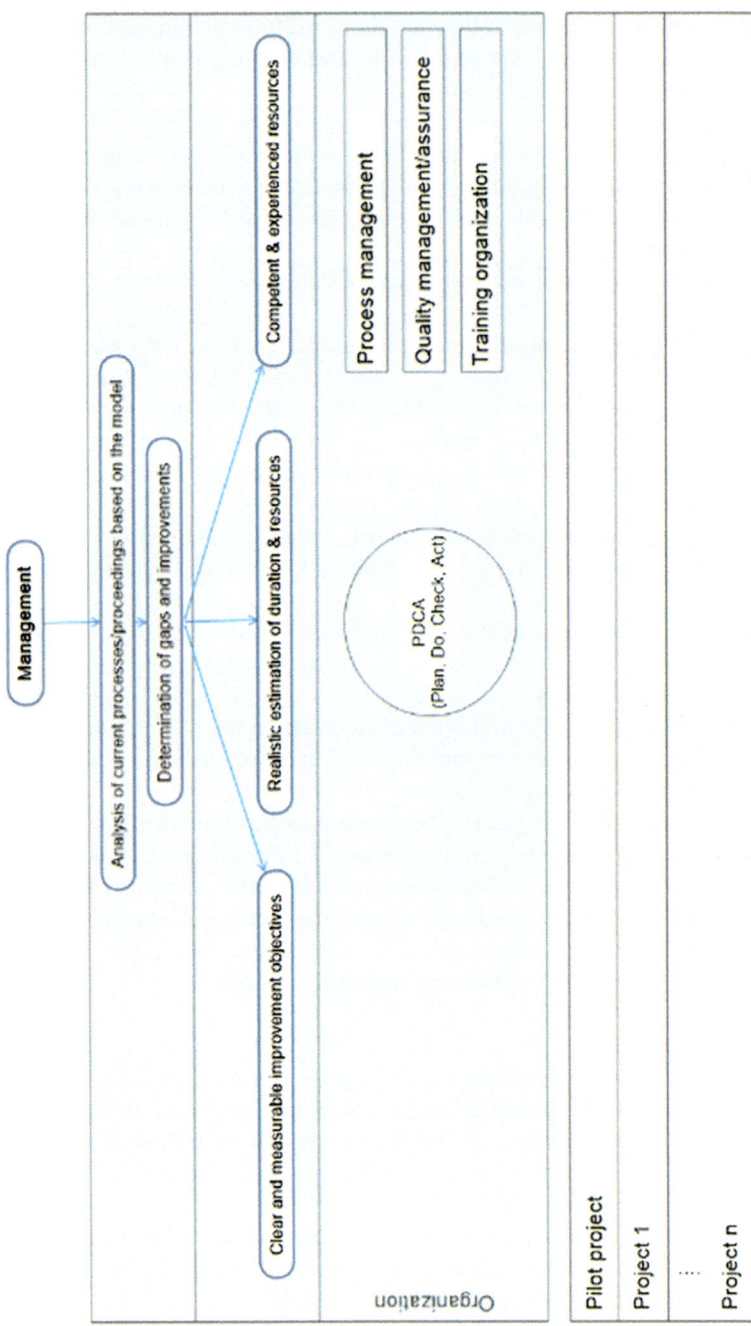

Fig. 4.25 ASPICE implementation driven by the organization

ID	Base Practice	Gap	Measures
SUP.8	BP1: Develop a configuration management strategy	Branch management is not fully described (the responsibility for merging branches is not defined).	Add responsibility in the strategy.
	BP2: Identify configuration items	Not all configuration items are defined.	Complete the configuration item list.
	BP3: Establish a configuration management system	---	---
	BP4: Establish branch management	It is not defined who is responsible for merging branches.	Add responsibility according to the updated strategy.
	BP5: Control modifications and releases	---	---
	BP6: Establish baselines	---	---
	BP7: Report configuration status	Currently no status is determined and reported.	Define, implement and report status report.
	BP8: Verify the information about configured items	---	---
	BP9: Manage the storage of configuration items and baselines	---	---

Fig. 4.26 Example of a gap analysis documentation for configuration management

Assigning competent and experienced resources

This is the Achilles' heel when implementing a new process. The leading employees must be organizational insiders: they must fully understand the new organization's purpose and related processes, have proven experience in the related development disciplines, and fully internalize the new process. Unfortunately, such employees are rare. Finding external experts can help if the organization understands what is needed. Once they are on board, the first step is to gain proper insight into the organization and its processes. Such a team can profit from the gap analysis before proceeding further.

A benefit of modern times is that resources can be distributed globally (partly due to resource problems in one's own country). Unfortunately, the fact that this causes cultural and planning problems is often ignored. Just imagine a team in Asia, another in the USA, and a European project manager.

Effectively assembling such "change" teams is always challenging. There is never enough time for everyone to collaborate (unless someone is willing to work outside regular hours). Also, cultural differences must be considered. For example, rough, direct feedback in the typical German manner can cause resentment and frustration. Global collaboration can be surprisingly expensive without suitable training tailored to cultural idiosyncrasies. The effort of managing an overly diverse team is often not sufficiently considered.

Definition of objectives

There must always be clear objectives at an organizational level. These should be in line with the business objectives specific to product development.

Simplified objectives could look like this (Fig. 4.27).

This is only an excerpt of possible objectives.

For example, if we consider customer satisfaction, potential objectives at the product level could include achieving the required product quality, meeting desired specifications, and ensuring on-time delivery (aligned with project milestones for intermediate deliveries and the final start of production). Detected defects also impact product quality, leading to additional costs.

In the same way, product quality depends on the competencies of the project members and, naturally, on the process quality. On-time delivery depends mainly on the required availability of the project members, but this is not the only challenge. In many projects, the effort often depends on estimations that are just rough "guestimates," and the project does not learn from them.

All in all, these objectives are often causally intertwined (Fig. 4.28).

Product quality can be broken down at the organizational level by defining additional objectives, such as maintainability, reducing technical debt, efficient resource usage, and software-specific goals (e.g., statement coverage, branch coverage, resource management, compiler warnings, MISRA [BF-034] warnings, etc.). These objectives can then be directly applied to individual projects, as is currently done in some organizations.

We will see more concretely how project-specific objectives can be derived from such organizational objectives in Sect. 4.3.5.5.

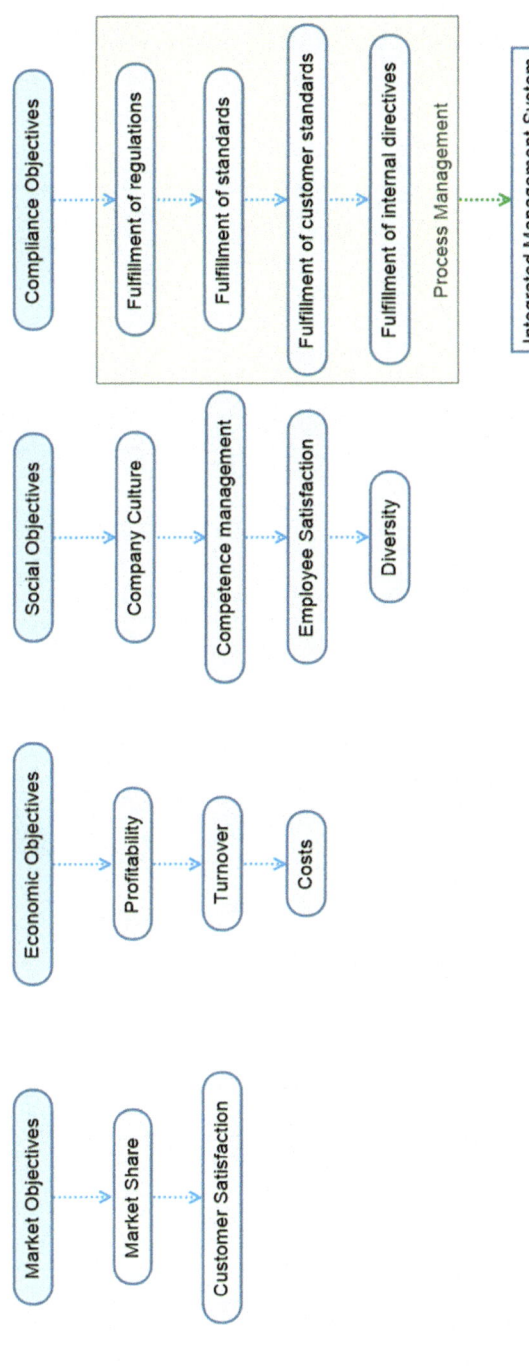

Fig. 4.27 Simplified business objectives

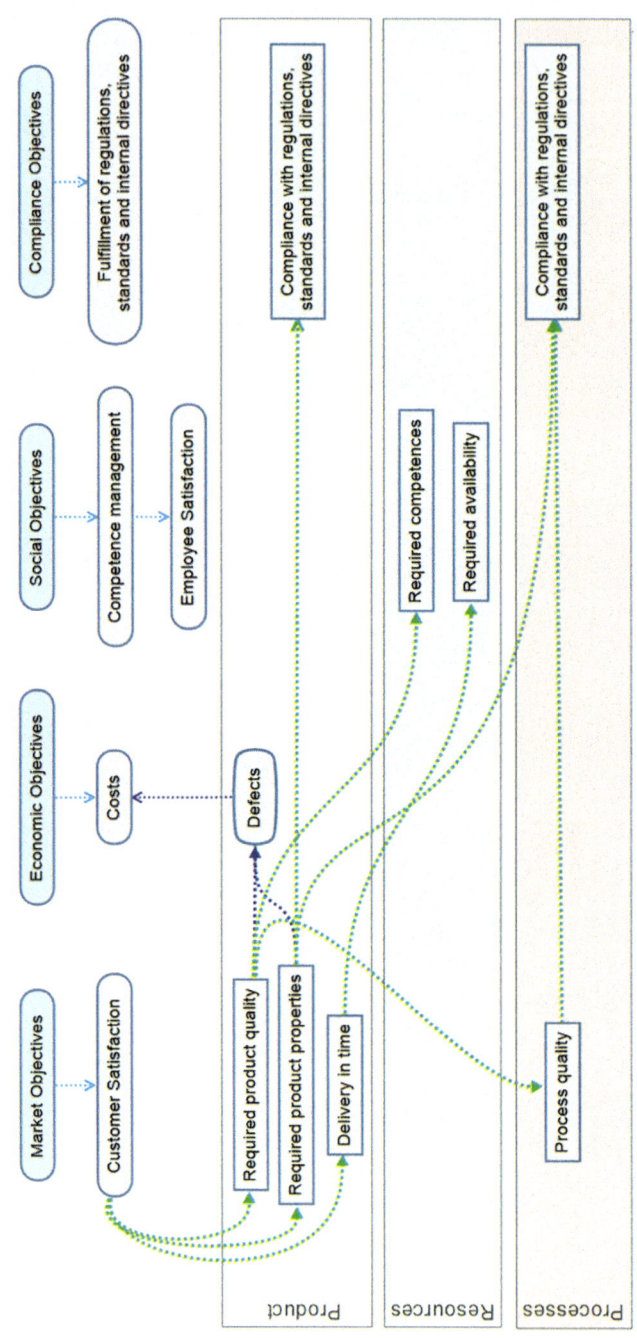

Fig. 4.28 Breakdown of objectives

Defects impact costs, delivery time, and resource allocation. Therefore, it would be prudent to learn from these defects and work towards reducing them over time.

The ASPICE VDA scope emphasizes quality assurance and problem-resolution management. These practices enable the identification and initiation of process improvement opportunities within the system. Another valuable approach is collecting employee feedback on potential improvements and incorporating it into the process.

Such a system could look like this in an ASPICE environment (Fig. 4.29).

On the other hand, the problems reported in SUP.9 (Problem Resolution Management) must be effectively "managed." This means that problem trends are visualized, and based on this trend, a prediction is made before the start of production (SOP). Thus, a minimal problem resolution rate can be determined, and a prediction can be made of whether every problem can be solved before SOP.

Such a trend analysis could look like this (Fig. 4.30).

At the same time, a simplified defect flow model is built around the processes of the V-model and the quality measures defined within engineering. Every engineering process includes expected quality assurance activities. On the left side of the V-model, these activities might involve required analyses or consistency checks (often performed during reviews). Additional quality assurance methods could include following coding guidelines for software or checking compliance with MISRA [BF-034] rules (Fig. 4.31).

Quality assurance activities focus on verifications and reviews on the right side of the V-model.

This simplified defect flow model is structured around two key elements: the source process, where the problem was initially generated, and the detection process, where the problem was eventually identified.

Figure 4.32 shows an example of a defect detected in software verification (SWE.6) and generated in software detailed design (SWE.3).

The following quality assurance activities apply:

- Consistency check (review) towards the software requirements (1)
- Consistency check (review) towards the software architecture (2)
- Review of the specification of the software unit verification measures (3)
- Review of the specification of the software component verification and integration verification measures (4)
- Review of the specification of the software verification measures (5).

The first two quality assurance activities should lead to the detection of possible mistakes in the software's detailed design.

In this case, the software detailed design and code must be corrected, and all tests up to software verification (SWE.6) must be repeated.

Thus, an objective could be to analyze the top three source processes and determine the root causes of the generated problems. Another objective could be to analyze why the quality measures between the source and the detection process didn't work and how these quality measures can be improved.

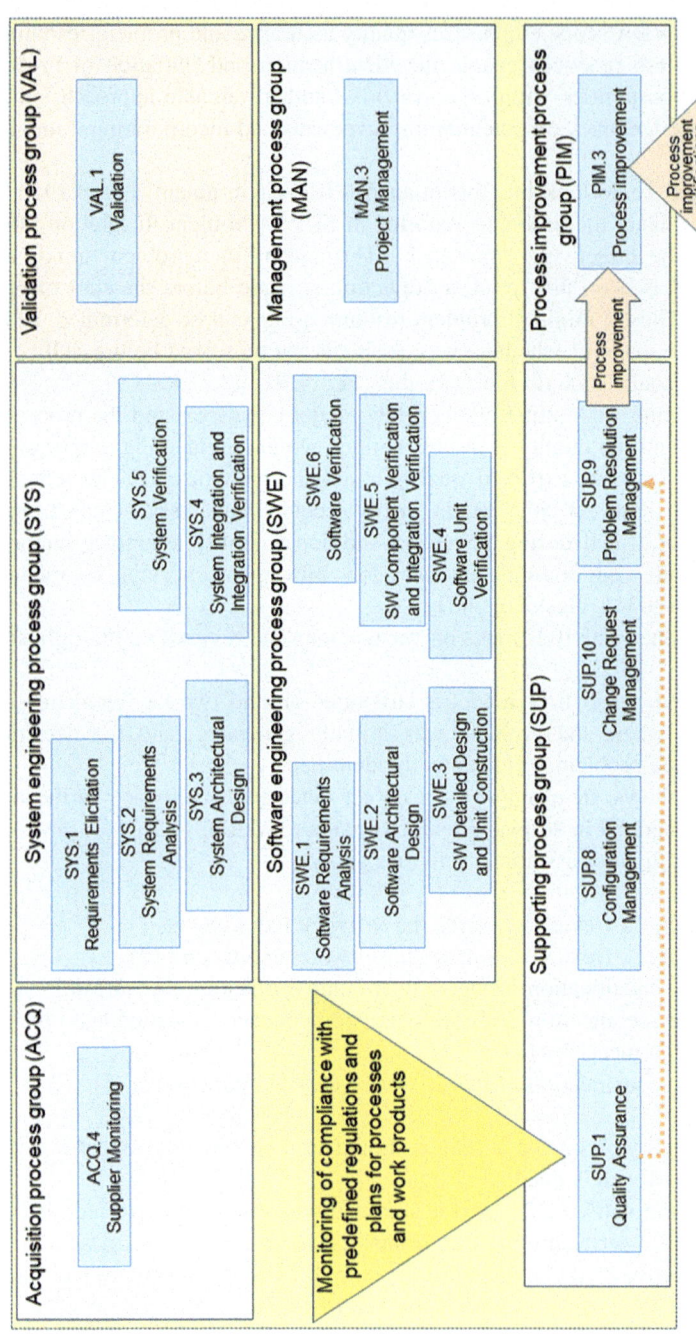

Fig. 4.29 Learning system

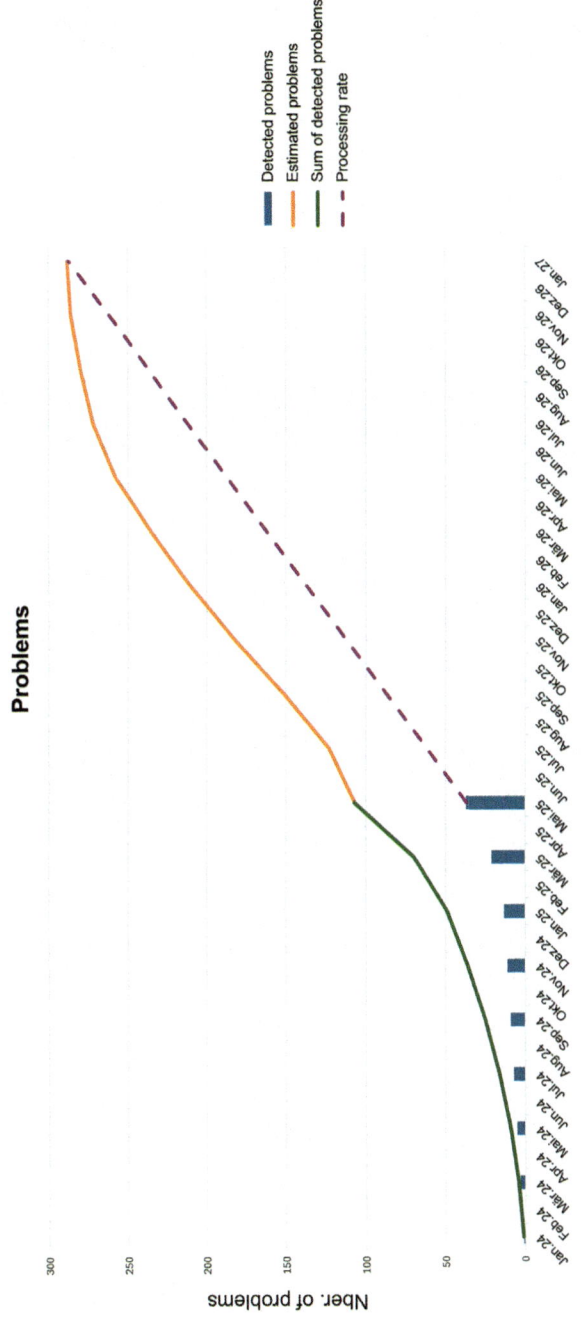

Fig. 4.30 Problem trend analysis

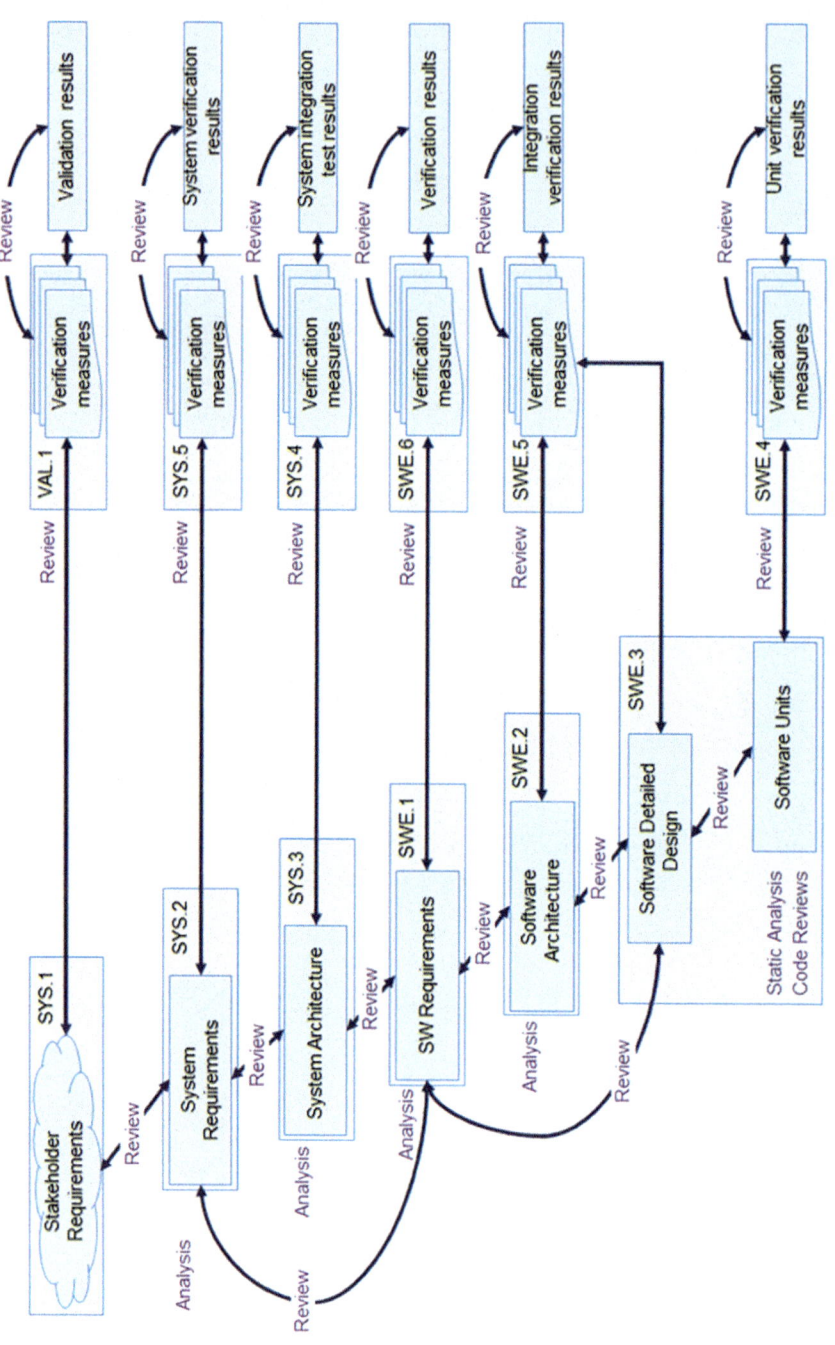

Fig. 4.31 Overview quality measures in the V-model

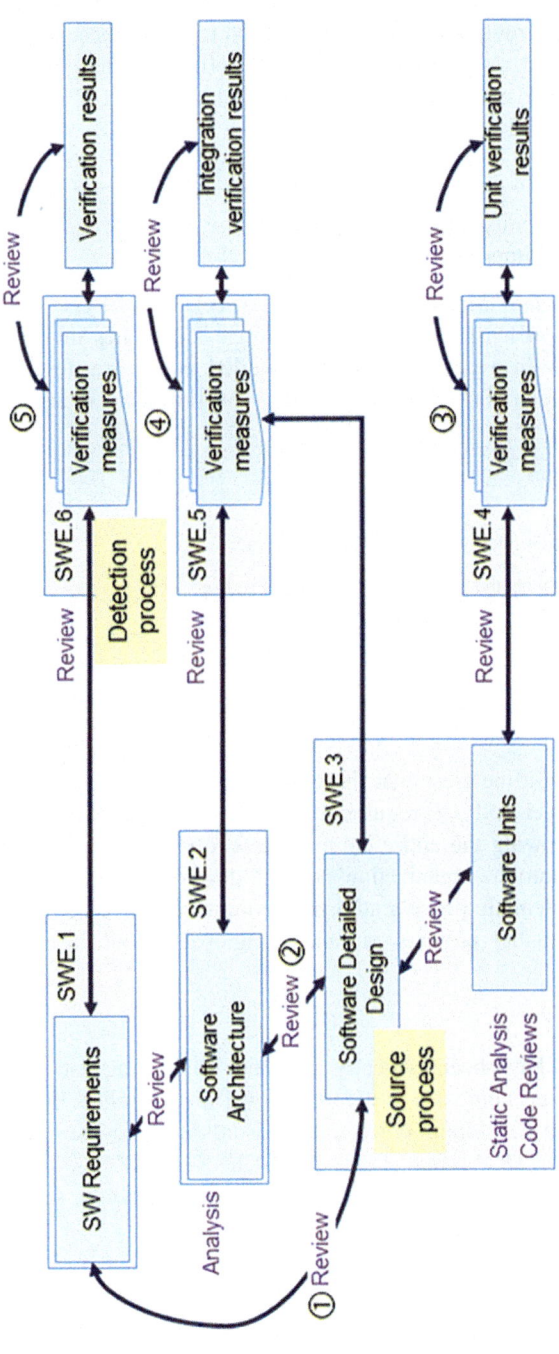

Fig. 4.32 Example of a defect with source and detection process

On the right side of the V-model, problems could also be caused by inappropriate test methods, test cases or conditions for performing the tests, and missing test cases. Again, the review would be the method to check these aspects.

This process could also be used for an IATF 16949 audit [BF-009]—a universally required standard. Indeed, the following chapters of IATF 16,949 require improvements based on nonconformities:

- 10 Improvement
  - 10.2 Nonconformity and corrective action
  - 10.3 Continual improvement.

In addition, there is the "Management Review" (Sect. 9.3 of IATF 16949), which must be conducted at least once a year. Section 9.3.2 defines the inputs for the management review in a generic manner. Detailed inputs are described in ISO 9001 [BF-010] under point (c) in Sect. 9.3.2. The following is an excerpt from these inputs:

"

(2) the extent to which quality objectives have been met

(3) process performance and conformity of products and services

(4) nonconformities and corrective actions

(5) monitoring and measurement results

(7) the performance of external providers".

At this stage, it should be clear that the goal should not be to enable just one or a few projects to meet ASPICE requirements but rather to set this as an organizational objective, allowing the entire company to evolve. In other words, ASPICE must be integrated into the organization's overall development system, with appropriate tailoring (systematic process adjustments) considered for different types of projects, such as smaller projects, variants, or those without customer maturity requirements.

Tracking

Once the measures have been correctly assessed, monitoring and tracking them enables a step-by-step approach to a certain level of capability. Planning, tracking, monitoring, and developing corrective measures in the event of deviations is nothing more than a PDCA cycle (Plan—Do—Check—Act).

However, if these measures have not been estimated correctly, this leads to the usual problems that can occur in a project, such as:

- Missed milestones
- Notorious overplanning
- Constant resource mismatch
- Inability to deliver on the agreed date.

For example, if achieving Capability Level 2 is required, the corresponding measures could be assigned to both Capability Level 1 (CL1) and Capability Level 2 (CL2). If milestones are set for when each capability level should be reached, progress could be tracked process by process. Once all measures have been defined, a burndown line for each capability level could also be established (Fig. 4.33).

Process Management: The Process "Multiverse"
In recent years, more and more standards and laws have been applied alongside models such as ASPICE (ISO 26262, ISO 21434, etc.).

To cope with the increasing number of standards and regulations, it is a good idea to establish integrated process management, i.e., to administer a single management system in which all possible requirements are included and incorporated. Unfortunately, another management system is frequently introduced when a standard comes into force (e.g., CSMS—Cyber Security Management System, ISMS – Information Security Management System, etc.). Developers rarely know which is applicable in which case.

Effective process management requires far-sighted planning and a deep understanding of organizational change management. We will not dive into the organizational change discipline here, as it is otherwise extensively covered in the literature (for example, in *Leading Change* by John P. Kotter [BF-011], with his change eight-stage process).

Another observation is that larger organizations often employ process developers who do not use the processes themselves. This usually leads to "good optics", where processes are theoretically excellent but have little practical value.

Another frequent observation in our practice is that processes are overdesigned, as they are a mixture of processes and guidelines. Process improvement here means that the process becomes increasingly detailed and complex. "More is better" seems to be the motto.

Furthermore, when process developers change within an organization, there is rarely a handover and appropriate training. That's highly unfortunate because a lack of understanding of the process landscape often leads to further deterioration in process quality.

We also observe that the tools used increase rather than reduce the development effort. Processes and tools are often not aligned, yet this alignment is a crucial success factor in process improvement.

We further observe that some organizations think *anyone* can design a development process. Improving a process requires professional experience, technical competence, high motivation, engineering freedom, and creativity to develop new, innovative products of the appropriate quality—and the same goes for process design.

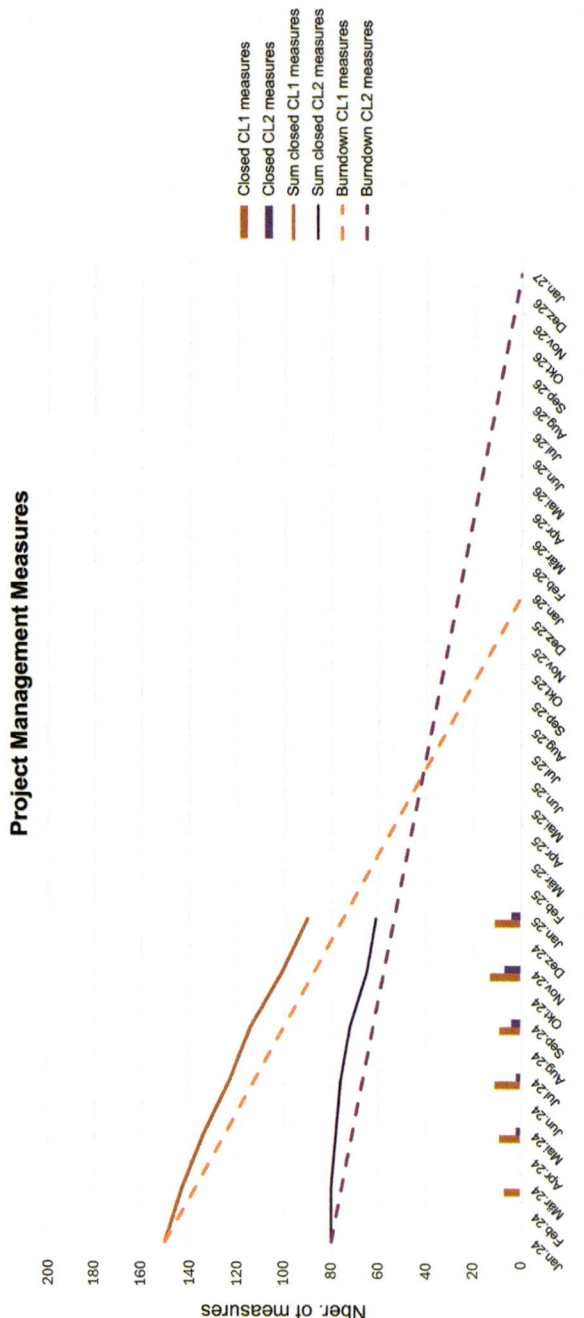

Fig. 4.33 Tracking example of CL1 and CL2 measures

Furthermore, clear goals must be defined for process improvement. These can be, for example:

- **Scope**

  The scope should define the processes to be integrated and how the process landscape should look (for example, see Fig. 4.34). Only development processes are often integrated, while management, business, and manufacturing processes should be addressed, leaving the interfaces for these processes poorly defined.
- **Architecture**

  Once the scope has been defined, the architecture must be set up, e.g., how many levels are allowed and how the processes are structured. Management processes can include marketing, business, human resources, and process management. Product development could contain all engineering processes. Supporting processes could include configuration management, change request management, problem resolution management, and quality management (Fig. 4.35).

The lowest level could be represented as an EITVOX (Fig. 4.36).

In addition, it is crucial to describe the connections between processes. Two processes are typically linked when they share a common work product. The linkage can be illustrated as follows (Fig. 4.37).

In this case, we speak of the *source* process, from which the corresponding work product for the next process is generated. Similarly, we speak of the *receiver* process, where the corresponding work product is inputted. One or more work products can form the interface between two processes.

It is also important to consider integrating fundamental aspects into the overall process landscape. Requirements could be a challenge; for example, is there a requirements management system that records all requirements (e.g., customer requirements, legal requirements, etc.) that categorizes them, sorts them, and makes them available to the corresponding engineering processes? Or are these steps integrated into the individual engineering processes?

Similarly, detailed planning and tracking aspects (see Sect. 4.3.5.5), which are required from the second capability level onwards, can be incorporated into each management, engineering, and support process, or consideration can be given to how this can be done at a higher process level. Such central planning could look like this in the project management process (Fig. 4.38).

Similarly, tracking, i.e., monitoring and any corrections to planning, could be mapped or integrated at this level. This would put both topics in one place. If it is a more complex system (system of systems), it could be structured similarly. The relevant specialist (usually a sub-project manager) is responsible for each specialist area. The final planning is coordinated in joint meetings with all planning roles.

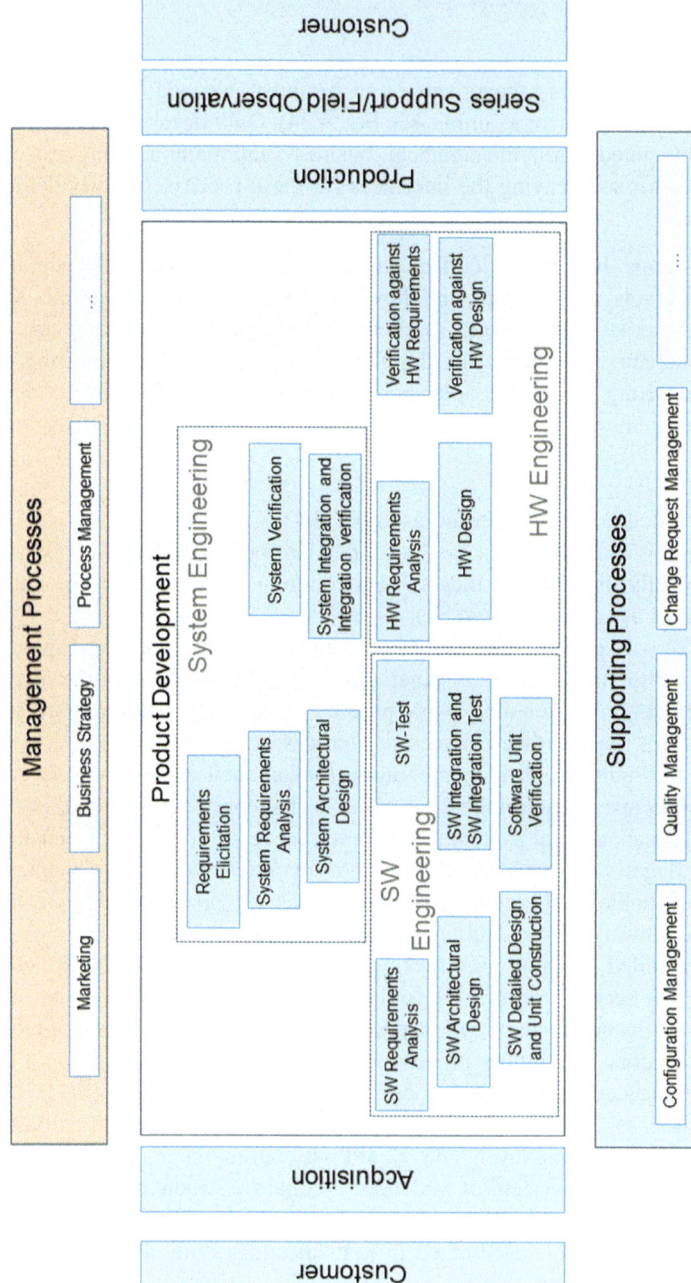

Fig. 4.34 Simplified process landscape

Process level	Content	Sub-content
Process level 1	• Process landscape	• Process groups
Process level 2	• Process groups	• Single processes (e.g., product development as a V-model)
Process level 3	• Single processes	• Main activities (top level of the process)
Process level 4	• Single activities	• EITVOX representation

Fig. 4.35 Example of a process structure

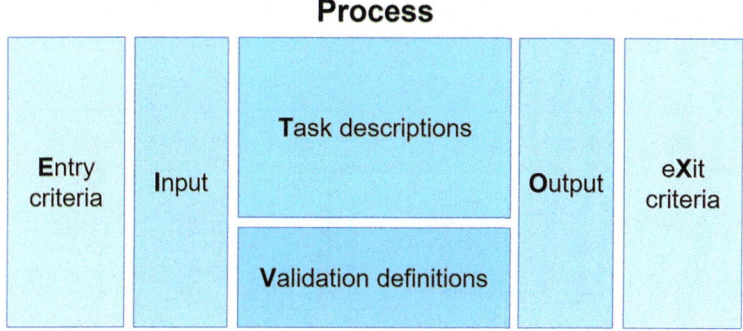

Fig. 4.36 Schematic representation of a process (EITVOX)

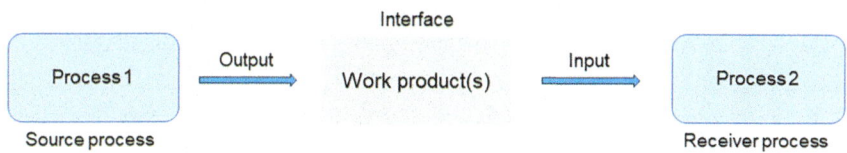

Fig. 4.37 Interaction between two processes

This also ensures the consistency of all individual plans with the overall project planning.

New topics, such as functional safety or cybersecurity, can be considered at this point. Once the requirements have been categorized accordingly, the engineering activities can be expanded depending on the categorization (an "add-on"). These are then implemented by means of corresponding strategies, concepts, and plans with detailed activities, allowing standard requirements to be met.

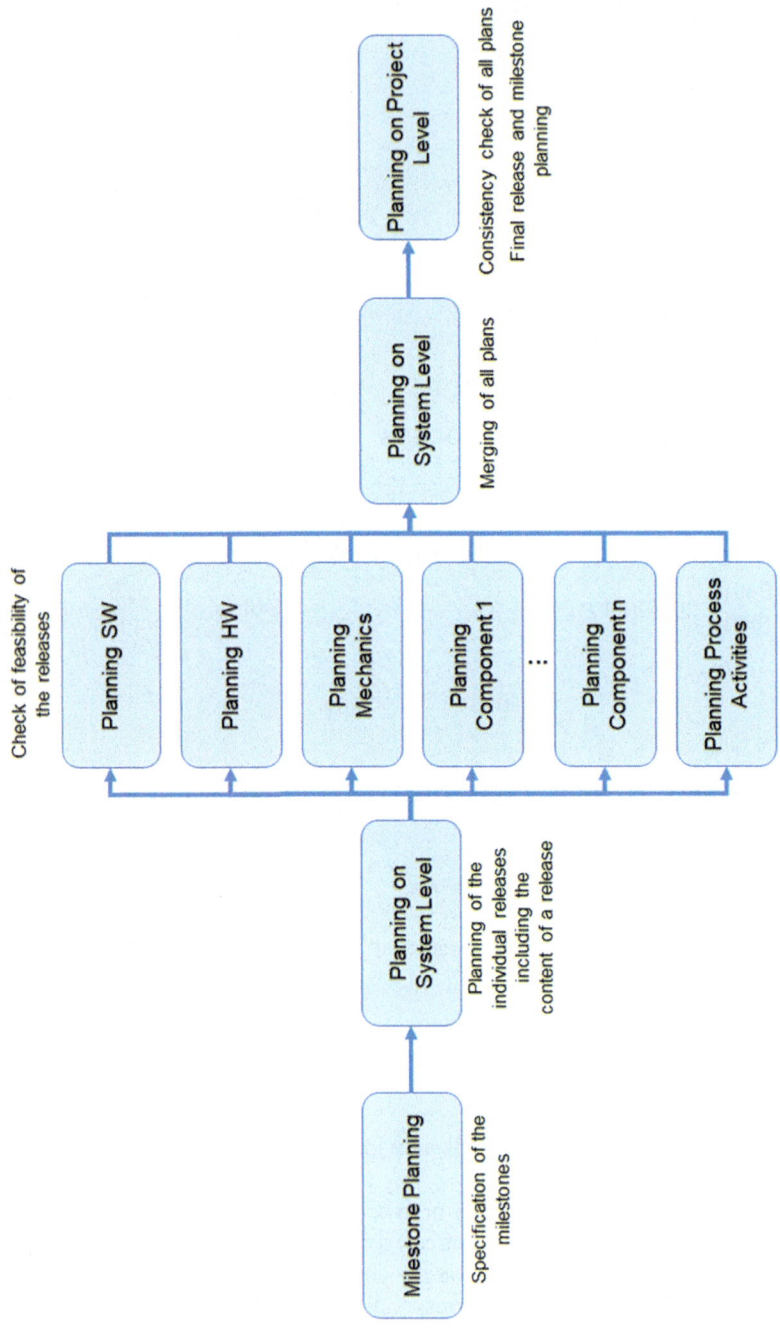

Fig. 4.38 Planning in project management

– Process detailing
  The fundamental issue here is to differentiate between what shall be done (process) and how it shall be done (guideline, method, tool description, etc.). As soon as both apply to a process, the process becomes enormous, and any updates and improvements require a lot of effort. Frequently, highly detailed solutions can be seen that overshoot the mark. For example, a configuration management process can consist of just a few activities, such as:
  – Define the configuration management strategy
  – Set up the configuration management system
  – Apply configuration management
  – Create baselines
  – Perform the baseline review
  – Report the status
  – Archive the project content at the end of the project.

That's usually sufficient.

The other aspect is that development processes don't need to reflect manufacturing processes in great detail. Manufacturing processes must be precisely described, as the repeatability of the product is a key aspect. These processes are usually written as work instructions, as each step is mandatory (we will return to the question of mandatory/non-mandatory later).

On the other hand, a development process should focus on critical process activities, such as a valid root cause analysis in the case of a defect (e.g., minimal documentation required) (Fig. 4.39).

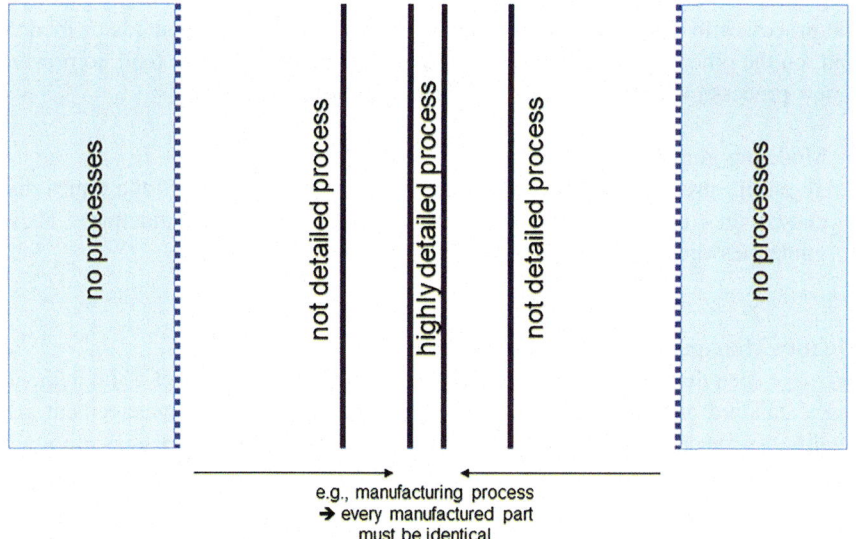

Fig. 4.39 Level of process detailing

The following aspects have to be considered:

- Reusable elements
  These can be sub-processes, roles, work products, and methods to maintain a lean structure and avoid expansive documentation.
- Trigger elements
  Depending on the process landscape, trigger elements can be integrated (e.g., process Y starts when process X has given the expected result). Examples could be a development project that can only start when a purchase order has been received from the sales process or a change request that can only be implemented when the sales process has received the order from the customer.
- Process mapping
  Users should not need to understand the exact mapping between defined processes and the standards or models they follow. Unfortunately, processes are often simply copied from the assessment model. Process developers often fail to recognize that this approach may not be best for the organization or projects and can lead to resistance.

The lack of critical thinking becomes apparent when the word ASPICE appears in every second paragraph, and the comment is that this is being done "because ASPICE wants it so." Such a style provides reluctant parties with a strong argument against working further on the process because copying the standards already appears "compliant." Often, these managers resort to various excuses, such as "insufficient resources," "overloaded staff," "the economic situation doesn't allow for this extra effort," "I need more people and additional support," and so on.

Depending on the modeling tool used for process design, a process expert can map the processes to ensure, on the one hand, the coverage of the standards or models and, on the other hand, that this coverage is maintained over time (e.g., to prevent a new process owner optimizing the process in the wrong place).

- Modeling guidelines
  To ensure processes remain comparable and understandable, all the topics discussed thus far should be documented in clear modeling guidelines. These guidelines should also be applied during process reviews.

Quality Management/Quality Assurance
We have seen that selecting the right resources is the Achilles' heel of implementing a standard or process model. This is equally true of quality management and quality assurance. It would be a mistake to assume that these two areas are interchangeable or that quality management experts can easily take on quality assurance

tasks. Quality management is focused on the achievement of internal or external regulations. It only addresses product quality superficially (for example, as part of a quality gate check).

In contrast, quality assurance aims to provide independent, objective verification that work products meet the requirements and that processes adhere to predefined standards and plans. The focus is not merely on formalities (e.g., ensuring the header of a work product is correctly filled out) but on evaluating the actual quality of the work products. The right resources are essential to do this effectively – such as a former software developer who can assess the code quality.

This dilemma can be solved in two different ways:

– Having an independent test group and assigning a test manager to this group ensures that the quality measures are performed correctly, as planned, and with the expected results. This would mean a work split between the test group and quality assurance.
– Process compliance checks are performed by quality assurers, who also check the work products related to management and supporting processes. The test manager checks the engineering work products and the quality measures performed in engineering.

The result of both activities can be resumed in a standard report, and the overall status can be considered for the release decision (Fig. 4.40).

This involves assigning resources with the appropriate qualifications and experience for quality assurance tasks. For instance, as previously mentioned, a former software developer would be well-suited to review and verify the quality of coded components or products.

The engineering processes always contain at least one quality measure (e.g., reviews).

Training
Implementing standards or process models means the existing processes are adopted. This involves the developers knowing and being trained to understand what has been changed and for what reasons (e.g., why a particular solution has been adopted). Unfortunately, organizations frequently believe that sending a mail to inform teams that a new version of a process has been introduced and listing the changes is sufficient. Suppose the new processes are first applied in a pilot project. In that case, this allows for quick training of the developers and checking with them to see if the process modifications can be applied quickly and efficiently. If this is not the case, training can be provided before the processes are rolled out.

4.3.5.5 Achieving Individual Capability Levels
Capability level 1 assesses whether the purpose of the process has been achieved. The process attribute PA1.1 "Process execution" is evaluated for this purpose. PA1.1 is based on achieving the process results, i.e., the purpose of the base

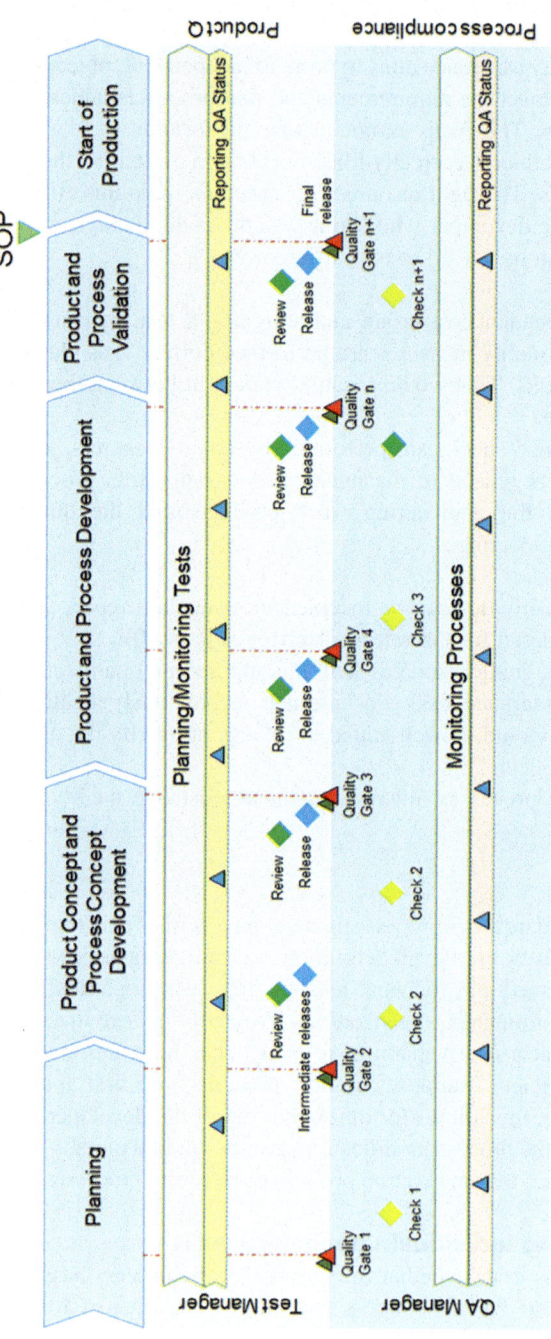

Fig. 4.40 Quality assurance split between QA Manager and Test Manager

practices (BPs) has been fulfilled, and the assigned work products have been created.

The fundamental prerequisite to achieving capability level 2 is that the process attribute PA1.1 "Process execution" has been assessed as "F" ("fully"). In addition, the process attributes PA2.1 "Management of process execution" and PA2.2 "Management of work products" are assessed.

In other words, it can be said that in addition to the base practices from capability level 1, requirements relating to project management, quality assurance, and configuration management are added for each process (see Fig. 4.41).

These relate to the respective process and not, for example, to overall project management, quality assurance, or configuration management.

An essential aspect of capability level 2 is the definition of process performance objectives. If the objectives are not defined, the planning, monitoring, and adjustment of process execution and the provision of resources cannot be completed.

As the focus is on the project, the objectives of the process performance shall be derived from the project objectives.

Figure 4.42 shows an example of an objective formulation at the highest level.

This formulation of objectives can be further detailed and could then look like this on a process level (Fig. 4.43).

Some of these targets can also be defined at the organizational level for individual products (e.g., error rates, permissible MISRA [BF-034] warnings, etc.). This facilitates the transition to the third maturity level later on.

Processes are linked on the right side in the overview of Fig. 4.43. This is intended to clarify which processes are influenced in principle by the concerned objectives. When deviations from these targets are identified, measures are defined to correct these deviations. This kind of objective allows monitoring and adjustments in the case of planning deviations. This is a typical capability level 1 objective.

However, this does not include any improvement of the applied processes. In this case, further objectives on capability level 2 shall be added, such as:

- On-time
  - Improvement of the estimations with a target not to exceed, for example, $\pm 15\%$
- On specification
  - Improvement of the delivered work products (requirements, architecture, design) with a target not to exceed a certain number of findings in the concerned reviews
- On quality
  - Improvement of the delivered work products with a target not to exceed a certain number of defects in the concerned verifications

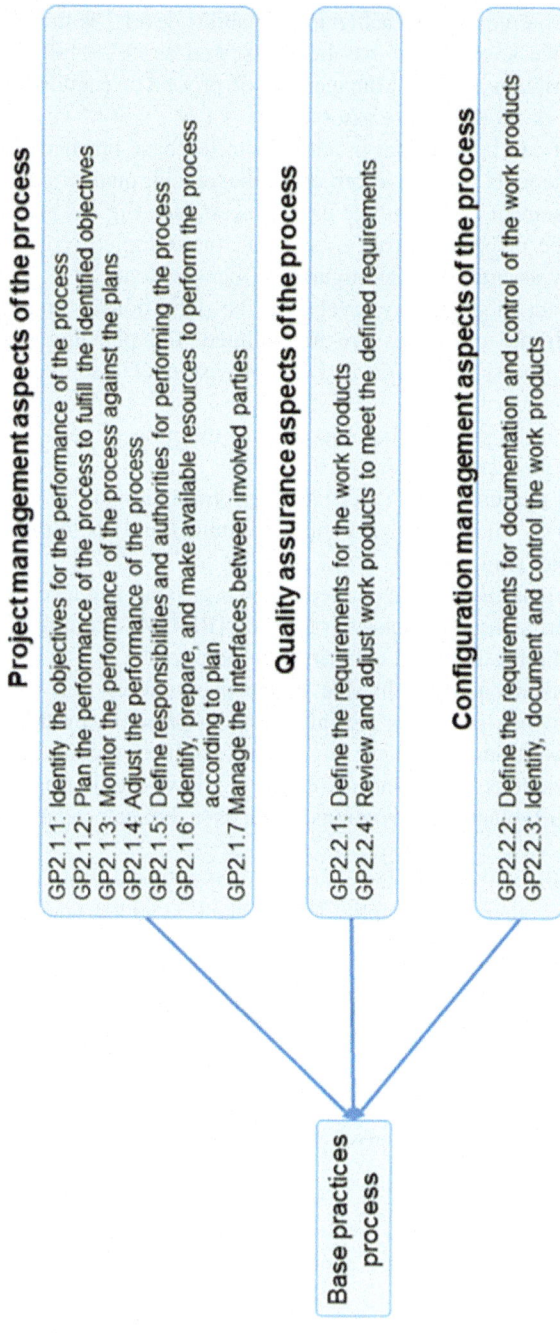

Fig. 4.41   Simplified presentation of the maturity level 2 requirements

Process	Project Objectives	Process Performance Goals	Monitoring	Reporting
1	"on time": all agreed milestones (int. & ext.) are achieved	All agreed milestones are on schedule	Milestone Plan	Status Report*
2	"on spec": all agreed requirements for a milestone are implemented	All requirements committed for a milestone are implemented	Status of requirements	Status Report*
3	"on quality": the expected quality for a milestone is achieved	All requirements committed for a milestone are implemented in the planned quality	Quality of the implemented requirements	Quality assurance report Status Report*
4	"on costs": target costs are achieved	The approved development budget is kept/met	Ratio planned/actual costs	Status Report*
5	Automotive SPICE® Capability Level 2 is achieved	Capability Level 2 is achieved	Gap Analysis Assessments	Gap Analysis Reports Assessments Reports

- These topics can also be covered using an earned value analysis. This can be used to measure and monitor the degree of completion of a project in comparison to the plan. This makes delays in deadlines visible, as well as the implementation of content and, ultimately, budget overruns.

Fig. 4.42 Example of a goal formulation

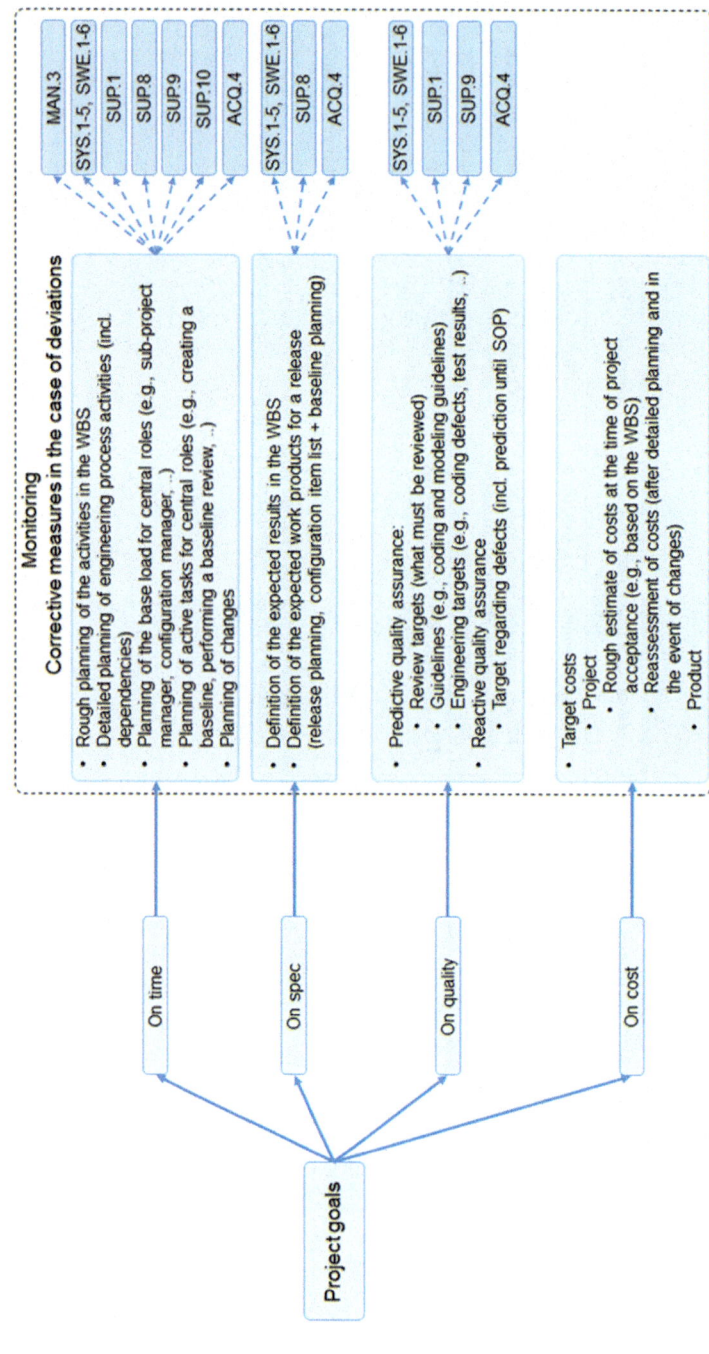

Fig. 4.43 Example of a more detailed formulation of objectives (without consideration of CL2 achievement)

Regarding the objectives for "on specification" or "on quality," the root cause analysis is based on the simplified defect flow model described in Sect. 4.3.5.4 can be used. If we take the example of findings in reviews, there could also be an issue with the quality of the reviews (lack of experience about the concerned topic or the review method). This could also be an objective for quality assurance as they have to ensure the quality of the work products (SUP.1.BP3: "Assure quality of work products").

This would allow lessons to be learned from the project and also, in some cases, at an organizational level. However, we have to pay attention to the fact that most lessons learned are not process topics. Often, the issue is that not everything is thought through in detail (due to the pressure to deliver on time, lack of experience, etc.), and it is also not checked with all involved stakeholders. Here, sensitization and how to spread this information (training, specialist meetings, data base, etc.) is the right approach.

Another important aspect of capability level 2 is the resources used. Every organization has defined a certain range of roles for different skill profiles. The expectation is that these roles in the project will be performed by employees with the necessary skills and experience. If this is not the case, an induction strategy must be planned (coaching, mentoring, training, etc.).

This is not so easy to check in an assessment, at least in Europe, as this is personal data and can only be viewed with the consent of the person concerned. Despite this, deviations can be found time and again. Either:

– measures have yet to be implemented or
– they can only be implemented later, or
– there are no accompanying measures (e.g., mentors) to enable the employees to carry out their tasks as required.

To achieve the third capability level, in addition to PA1.1, PA2.1, and PA2.2 (each rated "F"), PA3.1 "Process definition" and PA3.2 "Process deployment" must also be fulfilled (rated at least "L").

The third capability level requires a standard process, i.e., an organization-wide process that can be adapted according to the project's requirements. The organization must define bespoke guidelines that can be used to adapt the standard process to the project's requirements. This process is referred to as a "defined process" in the model.

Tables are often created for this with the different project categories or types and activities or work products that can be tailored.

A fundamental question now is the level of detail to which the processes are defined and which requirements they fulfill (for example, ISO 9000, ISO 9001, ISO 26262, etc.). The latter could be a measure of what can be tailored. The basis for tailoring would no longer be the project category but the requirements (standards, laws, maturity level, etc.) that must be met. This would result in mandatory and non-mandatory activities, work products, methods, and tools ("must"/"can" specifications).

This is often raised as a question within projects, which is understandable as the project team members do not know and do not need to know all the specifications. It is actually the task of the process developers and quality managers.

Unfortunately, precise documentation of specification coverage is often neglected, putting sustainable process development at risk (as discussed in Chapter 4.2). Database tools allow reference models to be stored and linked to processes, activities, tools, and methods. While this requires some initial effort and ongoing maintenance, it helps prevent inexperienced process developers from making unauthorized changes, avoiding the "Sisyphus effect" in process development.

4.3.5.6 Preparation for Assessments

In an assessment, evidence according to ISO/IEC 33002 is rated (see chapter 4.2.2: "b) each process identified in the assessment scope shall be assessed on the basis of objective evidence").

When planning the assessment, the scope is determined first, i.e., which processes are assessed (default set of reference processes of the VDA scope + plug-in processes + optional processes), whether it is an improvement assessment or a risk-based assessment, which product it concerns, which development scopes are available, which timeline is required, how many locations are involved, whether it will take place on-site or online, which assessors will participate, which documents will be made available in advance and which confidentiality agreements are required. The lead assessor and the sponsor coordinate this planning. The sponsor is usually the head of the division to which the assessed project is assigned.

The improvement assessment examines process implementation, while the risk-based assessment focuses on the quality of the developed product and the resulting risks.

As an assessment usually occurs in the middle of development, it is important to prepare for it and ensure that the required evidence is available in the appropriate quality. This means, for example, that 100% traceability must be available.

Traceability is defined or, respectively, required in ASPICE as follows (Fig. 4.44).

As you can see from this overview, traceability is a complex topic and often challenging to design in terms of tools. For example, different tools are generally used when creating architecture documents than in requirements management, and direct linking is often not easy to establish. In this case, an automated evaluation of traceability is also tricky. A remedy is often to add the IDs of the architecture elements to the requirements.

It is also essential to assess whether all requirements have been adequately addressed throughout the V-model, ensuring that all engineering processes are fully covered.

The corresponding process description must also be presented to prove the purpose of the practices. What is the corresponding activity that creates or revises the related work product? A well-documented purpose is needed to proceed with the rest of the process assessment. The answer must be well thought out and presented

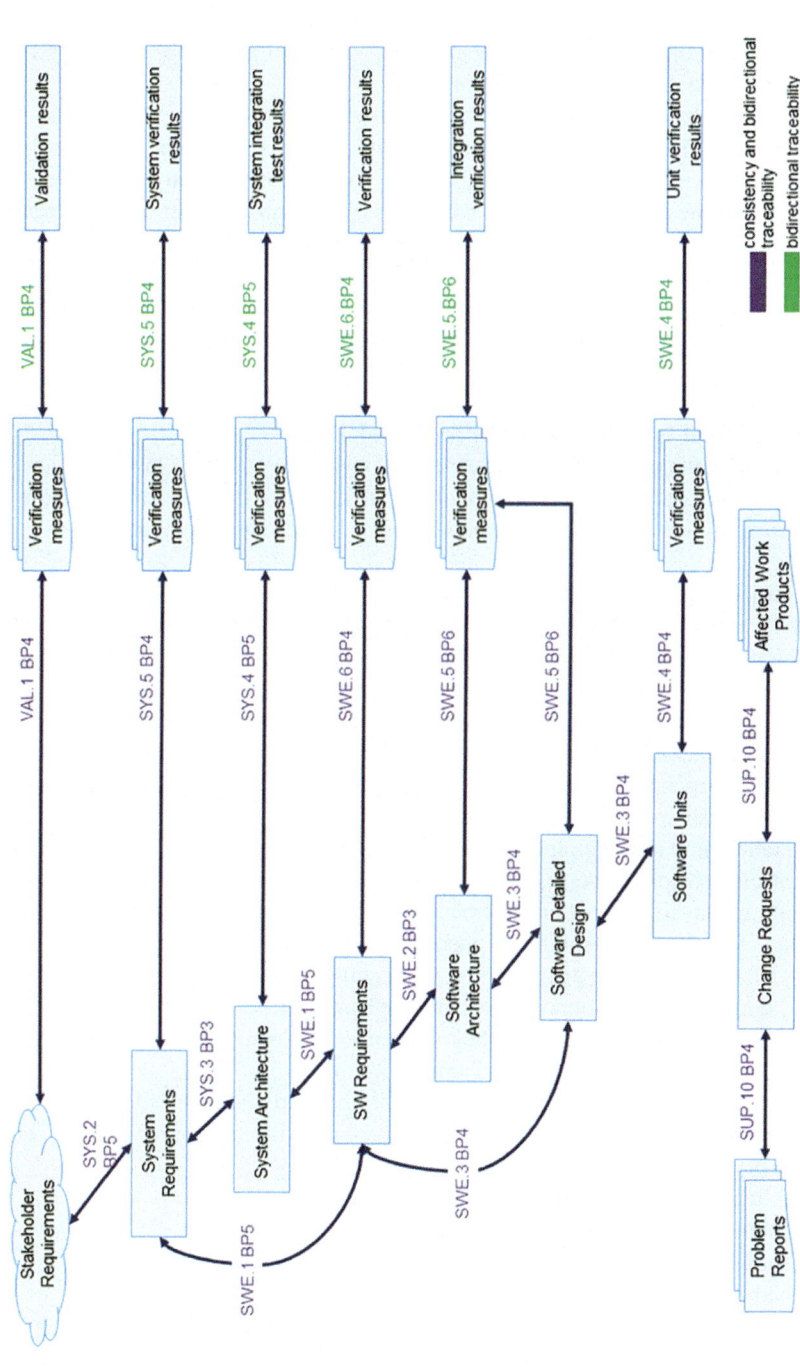

Fig. 4.44 Traceability according to PAM/PRM v4.0 (System/Software)

with confidence; otherwise, the discussion can become lengthy and confusing, creating unfavorable optics during an assessment, as the purpose cannot be proven or not all relevant work products can be shown. It is best to prepare well before the assessment to avoid such wasteful discussions.

We recommend scripting the assessment in advance to avoid these kinds of problems. The assessor is aware that the team typically prepares well for an assessment, and presenting a simplified script (e.g., slides showing individual process steps related to the practices) may be welcome. However, care must be taken with the work products to ensure that a so-called "golden sample" (a specially prepared, ideal example) is not solely presented. In such cases, the assessor should review additional work samples that they select independently. This can be challenging if the project is in a later stage of development, with only limited changes from the previous project.

4.3.5.7 Critical Issues at Capability Level 2

We have seen that a critical aspect of capability level 2 consists of planning, tracking, and, if necessary, taking corrective action. Certain elements are especially critical since project planning happens on several levels. These include:

- Key project milestones:
  - WBS (Work-Breakdown-Structure) was created at the start of the project. This is a basis for further detailing project activities (for example, in the sub-projects).
  - Release or feature planning. This is where the individual releases (deliveries) are planned, and the respective content/features of those releases are determined. Depending on the customer, a release's expected product maturity level is also specified for delivery.
  - Detailed activities. Complex projects involve thousands of detailed activities that can no longer be managed using simple spreadsheets. Many database applications for project management are also not designed for thorough planning. More modern solutions, such as JIRA with suitable extensions (such as JIRA Structure or "Big Picture"), are suitable here.
    Of course, the link between the upper planning level (milestone planning) and these activities must then be made easily recognizable, for example, by automatically adopting and updating milestones.
    A further topic is to limit the size of the activities in order to avoid micromanagement (for example, activities limited to a minimal size of one day).
- Other structuring aspects must also be taken into account:
  - Grouping of milestones and work packages by releases
  - Grouping by topic/employee
  - Load factor of the employees
  - Linking the individual activities
  - Estimated resources
  - Estimation assumptions

- Corrections of estimates and documentation of what has been corrected and why
- If corrections are necessary, it is essential to consider how to make these findings available to other projects or the organization (for example, in a dedicated database).

Lead management and supporting roles, such as project manager, sub-project manager, configuration manager, etc., are also typically used in the project. These roles are usually busy with frequent meetings, planning and tracking activities, and handling emails and other communications. Detailed planning could be required here, but it would be difficult to incorporate this level of detail into all available communication tools. For example, if the presenter shows longer-lasting activity, the assessor may request more detailed planning (e.g., activities longer than ten days must be broken down into smaller chunks).

One potential solution could be estimating and documenting this baseline workload for the year (e.g., in a spreadsheet), factoring in regular tasks and sporadic activities, such as unplanned meetings. If, for example, this basic load is 65%, this should also be visible in the detailed planning (e.g., with a resource availability of 35%). This thorough planning should demonstrate how baseline creation, review meetings, release approval, etc., are performed.

An example of such an estimation is shown in Fig. 4.45. It focuses on the role of a configuration manager.

Creating such detailed estimates can often be time-consuming and extensive, but assessors usually demand them unconditionally at the second capability level, and it is absolutely necessary to plan such management or supporting roles.

4.3.5.8 Costs of an Assessment

If the goal is to reach the second capability level, it makes sense to perform a gap analysis beforehand and only then continue with the official assessment. The gap analysis provides detailed insight into the project's status and any gaps that exist. Based on this, further assessment planning can be carried out.

The gap analysis could reveal significant or smaller gaps. Examples follow below.

Substantial gaps. There can be a whole spectrum of problems here, for example:

- Some processes are not defined
- Process interfaces are not complete
- Roles are not described at all or not described comprehensively.

This refers to organizational roles defined in the processes (e.g., project manager, developer, quality assurance representative, etc.). The ASPICE document does not describe any specific roles, but detailed role definitions are expected, and work will be carried out following them.

Estimation baseload of SUP.8 Configuration Manager	Assumptions & Risks	Hours per topic	Frequency per month	BL per year (2024) (nber.)	Year (2024) (hours)
Regular meetings	Participation in project meetings	1 meeting per week with a duration of 1,5 hours **Risks:** none	1,5	4		66
Non-regular meetings	Management meetings, meetings due to sporadic topics	2 meetings per week with a duration of 1 hour **Risks:** none	2	4		88
Baseline planning	Integration of the baselines into the configuration item list Update of the baseline planning	First baseline planning Update of the project-specific baseline planning for each baseline **Risks:** New topics could require more effort	64 2		4	64 8
Configuration management system	Setup and maintenance of the configuration management system(s) (incl. access rights)	Setup Maintenance **Risks:** New topics could require more effort	60 4	1	4	60 16
Training	Training of new project staff Training on new/improved configuration management topics	Training and initial support for new project members and for changed procedures **Risks:** Higher fluctuation in the project	2	1		22
Configuration management status report	Status report for each baseline	Status report before completion of the baseline **Risks:** Delay of artifacts may require additional baselines	8		4	32
Follow-up of open points	Tracking, contact in the event of missed deadlines, escalation	Tracking of open points assuming 24 measures with a duration of 30 minutes per month **Risks:** Delays in development could lead to more measures (greater focus on the completion of development activities)	12	1		132
Total (hours)					**488**

Fig. 4.45 Example of estimation of base load for a configuration manager

– Templates are not available or are incomplete
– Minimum expectations for work products are not defined
– Processes are not applied or applied differently than they are described.

In such a case, it can be assumed that it will take longer to achieve the targeted capability level (usually at least one year). This is because the identified gaps must be closed first, and only then can the required evidence be generated (by generating work products fulfilling the updated artifacts). This includes updates in the project schedule since those work products must be highly prioritized.

If the resulting work product is not of the expected quality, the related processes (engineering processes, SUP.1, SUP.9) must also be used to determine the causes of deficiencies and define corrective measures. A plan must also be drawn up and adhered to in order to rectify the defects—by the latest at SOP.

Closing the gaps becomes even more challenging when a customer assessment occurs during this time. In such cases, depending on the customer and the project's status, the customer may demand closer tracking (also called "boot camp" or on-site "task force," etc.) and become directly involved in lower-level project activities, including requiring the on-site presence of key project members. This is a proposition that is impossible to refuse. Even though such situations may feel like an embarrassing admission of failure, they occur regardless of the size or type of the supplier organization.

Minor gaps. These may be the following:

– Incomplete/incomprehensible strategies
– Small gaps in some processes
– Minor gaps in templates
– Gaps or inconsistencies in reporting
– Inconsistencies/insufficient timeliness of work products and planning
– Sluggish remedial measures (especially for critical bug fixes)
– Typical capability level 2 problems include incomplete proof of competence, inadequate resource planning, quality assurance measures, configuration management not implemented according to strategy, and adjustments/decisions not documented.

The same issues apply here as in the first case, but the gaps can be closed more quickly. However, as this is an ongoing project, it can be assumed that these gaps will still be difficult to close (unless additional and efficient resources are added to the project).

It's rare for project organizations to advance from the first to the second capability level in less than six months, though it does occasionally happen. Most organizations take much longer because of issues like inconsistent process improvement efforts, poor consulting, lack of motivation, or ongoing project overload—or, sometimes, all these factors combined.

As a rough estimate, the process designers and the project team members need 20 to 30 percent of their working time to close the process gaps and prepare for the assessment.

The gap analysis and the assessment will tie up the project team members for at least another two-and-a-half weeks. Some organizations also conduct so-called trial assessments to prepare the project team members for external assessments. Just as during formal audits, the team members must be trained to behave correctly in an assessment (for example, only answering questions about your role and area of responsibility, not opening up any sideshows, answering openly and honestly, not showing uncertainty, etc.)

A higher maturity level requires significant effort, but some organizations view ASPICE purely as a cost factor rather than an investment. Many organizations focus solely on achieving the required capability level for their most critical project (based on the number of units sold), often not integrating "lessons learned" into the overall process.

In addition to these preparation costs, there are the actual assessment costs—ASPICE assessors have a high price tag. As a rule, an assessment team of two or three experts is involved, and the presence of the actual project team members is required for the duration and preparation of the assessment.

Additionally, work products must be regularly and formally reviewed after reaching a specific level and establishing the quality assurance process. Some customers even mandate that all development processes be audited every two months.

These observations are consistent with the fact that only very few level 3 assessments are performed.

4.3.5.9 The Most Common Causes of Process Failure in Projects

We saw the difficulties in implementing the model in Sect. 4.3.5.4. This section will examine the most common causes of process failure in projects and organizations.

Corporate Culture

Complex projects require the assignment of numerous specialists in different areas. Conflicts of interest can arise between different line structures, regardless of their form (e.g., matrix versus project).

Sometimes, cumbersome multidimensional matrix structures are the norm. Whether such a structure is sustainable is a question for our dynamic automotive business. Wouldn't it be more effective for project team members to report directly to the project manager for the duration of the project without interference from line managers? Such questions should be taken seriously, as start-ups operating with smaller teams and presumably requiring less formality tend to be far more efficient and innovative.

Apart from the organizational structure, corporate culture is a significant success factor. What values does a company represent, and how sustainably does it act according to these values? What happens in stressful times? Is everything thrown aside in favor of fire-fighting, with the resulting shortcomings and costs unfairly shifted onto the employees?

What does an organization do to train, develop, and retain employees? Are there development programs that not only promote management careers but also specialists? Is there a so-called "expert career path"?

Organizations are often very complex and have long traditions. Navigating such a complex environment can lead to confusion and inefficiency without solid systems knowledge, analytical skills, and technological openness grounded in critical thinking. Developing such system specialists requires years of support and training.

The same applies to the quality of managers. Can they lead employees with skill and charisma? Can they deal with complex issues in a qualified manner? Do they have analytical skills? What target vision do they represent, and can they communicate it effectively to their employees? Can they offer their teams orientation and integrity? Or do they focus on unquestioning obedience to superiors and hypocritical rhetoric?

What about the evolving company culture? Are changes being implemented merely to maintain pressure, or is there a genuine effort to incorporate improvements with a clear understanding of their purpose? Corporate sustainability is crucial; without it, organizational development cannot continue. The key is to leverage strengths and experience rather than discard everything. A reliable foundation is essential, one that all employees can trust. While start-ups with no history to reconcile don't require this, experienced individuals can contribute their knowledge and provide direction. Strong vision can also offer guidance. Think of the founders of Apple, SpaceX, and Tesla—clear goals must be established before success can follow.

There are further aspects to consider. How does the company deal with internal criticism? Is it accepted or not tolerated at all? If it is not tolerated, employees will no longer use their brains over time—they will merely comply. Another critical aspect is the culture of handling errors. Is the focus on assigning blame, or is the goal to learn from mistakes collectively? In the worst-case scenario, a blame-oriented approach can result in the suppression of error reports, with far-reaching consequences.

Although we have previously discussed the somewhat dry and formal aspects of ASPICE, lively, upbeat, consistent corporate management with integrity is essential for meaningful process improvement. Proper use of the ASPICE model requires strong leadership. Without that, ASPICE is pointless.

The Importance of the Right Goals

Why is goal setting so crucial for effective process improvement? The answer lies in the meaningfulness of process improvement, as it results in meaningful goals. Goals are indispensable for reaching, monitoring, and tracking progress.

In our process improvement practice, we are experiencing two different "anti-patterns."

Anti-Pattern 1: The goals are well-defined but are not followed

Well-structured objectives are defined, e.g., 'I know we want to implement an ASPICE-compliant process (e.g., to improve quality)'. However, these goals are repeatedly overridden by operational objectives. The organization is in a so-called "firefighting" mode, focusing only on extinguishing fires. New fires occur, and the vicious circle never stops.

Anti-Pattern 2: The goals are not well-defined

The objectives are insufficiently defined and vary depending on customer pressure. Different experts/consultants support them during implementation, sometimes resulting in different implementation variants, which usually fail. Conclusion at the management level: ASPICE represents additional costs and does not move the project forward.

An anecdote from our practice once circulated about a quality manager who informed a working group of internal assessors that while ASPICE was crucial for the company, he had yet to learn how it could be implemented or within what time frame. Admitting such uncertainty requires a great deal of courage—though, with that kind of pay grade, it's a little easier to be candid.

It becomes apparent that well-defined goals alone are not enough. A goal must be followed by appropriate actions actively supported by management. The legendary "resources" sometimes play a decisive role in this.

The Challenge of Having the Right Resources

The most common problems are a need for more training and a lack of resources for such training. This lack of resources can be for several reasons, such as:

– Lack of *suitable* resources
– Improper effort estimation.

A lack of expertise and experience can present additional challenges for the project team, including coaching or quality assurance support. In some cases, software development seems fully compliant, only to discover sometime later that the actual software has yet to be developed.

Another anecdote illustrates the lack of qualification. Once, we asked a project lead whether coding was manual ("hand-coded") or used a model-based approach, and the puzzled project lead asked, 'What is modeling?'.

Issues arise when a project is supported by a "fake" organization whose task is to make an assessment "simply happen." Such a team is tasked with assessment preparation and participates in the assessment, pretending to be the actual project team. This quickly raises red flags during assessments, as the interviewees often scramble to find evidence for basic questions, revealing their lack of involvement in the daily project activities. In such cases, it's clear that the assessment is based entirely on "golden samples."

Another process design resource issue can involve using experts/coaches from different companies. This can result in cascading interfaces, possible loss of information at each interface, and a slowdown in the improvement progress (theoretical example in Fig. 4.46). In addition, greater project management effort is needed to manage all these interfaces. The same can apply at the engineering level when several engineering service providers are used for one development topic.

The Challenge with Consulting Services

In our experience, consultants are frequently selected based on hourly price ("best cost" phenomenon). However, knowledge and competence are not available at a low cost.

Another flawed consulting strategy is that consultants are not permitted to get involved in development activities. They are supposed to consult—not do the project work, because the actual development activities are supposed to be performed by "best cost" developers. This becomes critical if consultants themselves are not the"best cost" consultants. This might appear logical, but it causes more complex project work and, considering all factors, is most likely not saving anything. The question is how consultants can help efficiently if they are not allowed to have technical hands-on insight. If they do not see which problems arise from processes and tooling, how can they analyze the subject matter to find a more viable solution?

Instability in Line Management

Management can easily destabilize process improvement efforts in various ways. For instance, middle management—sometimes unflatteringly referred to as "princes"—may fail to support the clearly stated goals of their superiors, instead pursuing entirely different objectives within their own domains. Often, the result is prolonged inaction, with the standard excuse or "wild card" being the supposedly overwhelming resource demands of ASPICE.

Another aspect is that, for political or marketing reasons, tinkered and mikeshift solutions are requested to present a certain level of progress that could be more achievable.

The following is an example of how bringing unrest into a project can affect results (Fig. 4.47).

Two releases are planned in this example project. As delays in implementation have resulted in the first milestone—and, therefore, the start of series production—being late by at least eight months, the decision has been made to plan two interim deliveries. These are fundamental solutions from the previous project and new functions to carry out appropriate marketing measures. What needs to be addressed here is that the project is behind schedule, partly due to resource load. This means it will barely be possible to implement the interim deliveries without once again jeopardizing the start of series production and putting even more stress on the project.

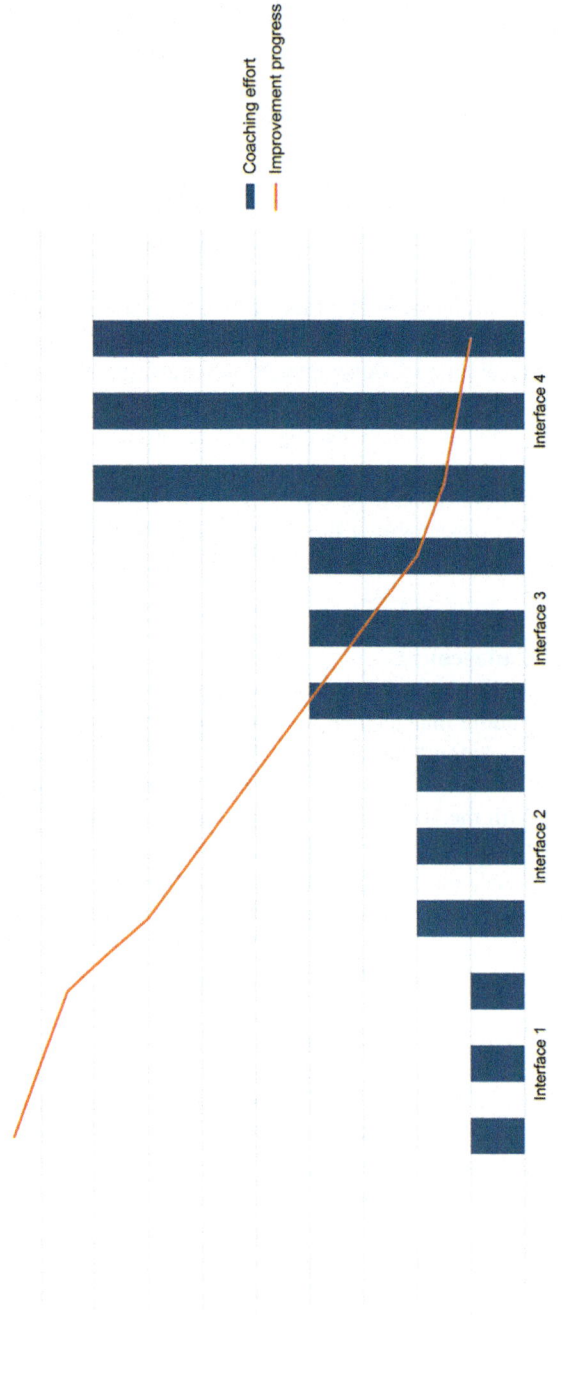

Fig. 4.46 Improvement progress

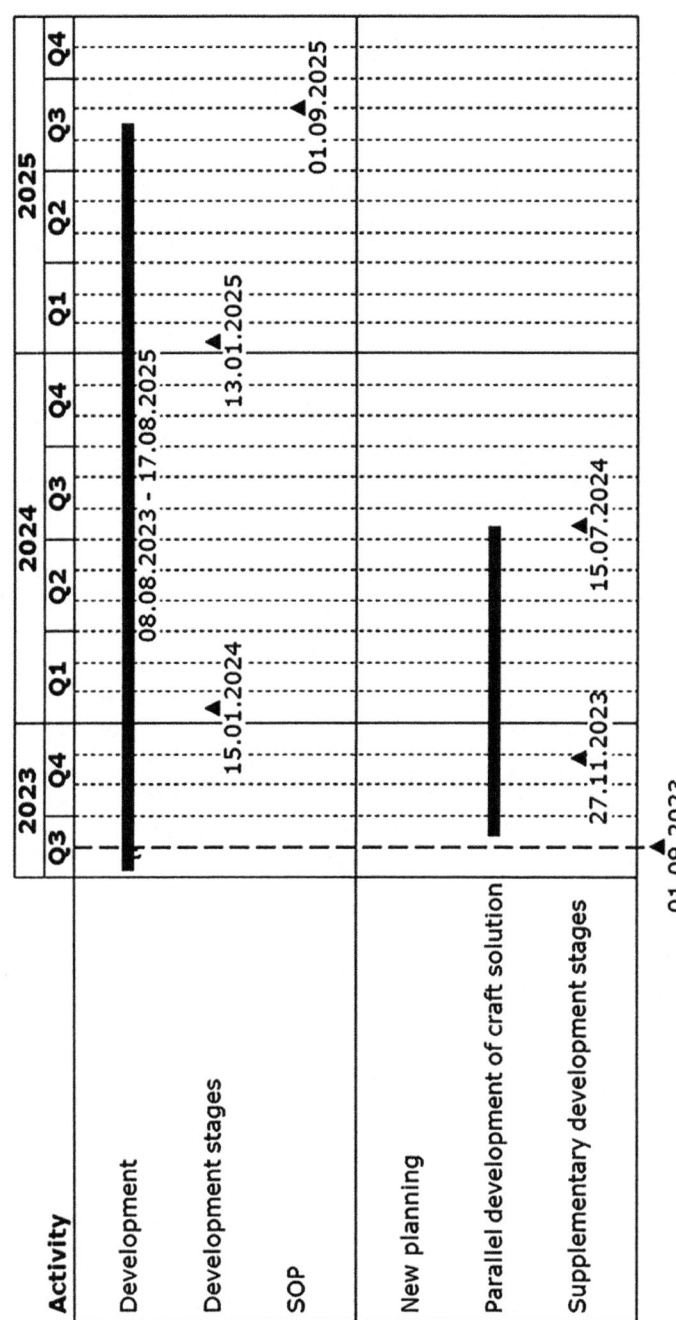

Fig. 4.47 Example of increased milestones in a time-critical project

Conclusion: in tight deadline situations, extending the project scope rarely helps; instead, it creates a higher risk, often resulting in resentment within the project team.

Sometimes, the decision is made to shorten development times to meet customer requirements (for example, because another supplier has dropped out). This usually means a significant increase in project risks that are rarely honestly evaluated. Depending on customer pressure, the expectation is still that the project will fully implement all processes, even if there is not enough time. In this instance, a decision would have to be made as to which process steps can be omitted with an acceptable level of risk to give the project the possibility of successful completion. Of course, this is only possible for functions not assigned to critical topics, such as functional safety, legal requirements, or cybersecurity.

Process Management

In Sect. 4.3.5.4, in the paragraph "Process Management: the Process "Multiverse,"" we have already described the main requirements for process management and how challenging it is to ensure consistent process improvement activities. Often, the most recently hired employee is tasked with process improvement. This can work in theory, but in most cases, it fails because the fresh hires don't know the corporate environment well enough to help.

One frequently observed phenomenon is that process designers are detached from the actual development team, and their solutions are ineffective. Instead, they continue with increasingly complicated and "heavy" processes. There's a common misconception that a complex process can mitigate risk, even when less experienced people are involved, but in reality, complexity often creates more problems than it solves.

Another issue is that there are no modeling guidelines for processes, which means every process can look different regarding the level of detail and the description method. In many cases, the process landscape is limited to development processes. As soon as interfaces to other processes come into play, problems can arise (e.g., purchasing, sales, etc.).

A proper *process strategy* is required for process development. There is often a lack of modeling guidelines here, reflected in both the structure and the quality of the processes.

All these weaknesses mean the processes are not implemented as described, inevitably leading to weaknesses in an assessment.

State-of-the-Art Development

Applying standards (e.g., ISO 9001, ISO 26262, and so forth) is considered state-of-the-art development. Contrary to occasionally expressed common belief, passing an ASPICE assessment does not guarantee such "state-of-the-art" quality. Instead, it is paramount to *implement* all relevant processes if they are based on all the appropriate standards (relevant for capability level 3).

If an organization does not internalize all the required standards, this leads to two problems:

- **Developers don't Know the state-of-the-art expectations**
This represents a risk on the product development level, especially if it is a complex product (e.g., a system of systems) with new features (e.g., advanced safety and drivability features)
- **Consistency Issues**
A consequence of the problem above is that releases/deliveries are not planned as a consistent set of work products (e.g., according to a consistently defined baseline). This leads to all kinds of issues, such as interfaces not working correctly, features performing only partially or not at all, a multitude of defects, etc.

4.3.5.10 Key Differences Between PAM/PRM v4.0 and PAM/PRM v3.1

PAM/PRM v4.0 includes new reference processes, while other processes that were included in PAM/PRM v3.1 have been dropped (for example, SUP.2, SUP.4, SUP.7, ACQ.3, ACQ.2, or ACQ.11).

In principle, there are three significant changes in v4.0 compared to version 3.1:

- **New processes for hardware engineering** (HWE.1-4), **machine learning** (MLE.1-4), and **validation** have been added, integrating the hardware engineering and machine learning models that previously existed as independent process reference models.
- **The ASPICE "strategies" are no longer included** in capability level 1. Instead, they are integrated into capability level 2.
"Identify the objectives for the performance of the process" was changed to "Identify the objectives and **define a strategy** for the performance of the process." In this context, the new notes are also more helpful than the previous ones.
Note 1 gives the following examples of process execution targets: budget targets, delivery dates to the customer, test coverage, and process throughput time (see Fig. 4.42: Example of an objective formulation).
Note 3 adds that the process execution strategy must not necessarily be documented individually for each process but can be described in the project manual or as part of a common test strategy document. Both notes provide better orientation for the projects, as the previous definition was confusing.
- **The training model has changed** because there is now an introductory training course for process experts; this is the basis for training as a Provisional Assessor and Competent Assessor. This training is either generic or specific for the model supplements (such as HW or mechanics).

In addition to these changes, there are also changes in terminology and individual practices, for example:

- **Verification Criteria**:
  There was a separate base practice (SYS.2, SWE.1) for this in the requirements processes (SYS.2.BP5, SWE.1.BP5). This could mean that verification criteria must be documented separately from the requirements. However, they are part of a requirement, the verifiability of which is now described in SYS.2.BP1 and SWE.1.BP1. The specific base practices for the verification criteria have been dropped (BP5). Note 2 also contains the addition that verification criteria are included in the requirements set.
- **Traceability and Consistency**:
  In PAM v3.1, there were separate base practices for both traceability aspects. These are now described as one base practice, as both are required to ensure consistency. For example, you can use an automatic evaluation to check that each stakeholder requirement is linked to at least one system requirement. In addition, a review ensures that the implementation of the system requirements fully covers the content of the stakeholder requirements.
- **Application Parameter**:

  In PAM v3.1, application parameters were described as part of the system requirements (Note 1). These are omitted entirely in v4.0 as they are part of the verification (i.e., SYS.4, SYS.5) and, in this context, are defined by the product lifecycle (for example, the parameterization of a specific product variant).
- **Terminology for "Measure" and "Metric"**:
  Previously, these two terms were not used consistently. A uniform definition has now been identified:
  - quantitative measurement—metric
  - action plan—measure
  - act operationally—action
- **Architecture Design**:
  In PAM v3.1, architecture design was described in three base practices in SYS.3 and SWE.2, for example:
  - SYS.3.BP1: Develop system architectural design
  - SYS.3.BP3: Define interfaces of system elements
  - SYS.3.BP4: Describe dynamic behavior.

These practices were summarized in two base practices in PAM 4.0:

- SYS.3.BP1: Specify static aspects of the system architecture
- SYS.3.BP2: Specify dynamic aspects of the system architecture.

The base practice BP2 ("Allocate system requirements") has been omitted as the traceability and consistency check now ensures this aspect.

Similarly, the BP5 for evaluating alternative solutions has also been dropped. The new base practice 3 (Analyze system architecture) covers this point with the following passage: "Document a justification for the decision regarding the system architecture."

4.3.5.11 ASPICE Conclusion

Every assessment model always stands or falls with the following factors:

- Incorrect estimation of the required change
- Insufficient support for change (sustainable, reasonable, and organization-wide goals, sufficient resources, etc.)
- Insufficient analysis of the effects of the change on the organization (mostly retention of current structures and further splitting of responsibilities with the addition of further managers)
- Theoreticians, not users, evaluate and describe the necessary changes (processes including tools)
- Lack of sustainable and holistic process development approaches
- Inadequate support of processes through suitable tools.

ASPICE has several advantages over CMMI. Firstly, ASPICE explicitly requires the V-model, whereas this was only an option in CMMI. While this freedom of choice could be seen as an advantage, in ASPICE, the question is eliminated, which saves us some discussion.

In addition, ASPICE offers potential cost savings compared to CMMI, especially at the second maturity level. In ASPICE, only one project must be assessed, while in CMMI, the entire organization must be appraised.

Furthermore, unlike CMMI, the ASPICE reference model is tailored to the automotive world and is not described generically. However, since all critical projects above a certain size and those with a high degree of innovation or importance are now required to reach a certain maturity level, costs are skyrocketing, partly due to the abovementioned reasons.

The increasing complexity and demands of automotive manufacturers on their suppliers will require more balanced approaches to ASPICE in the future. Because of the current challenges in our industry, developing effective and sustainable implementation strategies for ASPICE is becoming increasingly important, with concrete solutions being proposed in Chap. 6 (CORE SPICE). A key question remains about how in-depth and detailed ASPICE should be applied to ensure the competitiveness of vehicle manufacturers, particularly in the face of international competition, many of whom do not demand ASPICE compliance.

4.4   Security and Safety

4.4.1   Cybersecurity

With the growth of electronic systems, interest in manipulation possibilities has also risen. The rise in popular tuning measures is consistent with this trend. Tuning refers to modifications to the vehicle that are not originally part of the vehicle's design. Both OEMs and some tuners have supported such measures. A good example is the increased performance of common-rail diesel systems with an inexpensive tuning kit. A simple resistor is installed between the control unit and the rail pressure sensor that simulates a lower pressure for the control unit, meaning more diesel is injected and more power is generated. With modern diesel injection systems, however, it is questionable how long the particle reduction system, for example, will be able to handle this.

The introduction of wireless (e.g., transponder keys) and connectivity (diagnostic interface, Bluetooth, WiFi, eSIM, over-the-air (OTA) software updates, V2X, and so on) has forced the industry to focus on cybersecurity risks.

A 2015 Jeep Cherokee hack [BF-016] is an example of a security risk that turned into a problem. Security researchers managed to remotely hack into a Jeep Cherokee and take control of various functions, including the transmission and brakes. Although no accident is known to have occurred in connection with this software weakness, it did demonstrate potential vulnerabilities in the vehicle software.

Statistical evidence of cybersecurity hazards demonstrates the severity of this problem. In Fig. 4.48 (reproduced with permission from Upstream Security Ltd), the *Upstream Security—Global Automotive Cybersecurity Report 2020* [BF-033] shows a sharp increase in cybersecurity incidents between 2010 and 2019.

Figure 4.49 (reproduced with permission from Upstream Security Ltd) illustrates the distribution of the main attack vectors according to *Upstream Security— Global Automotive Cybersecurity Report 2022* [BF-031].

Figure 4.50 (reproduced with permission from Upstream Security Ltd) illustrates the most critical threats and vulnerabilities according to *Upstream Security—Global Automotive Cybersecurity Report 2022* [BF-031].

Based on the forecast for the first half of 2023 from the *Upstream H1'2023 Automotive Cyber Trend Report* [BF-032], data leaks (including personal data) will increase. This is also due to the increasing number of attacks on backend servers (including APIs), which enable access to more data and can impact entire fleets.

Organizations have emerged to counteract this, and databases have been set up to document and share knowledge about known attack scenarios. These include the following databases:

- **NIST Vulnerability Database (NVD)** [BF-017]:
  The NIST (National Institute of Standards and Technology) promotes innovation and US industrial competitiveness by advancing the science of measurement, standards, and technology. The NVD includes datasets referencing safety checklists, safety-related software defects, misconfigurations, product names, and impact metrics.
- **Common Attack Pattern Enumeration and Classification (CAPEC™)** [BF-018]:
  CAPEC™ provides a comprehensive dictionary of known attack patterns used by hackers.

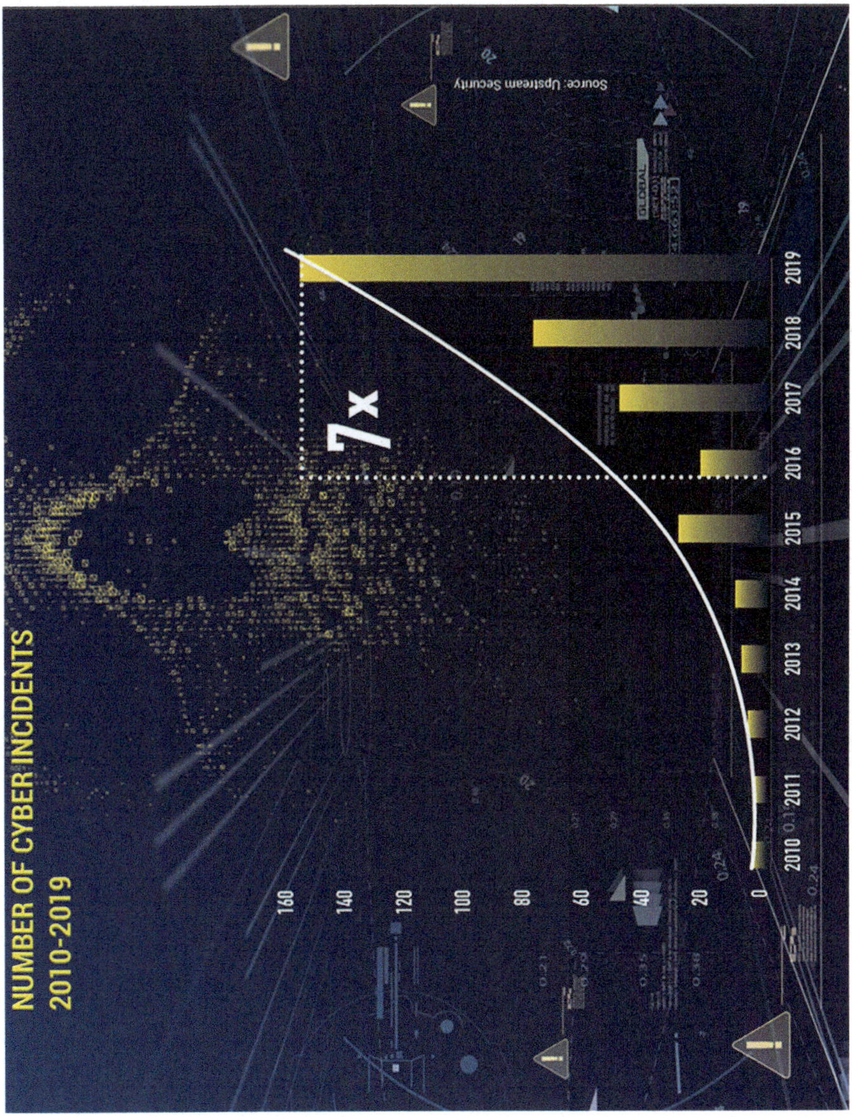

Fig. 4.48 Number of cybersecurity incidents 2010–2019

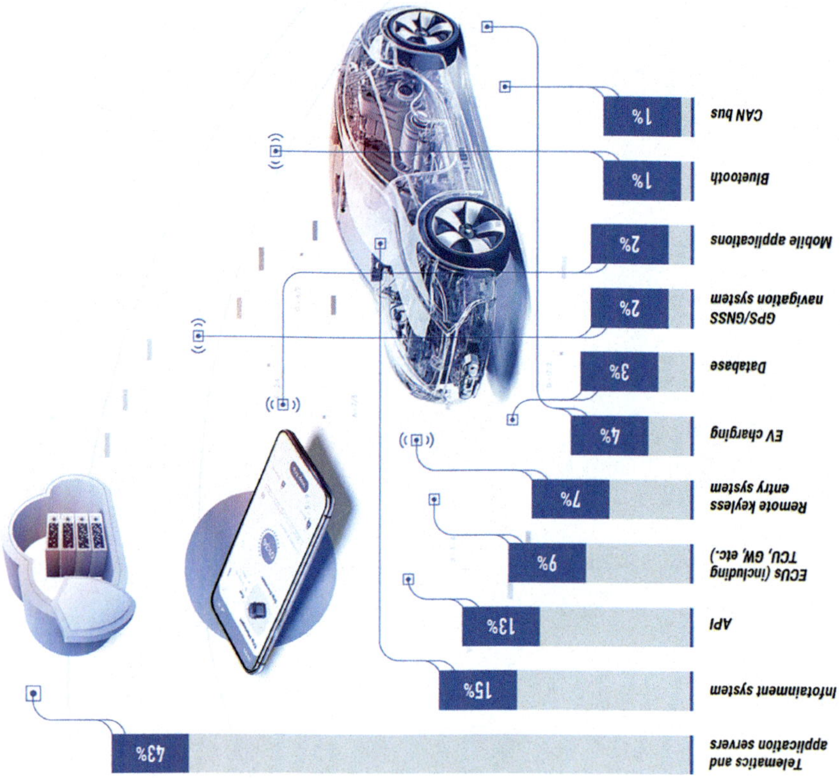

CAN bus	1%
Bluetooth	1%
Mobile applications	2%
GPS/GNSS navigation system	2%
Database	3%
EV charging	4%
Remote keyless entry system	7%
ECUs (including TCU, GW, etc.)	9%
API	13%
Infotainment system	15%
Telematics and application servers	43%

Fig. 4.49 Attack vectors in the automotive area

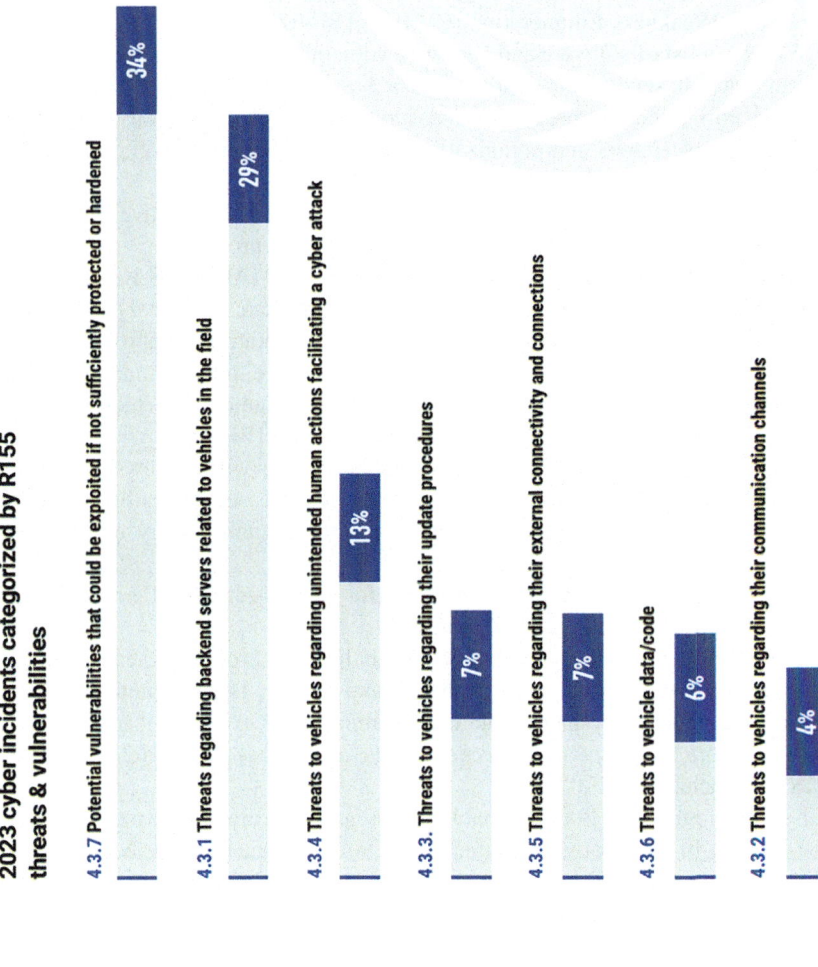

Fig. 4.50 Threats and vulnerabilities

- Common **V**ulnerabilities and **E**xposures (CVE®) [BF-019]:
  Publicly disclosed cybersecurity vulnerabilities are identified, defined, and cataloged. Each vulnerability in the catalog has a CVE entry. The vulnerabilities are published by organizations worldwide that have partnered with the CVE program.
- MITRE [BF-021]:
  MITRE was founded as a non-profit organization to promote national security in new ways and to serve the public interest as an independent advisor. One of the main areas of focus is cybersecurity, which includes ATT&CK® [BF-022], a knowledge database that documents hacking techniques based on real-world observations.
- Common **W**eakness **E**numeration (CWE™) [BF-020]:
  CWE™ is a list of software and hardware vulnerabilities developed by the CWE community. It serves, for example, as a basis for identifying vulnerabilities and for mitigation and prevention measures. The CWE community consists of individual researchers and organizations such as Apple, Microsoft, and NIST.
- **A**utomotive **S**ecurity **R**esearch **G**roup (ASRG) [BF-029]:
  The Automotive Security Research Group is a non-profit initiative to promote the development of security solutions for automotive products.
- **Auto**motive **I**nformation **S**haring and **A**nalysis **C**enter (AUTO-ISAC) [BF-030]:
  AUTO-ISAC is an industry-driven community that shares and analyzes insights on emerging vehicle cybersecurity risks and collaboratively improves vehicle cybersecurity capabilities across the global automotive industry, including light and heavy-duty vehicle OEMs, suppliers, and the commercial vehicle sector.
- European Union Agency for Cybersecurity (ENISA) [BF-027]:
  ENISA contributes to the EU's cyber policy and works with its key stakeholders to strengthen trust in the connected economy, increase the resilience of the Union's infrastructure, and ultimately ensure the digital security of European society and citizens.
- **BSI** (**B**undesamt für **S**icherheit in der **I**nformationstechnik), German federal office for information security [BF-028]:
  The BSI shapes information security in digitalization through prevention, detection, and response for the state, economy, and society. The automotive sector is also listed under "Companies and organizations"; for example, the *Automotive sector situation report – cyber security in the automotive sector 2022/2023*.
- **ENX** Association [BF-075]:
  ENX was founded in 2000 to enable and simplify secure collaboration. It is an association of automotive manufacturers (Volkswagen, Renault, Nissan, PSA, Fiat, etc.), suppliers (Bosch, Continental, Magna, etc.), and national automotive associations like VDA, SMMT, and ANFAC.

NIST has also published various frameworks covering automotive infrastructure topics, such as *NIST IR 8473—Cybersecurity Framework Profile for Electric Vehicle Extreme Fast Charging Infrastructure* [BF-023] and *Cybersecurity Framework Profile for Connected Vehicle Environments* [BF-024].

In addition, international regulations have emerged, such as *UN Regulation No. 155 Uniform provisions concerning the approval of vehicles with regards to cyber security and cyber security management system* [BF-058] and *UN Regulation No. 156 Uniform provisions concerning the approval of vehicles with regards to software updates and software updates management system* [BF-059] of March 4th 2021.

The requirements from *UN Regulation No. 155*, which have been adopted at the European level in ECE/TRANS/WP.29/2020/79, contain the following:

- A Cyber Security Management System (CSMS) must be introduced and established.
- A licensing authority must check this.
- The approval authority must issue a certificate of conformity.
- The OEM applies for a Certificate of Compliance for CSMS from the relevant authority. Documents describing the CSMS must also be submitted. The purpose is to demonstrate that the OEM has suitable processes to meet the regulations' requirements. It should be noted that this applies to the development, production, and post-production phases.
- If the test is positive, the authority issues a three-year certificate of conformity for the CSMS.
- The authority reserves the right to check compliance with the requirements of the regulation at any time and to withdraw the certificate of conformity if necessary.
- The OEM must inform the authority of any changes that affect the relevance of the certificate of conformity.
- Based on this information, the authority decides whether a new inspection is required.
- For a vehicle's type approval, the OEM must identify the critical elements of the vehicle type and carry out a risk assessment. The supplier-related risks must also be determined. For all identified risks, the OEM must implement mitigation measures, covering all mitigation measures from Annex 5 Parts B and C of the regulation relevant to these identified risks. If the measures from Annex 5 are not appropriate or sufficient for the identified risks, the OEM must ensure that other suitable measures have been implemented.
- Furthermore, the OEM must monitor the findings on new risks (monitoring, for example, through feedback from the field or new entries in corresponding databases).
- If new risks arise, appropriate mitigation measures must be determined and introduced. The corresponding requirements for updating the software and its management are defined in regulation no. 156.

This regulation has been mandatory for all new vehicle types since July 7th, 2022.

The following section describes how to introduce and test a CSMS.

4.4.1.1 Cybersecurity Standards

In 2016, the SAE published the *Surface Vehicle Recommended Practice J3061— Cybersecurity for Cyber-Physical Vehicle Systems* [BF-025], which described production, operation, and customer service and included process implementation for all cybersecurity-relevant processes.

SAE J3061 was replaced by ISO/SAE 21434 [BF-026] in 2021. It was later followed by ISO/PAS 5112 (2022) as a guideline for auditing cybersecurity engineering and ISO 24089 (2023) as a specification for the engineering of software updates (Fig. 4.51).

In the following section, we detail the ISO/SAE 21434 cybersecurity standard.

4.4.1.2 ISO/SAE 21434 (Cybersecurity)

ISO/SAE 21434 for cybersecurity consists of 15 clauses. Clauses 1 to 4 describe informative topics such as scope, normative references, terms, definitions, and general considerations.

The clauses are broken down as follows:

- Organizational clauses:
  - Clause 5—Organizational cybersecurity management: includes cybersecurity management and the specification of organizational cybersecurity policies, rules, and processes.
  - Clause 7—Distributed cybersecurity activities: includes supplier capability, requests for quotations, and the work split with suppliers.
  - Clause 8—Continual cybersecurity activities: includes all activities required to secure vulnerability management until the end of cybersecurity support.
- Project and phase-specific clauses:
  - Clause 6—Project-dependent cybersecurity management: includes cybersecurity management and security activities at the project level.
  - Clause 9—Concept phase: includes activities that define cybersecurity risks, objectives, and requirements for a product.
  - Clause 10—Product development phase: includes the activities for defining cybersecurity requirements, their implementation, and verification.
  - Clause 11—Cybersecurity validation: includes the cybersecurity validation of an object (item) at the vehicle level.
  - Clause 12—Production: includes the aspects of manufacturing and assembling an element or component.
  - Clause 13—Operations and maintenance: includes responding to cybersecurity incidents and updates to an element or component.
  - Clause 14—End of cybersecurity support and decommissioning: includes communication on the termination of support.
  - Clause 15—Threats analysis and risk assessment methods: assets that have a value or contribute to a value are determined, and the corresponding risks of a threat to these assets are assessed. Depending on the risk assessment,

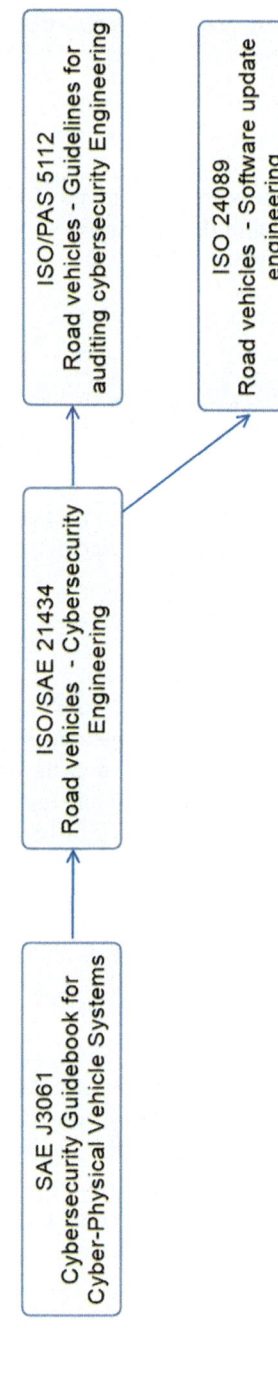

Fig. 4.51 Overview of cybersecurity standards in the automotive sector

measures are determined. The seven steps described here are usually carried out in a TARA (**T**hreat **A**ssessment and **R**isk **A**nalysis). Another method for risk assessment is MoRA (**Mo**dular **R**isk **A**ssessment for the Development of Secure Automotive Systems), which offers a risk-oriented, iterative, and modular approach (Fig. 4.52).

The following keywords and cross-references are assumed in the standard (Fig. 4.53).

The clauses from Clause 5 onwards are structured as follows (Fig. 4.54).

The following identifiers are used in the standard:

– RQ for a requirement
– RC for a recommendation
– PM for permission
– WP for a Work Product.

This identifier is supplemented by two numbers separated by hyphens, for example, "RQ-05–01." The first number indicates the corresponding clause, and the second is the order of the provisions or work products in this clause.

In the following, we will look at the requirements/recommendations of the individual clauses.

4.4.1.3 Organizational Clauses

The following organizational clauses are used.

Clause 5—Organizational Cybersecurity Management
Management must have a cybersecurity strategy and the processes, rules, tools, and resources to implement it. These must be available until the product is decommissioned. The extent to which these procedures can be integrated into existing processes should also be examined. This is optional for the management system, but it certainly makes sense to implement and maintain an integrated management system. Other topics include promoting a corresponding cybersecurity culture and auditing processes.

The following requirements/recommendations are made.

Cybersecurity governance
Five requirements (RQ-05-01 to RQ-05-05) are used to define the expectations of a cybersecurity policy and the rules and processes to implement the requirements of this standard. This also implies that responsibilities and authority are defined to achieve and maintain cybersecurity, and the right resources are assigned and communicated. Finally, the organization shall check which cybersecurity disciplines are affected and ensure that communication with these disciplines is established

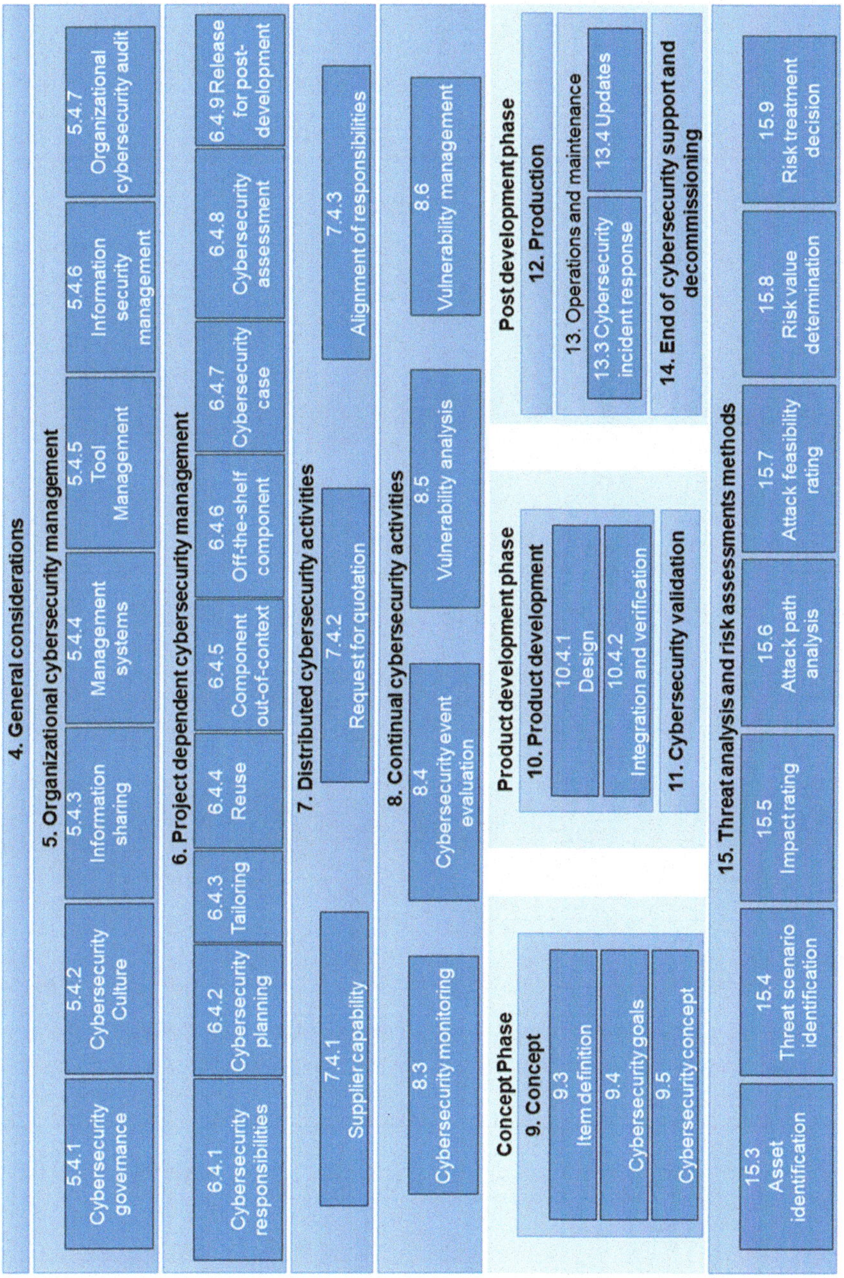

Fig. 4.52 Overview of clauses of the standard

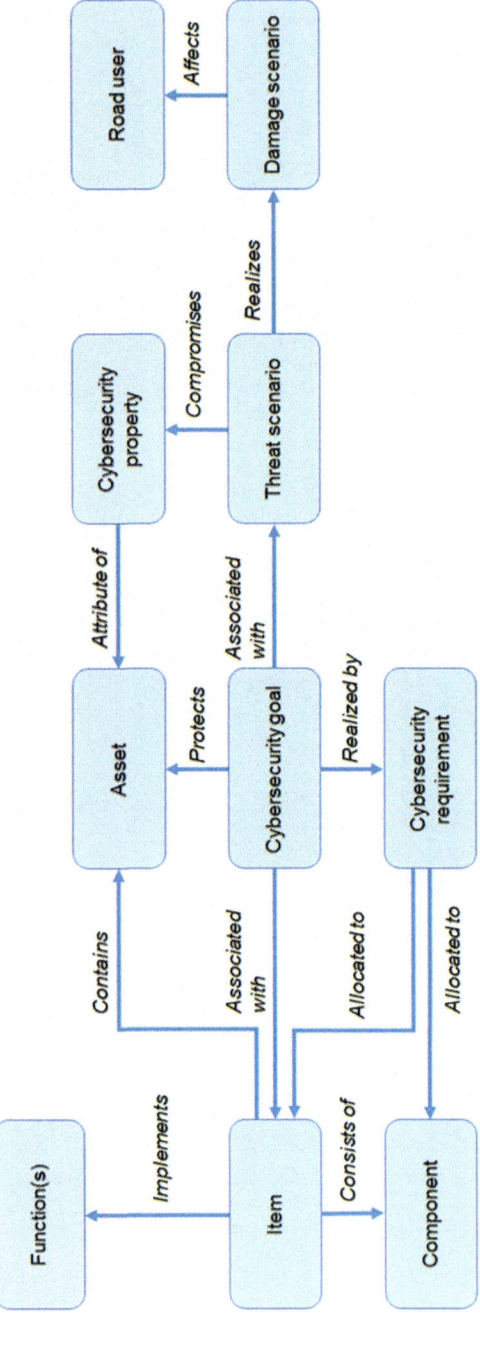

Fig. 4.53 Keywords and cross-references

Fig. 4.54 Structure of the
clauses in ISO/SAE 21434
(from clause 5 on)

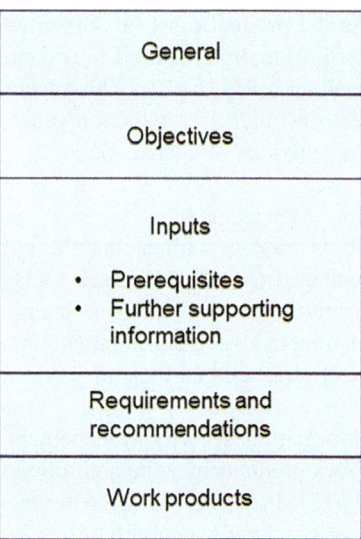

General

Objectives

Inputs

- Prerequisites
- Further supporting
  information

Requirements and
recommendations

Work products

and maintained. This also includes the integration of cybersecurity into existing
processes.

Cybersecurity culture

Three requirements (RQ-05-06 to RQ-05-08) focus on establishing a strong cybersecurity culture, a continuous improvement process, and assigning responsibility
with the right competencies.

Information Sharing

Organizations already control information sharing, but cybersecurity is even more
important when controlling it. This includes when sharing is required, permitted, or
forbidden, and it differs depending on whether it is internal or external (RQ-05–09)
or with other parties (RQ-05-10).

Management Systems

The organization shall set up and maintain a quality management system following
international standards (RQ-05-11). IATF 16949 [BF-009], in conjunction with
ISO 9001 [BF-010], is cited as an example of such a quality management system.
    This quality management system shall address the following:

– Change management
– Documentation management
– Configuration management
– Requirements management.

As the product must be supported until decommissioning or the end of support, the information required for maintaining its cybersecurity in the field must also be available (RQ-05–12). This means the cybersecurity management system should also include the production processes (RQ-05-13), which are needed to support the activities of Clause 12.

Tool Management

Tools used or influencing the cybersecurity of an item or component shall be managed (RQ-05-14). Such tools can be, for example, modeling tools and verification tools during development, flashing, and diagnostic tools in the field. This also includes the development environment needed to implement remedial actions (software build and testing environment; RC-05-15).

Information Security Management

Work products must be controlled so that no alteration or deletion is possible (RQ-05-16). In an ASPICE environment, work products are defined as configuration items and are part of a baseline. Baselines are typically frozen, which means they are stored on special servers or databases where any modification is no longer possible.

Organizational cybersecurity audit

The last requirement (RQ-05-17) of Clause 5 concerns the performance of an audit to judge whether the objectives of this standard have been achieved.

Work products

The following work products are defined in this clause (Fig. 4.55).

Clause 7—Distributed Cybersecurity Activities

This clause defines the requirements regarding the selection of suppliers, requests for quotations, and the job split between partners.

Supplier capability

Car manufacturers usually have requirements for their suppliers (e.g., IATF 16,949 certified). Additionally, they perform audits to ensure that a new manufacturing plant or supplier is covering their requests. The suppliers are listed in a supplier pyramid based on these results and on the supplied services or products (e.g., costs, quality, on-time delivery, defects, etc.).

The first part of Clause 7 describes similar requirements, e.g.,

- Evaluation of the capability to develop and ensure post-development activities as required by this standard (RQ-07-01)
- Have a proof of cybersecurity capability (RC-07-01).

Both requirements can be supported by audits proving that a cybersecurity management system has been implemented and is maintained.

Work product	Requirements/ Recommendations
WP-05-01 Cybersecurity policy, rules and processes	RQ-05-01 RQ-05-02 RQ-05-03 RQ-05-04 RQ-05-05 RQ-05-06 RQ-05-07 RQ-05-08 RQ-05-09 RQ-05-10
WP-05-02 Evidence of competence, awareness management and continuous improvement	RQ-05-08
WP-05-03 Evidence of the organization's management system	RQ-05-11 RQ-05-12 RQ-05-13 RQ-05-16
WP-05-04 Evidence of tool management	RQ-05-14 RC-05-15
WP-05-05 Organizational cybersecurity audit report	RQ-05-17

Fig. 4.55 Overview of work products/requirements for organizational clauses

Request for quotation

The basis for each supplier contract is a request for quotation (RFQ). This shall stipulate that this standard is adhered to and that the supplier will take on the defined cybersecurity responsibilities, goals, and requirements relevant to the item or component for which the supplier is asked to quote (RQ-07-03).

Alignment of responsibilities

The request is defined roughly for the quotation request and shall be refined before starting the activities. This is usually done in a cybersecurity interface agreement (RQ-07–04) which defines:

- The responsibilities on both sides (customer and supplier)
- The activities to be performed and to whom they are assigned
- Tailoring of activities, if applicable
  The work products and information to be shared/distributed. Information can include vulnerabilities, findings, data handling, and communications channels/ methods used.
- The milestones where the defined work products are expected to schedule the needed activities (including the end of the cybersecurity support).

Most of this information can and shall be documented in a RASIC (an acronym for Responsible, Approve, Support, Inform, Consult) (RC-07-08).

Work product	Requirements/Recommendations
WP-07-01 Cybersecurity interface agreement	RQ-07-04 RQ-07-05 RQ-07-06 RQ-07-07 RQ-07-08

Fig. 4.56 Overview of work products/requirements for distributed cybersecurity activities

Naturally, the cybersecurity interface agreement shall be signed by both sides before starting the activities (RQ-07-05).

Suppose the requirements are not precise or feasible or contradict other cybersecurity requirements or requirements from different disciplines. In that case, either the customer or the supplier must communicate this to discuss and decide on further action to solve the issue (RQ-07-07).

If a vulnerability is identified, the customer and the supplier will commit to the risk treatment decision and related activities (RQ-07-06).

Work products
The following work products are defined in this clause (Fig. 4.56).

Clause 8—Continual Cybersecurity Activities
This involves monitoring, evaluating, and analyzing cybersecurity information and determining the necessary measures. This monitoring must be carried out until decommissioning, which means a dedicated group in the organization must carry out this monitoring on an ongoing basis. During development, the information from this monitoring is used in the project to ensure that all possible threats are considered. Similarly, at the end of the development project, the implemented concept must be handed over to this group so that new threats can be correctly assessed and necessary measures integrated into the existing product. The requirements for this handover are described later as the release for post-development (RQ-06-33 and RQ-06-34).

Based on the standard, the activities and the used or generated work products look like this (Fig. 4.57):

Cybersecurity monitoring
Monitoring is based on selected sources (RQ-08-01). These can be both internal and external sources: information from the field and information from corresponding databases such as MITRE ATT&CK® [BF-022] can be used for monitoring (Fig. 4.58).

This information can have different impacts on an item or component. For this reason, it is essential to install a triage of this information, which is done based on defined triggers (RQ-08-02).

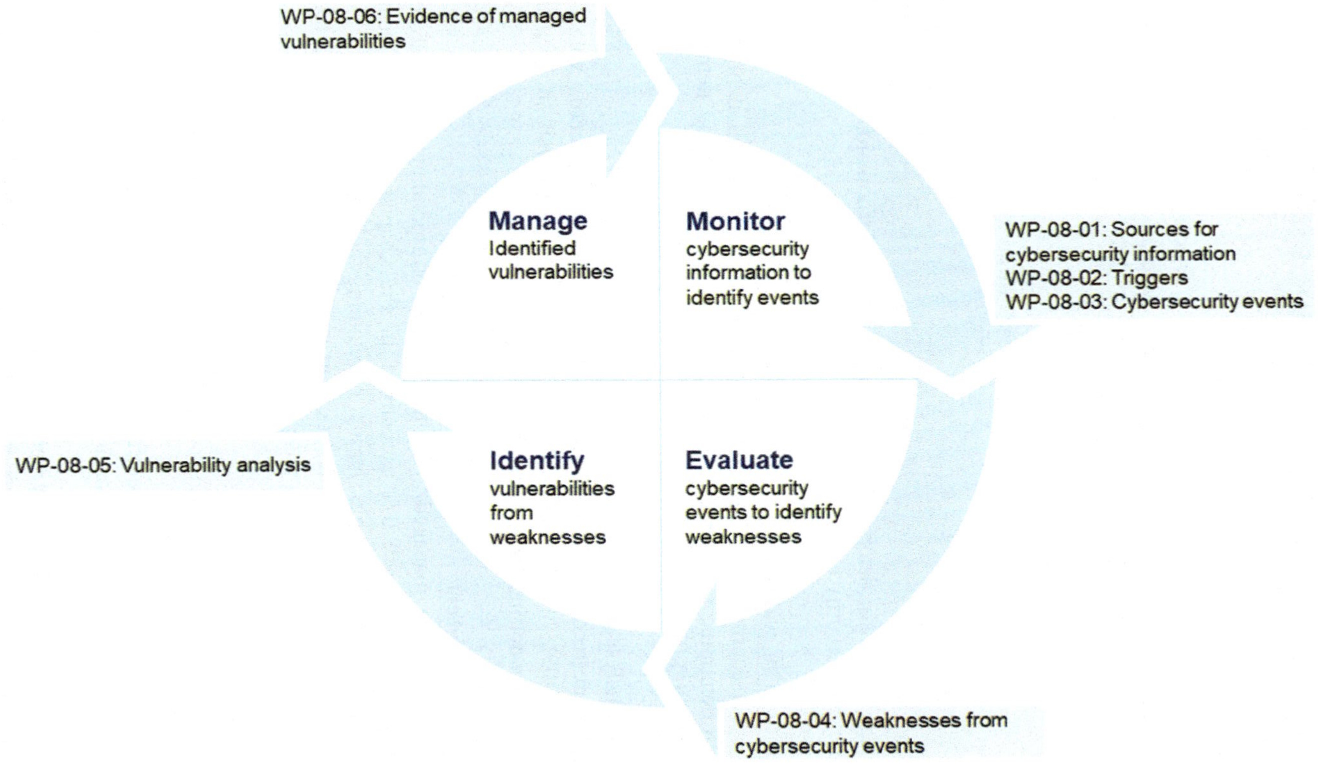

Fig. 4.57 Continual cybersecurity activities

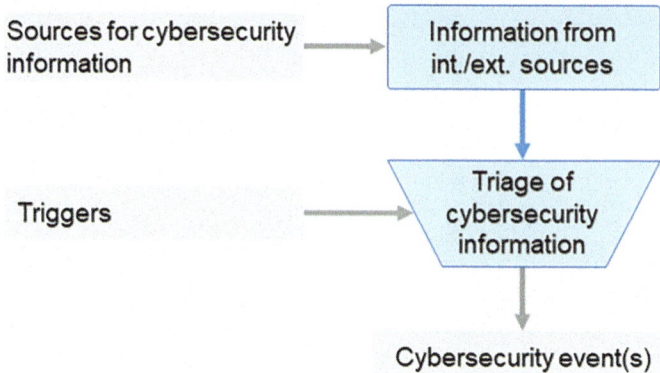

Fig. 4.58 Cybersecurity monitoring

This triage allows for determining if the cybersecurity information consists of one or more cybersecurity events (RQ-08-03).

Cybersecurity event evaluation
Once the cybersecurity information has been assigned to an event, it must be evaluated to determine whether this event leads to weaknesses in an item and/or component (RQ-08-04).

Vulnerability analysis
If weaknesses are recognized, an analysis must be performed to identify vulnerabilities (RQ-08-05). A rationale must be documented if this analysis shows that the weakness does not induce a vulnerability (RQ-08-06).

Vulnerability management
Vulnerabilities must be managed. This means the corresponding risk must be assessed and treated (RQ-08-07) as defined in the chapter "Risk treatment decision," e.g., avoiding, reducing, sharing, or retaining the risk.

If a risk treatment decision requires a cybersecurity response (RQ-08-08), the chapter "Cybersecurity incident response" shall be applied.

Work products
The following work products are defined in this clause (Fig. 4.59).

4.4.1.4 Project and Phase-Specific Clauses
The following project-specific clauses are used.

Clause 6—Project-dependent cybersecurity management
The main activities of project-dependent cybersecurity management are listed in Fig. 4.60.

Work product	Requirements/ Recommendations
WP-08-01 Sources for cybersecurity information	RQ-08-01
WP-08-02 Triggers	RQ-08-02
WP-08-03 Cybersecurity events	RQ-08-03
WP-08-04 Weaknesses from cybersecurity events	RQ-08-04
WP-08-05 Vulnerability analysis	RQ-08-05 RQ-08-06
WP-08-06 Evidences of managed vulnerabilities	RQ-08-07

Fig. 4.59 Overview of work products/requirements for continual cybersecurity activities

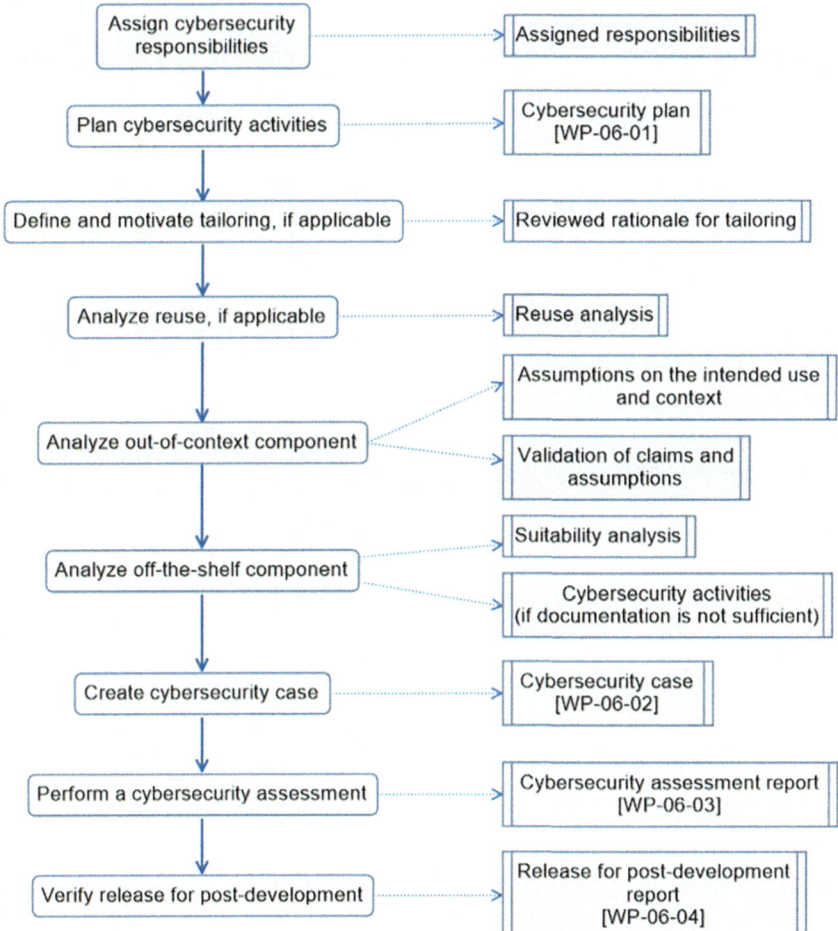

Fig. 4.60 Activities and work products for project-dependent cybersecurity management

Cybersecurity responsibilities

The project's cybersecurity activities must be assigned and communicated (RQ-06-01) based on the organizational definitions (see RQ-05-03).

Cybersecurity planning

The first step is to analyze the item or component to determine whether it is cybersecurity-relevant, whether it is a new development or reuse, and whether tailoring is applicable (RQ-06-02).

Based on this information, cybersecurity activities can be planned. These activities can be integrated into a specific cybersecurity or project plan (RQ-06-05). In both cases, the objectives of the activities, the dependencies from other activities, the required resources to perform the activities, the start and end dates, and the expected work products must be documented (RQ-06-03). To determine the start and end date, both the duration and, naturally, the possible load of the resource (100% or less) must be estimated.

As required in project management, a responsible person must be assigned to set up and maintain the cybersecurity plan (every automotive plan must be up to date on the day of its release) and monitor progress (RQ-06-04). This assignment is done based on organizational governance (RQ-05-03) and (RQ-05-04).

The cybersecurity plan shall also structure the activities based on the concept and development phases (RQ-06-06). It must be updated if changes or refinements of some activities are needed (RQ-06-07.

Suppose an analysis based on the chapter "Risk value determination" leads to a risk value 1 for a threat scenario. In this case, conformity with the chapters "Cybersecurity concept", "Product development", and"Cybersecurity validation" can be omitted (PM-06-08).

The planning and the created work products are subject to configuration, change, requirements, and documentation management (RQ-06-11, RQ-06-12).

The expected work products must be updated and maintained until the release for post-development (RQ-06-09).

If a supplier is involved, the customer and supplier must set up and maintain a cybersecurity plan (RQ-06-10) based on the committed job split defined in the cybersecurity interface agreement (see Clause 7).

Tailoring

Cybersecurity activities can be tailored (PM-06-13). If this is the case, a rationale explaining why tailoring that still covers the standard's objectives is appropriate and must be documented and reviewed (PM-06-14).

Reuse

Often, items or components are reused in new or other systems. In this case, an analysis must be performed to determine if modifications are necessary, if the operational environment changes, or if there are information changes (RQ-06-15). In addition to these aspects, the implications of the changes must be evaluated (e.g., the validity of the cybersecurity claims and assumptions, affected or new required work products, and supplementary required cybersecurity activities) (RQ-06-16). Finally, the analysis must also show if the cybersecurity requirements of the item or component in which the integration takes place are fulfilled and if the available documentation is sufficient (RQ-06-17).

Component out-of-context
A component can be developed by assuming a context and an intended use (e.g., without a concrete application). Using such components, the assumptions on the intended use and context must be checked and documented (RQ-06-18), and the cybersecurity requirements shall be adapted to the updated assumptions (RQ-06-19). When integrating an out-of-context component, the cybersecurity claims and assumptions must be validated (RQ-06-20).

Off-the-shelf component
An off-the-shelf component is one that can be used without modification, e.g., an open-source software component. Integrating such a component also requires checking if the assigned cybersecurity requirements can be fulfilled, if it is suitable for the intended use and context, and if the existing documentation must be updated (RQ-06-21). If so, the required cybersecurity activities must be determined and performed (RQ-06-22).

Cybersecurity case
A cybersecurity case must be created to provide an argument for the degree of cybersecurity achieved by the item or component (RQ-06-23).

Cybersecurity assessment
For any item or component, a decision regarding the performance of a cyber-security assessment must be made, and the rationale for this decision must be documented (RQ-06-24) and reviewed (RQ-06-25). For the assessment itself, the following topics must be defined:

- The assessment must evaluate the cybersecurity of the component or item (RQ-06-26)
- A responsible person shall be assigned to plan and perform the assessment (RQ-06-27)
- This person shall have access to the relevant information and tools as well as to the developers (RQ-06-28)
- The evaluation shall be based on the requirements of this standard (RQ-06-29)
- The assessment scope must be defined (e.g., work products to be checked, cybersecurity activities, etc.) (RQ-06-30)
- A report must be created recommending the acceptance, conditional approval, or rejection of the cybersecurity of the component or item (RQ-06-31)
- If a conditional acceptance is recommended, the report shall also describe the supplementary conditions for acceptance (RQ-06-32).

Release for post-development
The following conditions must be fulfilled:

- The cybersecurity case, the cybersecurity assessment report (if applicable), and the cybersecurity requirements for post-development must be available (RQ-06-33).
- In addition, the argument for cybersecurity provided by the cybersecurity case is convincing, the cybersecurity case is confirmed by the cybersecurity assessment (if applicable), and the cybersecurity requirements for the post-development phases have been accepted (RQ-06-34).

Work products
The following work products are defined in this clause (Fig. 4.61).

In addition to the project-specific clauses, phase-related clauses are used that are assigned to three different phases:

- Concept phase
- Product development phase
- Post-development phases.

Clause 9—Concept (Concept Phase)
This is where the expectations of the items, the cybersecurity goals, and the cyber-security concept are defined. The process for determining the work products can be represented as follows (Fig. 4.62).

Item definition
The item shall be defined by the item boundary, the item functions, and the preliminary architecture (RQ-09-01). Furthermore, the cybersecurity-relevant operational environment of the item shall be determined (RQ-09-02).

Work product	Requirements/Recommendations
WP-06-01 Cybersecurity plan	RQ-06-01 to RQ-06-09 RQ-06-10 to RQ-06-19 RQ-06-20 to RQ-06-22
WP-06-02 Cybersecurity case	RQ-06-23
WP-06-03 Cybersecurity assessment report	RQ-06-24 to RQ-06-29 RQ-06-30 to RQ-06-32
WP-06-04 Release for post-development report	RQ-06-33 RQ-06-34

Fig. 4.61 Overview of work products/requirements for project-dependent cybersecurity management

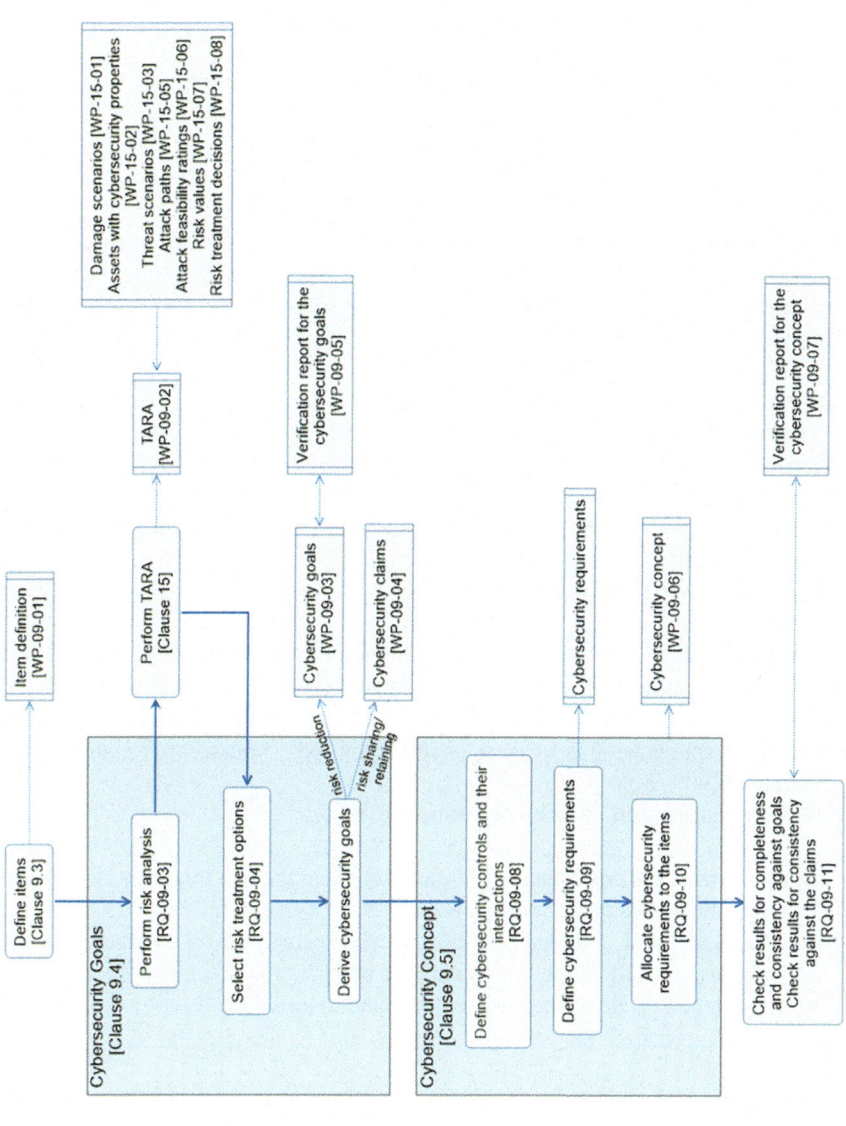

Fig. 4.62 Procedure for determining goals and concept

Work product	Requirements/ Recommendations
WP-09-01 Item definition	RQ-09-01 RQ-09-02

Fig. 4.63 Overview of work products/requirements for items definition

Work products
The following work products are defined in this sub-clause (Fig. 4.63).

Cybersecurity goals
Based on the item definition, a threat and risk analysis must be performed (as defined in Clause 15—Threat Analysis and Risk Assessment Methods). The proposed method is the TARA, which consists of seven steps and leads to the following work products:

- Damage scenarios [WP-15-01]
- Assets with cybersecurity properties [WP-15-02]
- Threat scenarios [WP-15-03]
- Impact ratings with associated impact categories [WP-15-04]
- Attack paths [WP-15-05]
- Attack feasibility ratings [WP-15-06]
- Risk values [WP-15-07] (RQ-09-03)

Further outputs of the TARA are the risk treatment decisions [WP-15-08], for which the risk treatment options must be defined (RQ-09-04). If the risk treatment decision is to reduce the risk, cybersecurity goals must be defined (RQ-09–05). If the risk treatment decision is to share or retain the risk, cybersecurity claims must be defined (RQ-09-06).

Finally, a verification must be performed to ensure:

- The correctness and completeness of the TARA and the risk treatment decisions, as well as the consistency between both
- The correctness and completeness of the cybersecurity goals and the cybersecurity claims, as well as the consistency of both the risk treatment decisions
- The consistency of all cybersecurity goals and claims (RQ-09-07).

Work products
The following work products are defined in this sub-clause (Fig. 4.64).

Work product	Requirements/ Recommendations
WP-09-02 TARA	RQ-09-03 RQ-09-04
WP-09-03 Cybersecurity goals	RQ-09-05
WP-09-04 Cybersecurity claims	RQ-09-06
WP-09-05 Verification report for cybersecurity goals	RQ-09-07

Fig. 4.64   Overview of work products/requirements for cybersecurity goals

Cybersecurity concept

The first step is to define the cybersecurity controls and their interactions so they fulfill the cybersecurity goals and consider the dependencies between the item's functions and the cybersecurity claims (RQ-09-08).

The standard defines cybersecurity controls as measures that modify risk. Risk management has to be performed during a product's whole lifecycle, i.e., in all cybersecurity phases (Fig. 4.65).

Regarding risk management, the standard relates to ISO 31000:2018, Risk Management—Guidelines [BF-077].

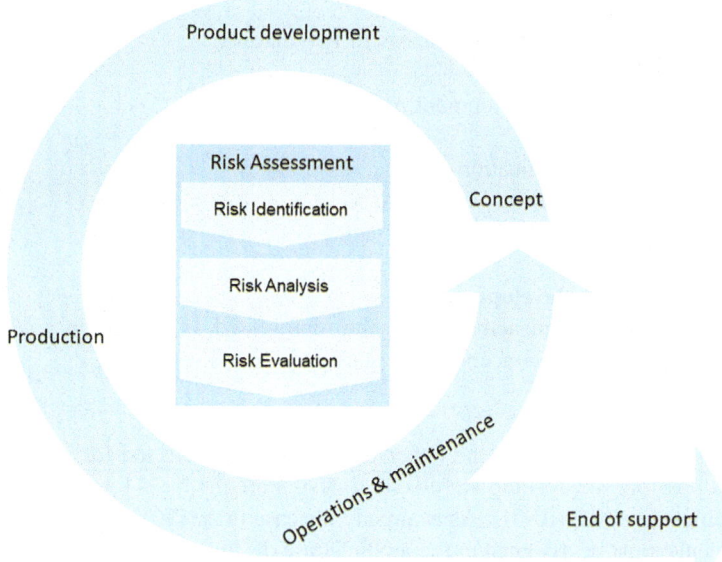

Fig. 4.65   Cybersecurity risk management in all phases

Work product	Requirements/ Recommendations
WP-09-06 Cybersecurity concept	RQ-09-08 RQ-09-09 RQ-09-10
WP-09-07 Verification report for the cybersecurity concept	RQ-09-11

Fig. 4.66 Overview of work products/requirements for the cybersecurity concept

The next step is to define the cybersecurity requirements and the requirements for the operational environment (RQ-09-09). The last step is to allocate the cybersecurity requirements to the item (RQ-09-10).

These results must all be checked for completeness, correctness, and consistency with the cybersecurity goals and claims (RQ-09-11).

Implementing the cybersecurity concept ensures that the cybersecurity objectives are achieved. The activities for implementing the cybersecurity concept are included in the cybersecurity plan.

Work products
The following work products are defined in this sub-clause (Fig. 4.66).

Product Development Phase
The product development phase consists of two clauses:

- Clause 10—Product Development, which includes:
  - Design, and
  - Integration and verification
- Clause 11—Cybersecurity Validation.

Clause 10—Product Development
The product development activities can be represented as follows in a V-model (which also includes the concept as the starting point) (Fig. 4.67).

Design
Higher levels of architecture, the cybersecurity controls selected for implementation, and possibly the existing architectural design are the basis for the cybersecurity specifications (RQ-10-01). As is already the case in ASPICE, the cybersecurity requirements must be assigned to the architectural design components (RQ-10-02).

As cybersecurity must be maintained through the complete production cycle and until the end of the cybersecurity support, procedures to ensure this must

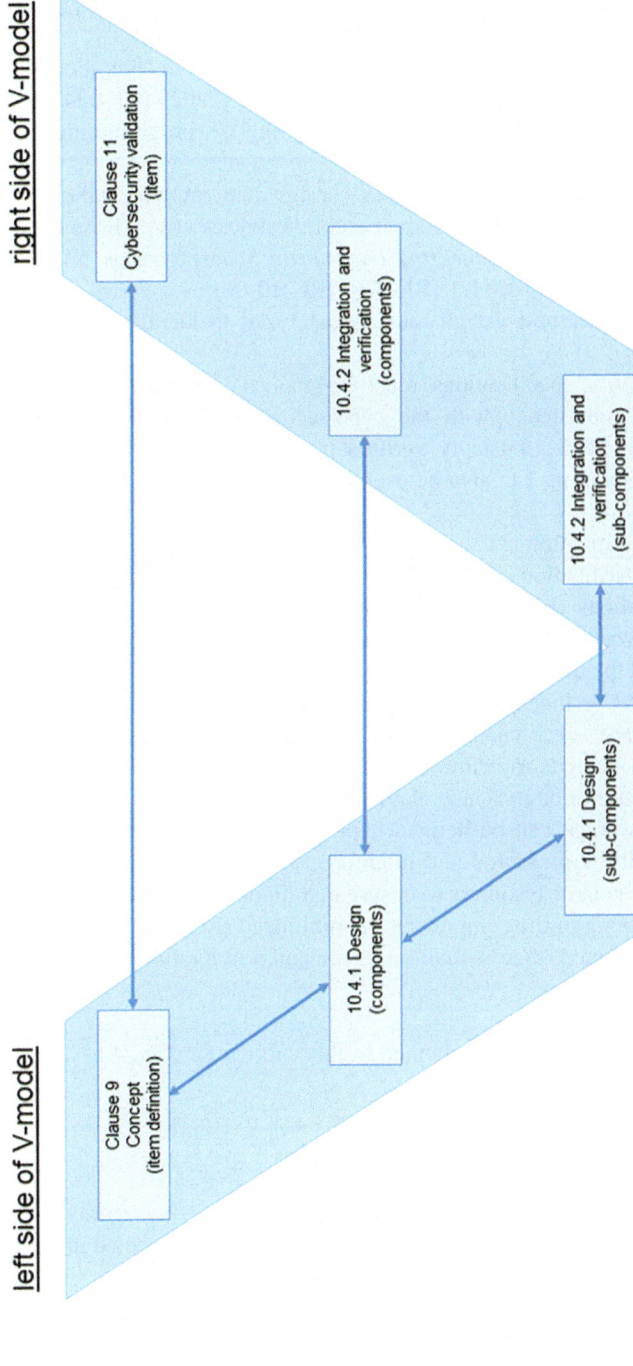

Fig. 4.67 Example of development activities in the V-model (ISO/SAE 21434)

exist (RQ-10-03). If programming notations or languages are used to specify or implement cybersecurity, these notations or languages shall fulfill minimal requirements (e.g., understandable and unambiguous definitions in semantics and syntax, support for structured constructs, modularity, etc.) (RQ-10-04).

If not all criteria can be addressed by the language itself, design, modeling, and programming guidelines shall be applied (e.g., MISRA C:2023 [BF-034], or CERT C Coding Standard [BF-035] for secure coding in "C" programming language (RQ-10-05).

In the same way, principles of trusted design and implementation shall be applied to minimize or avoid the introduction of weaknesses (e.g., Ross R, McEvilley M, Winstead M (2022) *Engineering Trustworthy Secure Systems*, NIST Special Publication NIST SP 800-160v1r1 [BF-076] (RC-10-06).

Finally, the architectural design must be analyzed to identify possible weaknesses (RQ-10-07).

The cybersecurity specifications must be analyzed to ensure completeness, correctness, and consistency with the cybersecurity specifications from higher architecture levels. This is usually verified in a review, but other methods like simulation or prototyping can also be used (RQ-10-08).

Integration and verification

Integration and verification aim to verify that the components fulfill the cybersecurity specifications (RQ-10-09). The activities required for this need at least to consider the cybersecurity specifications, the resources needed for the specified functionality, and the defined modeling, design, and coding guidelines (RQ-10-10). Like ASPICE, the verification methods must be determined (e.g., requirements-based test, interface test, dynamic and/or static analysis, resource usage). In the case of tests, the methods to define the test cases shall be specified (e.g., analysis of requirements, use of equivalence classes or boundary values). The test environment and test coverage can be defined depending on the project phase, and the test coverage shall be measured and monitored to ensure sufficient test activities (RQ-10-11). One goal of testing is to ensure that unidentified weaknesses and vulnerabilities remaining in the component are minimized (RC-10-12). If this kind of testing is not performed, then it shall be documented in a rationale (RQ-10-13).

Work products

The following work products are defined in this sub-clause (Fig. 4.68).

Clause 11—Cybersecurity Validation

Cybersecurity Validation

Validation is performed on the vehicle level against the cybersecurity goals and claims (RQ-11-01). Validation activity selection must be documented in a rationale (RQ-11–02). The following validations can be performed:

Work product	Requirements/ Recommendations
WP-10-01 Cybersecurity specifications	RQ-10-01 RQ-10-02
WP-10-02 Cybersecurity requirements for post-development	RQ-10-03
WP-10-03 Documentation of the modelling, design or programming languages and coding guidelines	RQ-10-04 RQ-10-05
WP-10-04 Verification report for the cybersecurity specifications	RQ-10-08
WP-10-05 Weaknesses found during product development	RQ-10-07 RC-10-12
WP-10-06 Integration and verification specification	RQ-10-10
WP-10-07 Integration and verification report	RQ-10-09 RQ-10-10 RC-10-12

Fig. 4.68 Overview of work products/requirements for product development

- Review of the work products of the cybersecurity concept and product development
- Penetration testing
  The goal of the test is to demonstrate the achievement of the cybersecurity goals. Suppose external sources for cybersecurity information are used. In this instance, the coverage of the selected threats and attack paths from these sources by the single tests shall also be documented in the penetration test report.
- Review of all managed risks identified in the previous phases.

Work products
The following work products are defined in this sub-clause (Fig. 4.69).

Post-Development Phases
The post-development phases consist of three different clauses:

Work product	Requirements/ Recommendations
WP-11-01 Validation report	RQ-11-01 RQ-11-02

Fig. 4.69 Overview of work products/requirements for cybersecurity validation

- Clause 12—Production
- Clause 13—Operations and Maintenance, which includes:
  - Cybersecurity incident response
  - Updates
- Clause 14—End of Cybersecurity Support and Decommissioning

Clause 12—Production

This clause relates primarily to the security and access of production facilities.

A production control plan is created to ensure that the cybersecurity requirements for post-development are applied ([RQ-12-01). Naturally, this requires that the information needed to set up the plan is transferred from development to production and post-development, implying also that the production personnel are trained in cybersecurity and have the right mindset.

The production control plan includes activities to apply the cybersecurity requirements for post-development, configuration of the production equipment, cybersecurity controls to prevent dangerous alterations during production (e.g., physical access control to servers, logical controls applying cryptographic techniques), and methods to confirm that the cybersecurity requirements for post-development are covered (RQ-12-02).

Finally, this plan must be implemented (RQ-12-03).

Work products

The following work products are defined in this sub-clause (Fig. 4.70).

Clause 13–Operations and Maintenance

This clause deals with responding to incidents and updating functions in the operations and maintenance phase.

Cybersecurity Incident Response

During the serial life of each cybersecurity incident, a cybersecurity incident response plan must be set up (RQ-13–01) containing the following actions (Fig. 4.71).

The cybersecurity incident response plan must be implemented (RQ-13-02).

Work product	Requirements/ Recommendations
WP-12-01 Production control plan	RQ-12-01 RQ-12-02

Fig. 4.70 Overview of work products/requirements for the production control plan

Fig. 4.71 Cybersecurity
incident response plan

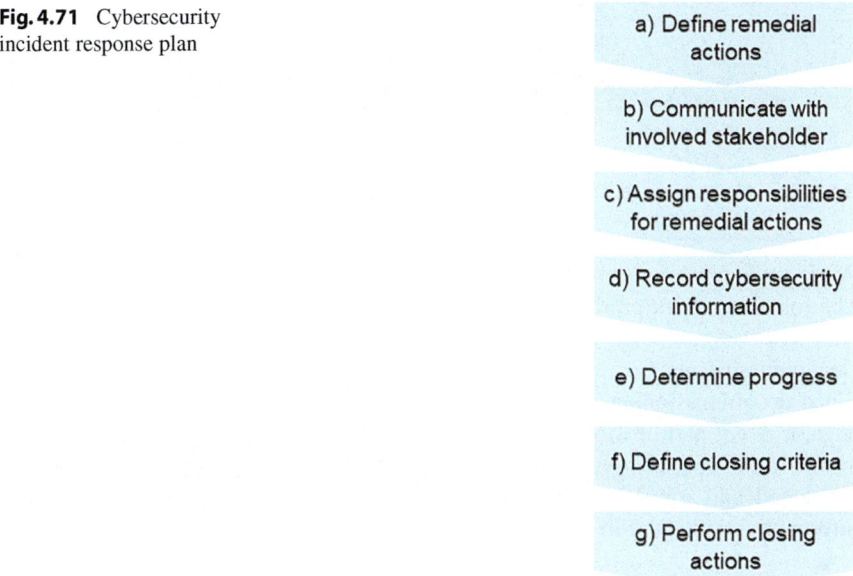

a) Define remedial
actions

b) Communicate with
involved stakeholder

c) Assign responsibilities
for remedial actions

d) Record cybersecurity
information

e) Determine progress

f) Define closing criteria

g) Perform closing
actions

Work product	Requirements/ Recommendations
WP-13-01 Cybersecurity incident response plan	RQ-13-01

Fig. 4.72 Overview of work products/requirements for the cybersecurity incident response plan

Work products
The following work products are defined in this sub-clause (Fig. 4.72).

Updates
To be able to perform updates in vehicles, update-related capabilities (e.g., over-the-air updates) must be developed based on this standard (RQ-13-03). The updates must comply with UN Regulation No. 156 Software update and software update management system [BF-059].

Work products
None (the work products result from other clauses).

Clause 14—End of Cybersecurity Support and Decommissioning

End of cybersecurity support
The organization must define a process to communicate to their customers the end of cybersecurity support for an item or component (RQ-14-01).

Work product	Requirements/ Recommendations
WP-14-01 Procedures to communicate the end of cybersecurity support	RQ-14-01

Fig. 4.73 Overview of work products/requirements at the end of cybersecurity support

Work products
The following work products are defined in this sub-clause (Fig. 4.73).

Decommissioning
Since decommissioning can occur without the concerned organization's knowledge, it is not part of this standard. However, cybersecurity requirements must be available for decommissioning (RQ-14–02). If, for example, a vehicle is decommissioned and a cybersecurity-relevant ECU is resold separately, the personal information contained in the ECU must not be recoverable.

Work products
None.

Clause 15—Threat Analysis and Risk Assessment Methods
A threat analysis and risk assessment are performed to determine to what extent a road user can be affected by a threat scenario. For this, a formal and disciplined approach is needed; the TARA is referenced in the standard as such an approach. The TARA can be performed with regular office tools (e.g., Excel), but a specific tool allows better reuse or rework and easier structuring and overview. One example of such a tool is Ansys Medini Analyze for Cybersecurity [BF-036].
    The TARA consists of a total of seven steps (Fig. 4.74).
– Step 1: Asset identification
    The first step is to determine which damage scenarios are possible (RQ-15-01) and which assets with cybersecurity properties can be affected by these scenarios (RQ-15-02).

Work products
The following work products are defined in this sub-clause (Fig. 4.75).

– Step 2: Threat scenario identification
    The next step is to determine the threat scenarios for the identified assets, the compromised properties of these assets, and the cause of these compromises (RQ-15-03).
    Different threat modeling methods can be used:
    – EVITA [BF-038, BF-039]
    – TVRA [BF-040]

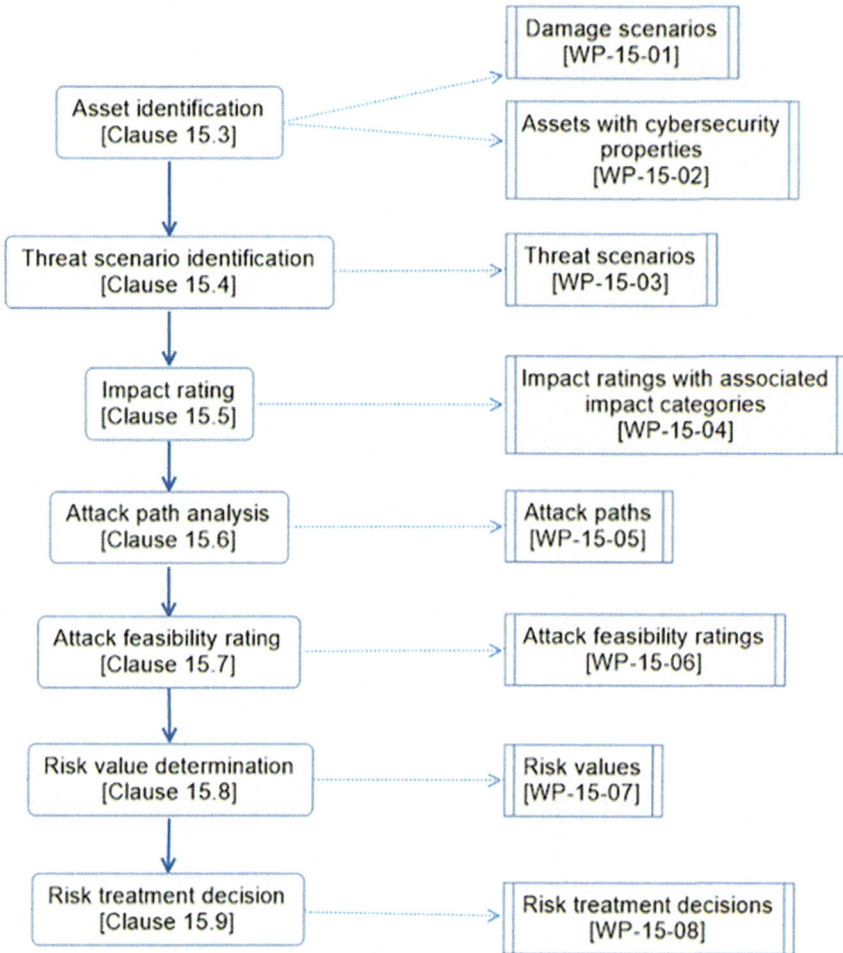

Fig. 4.74 Activities and work products of a TARA

Work product	Requirements/ Recommendations
WP-15-01 Damage scenarios	RQ-15-01
WP-15-02 Assets with cybersecurity properties	RQ-15_02

Fig. 4.75 Overview of work products/requirements for the identification of assets

- PASTA [BF-041]
- STRIDE (*Spoofing, Tampering, Repudiation, Information Disclosure, Denial of Service, Elevation of Privilege*).

Work products
The following work products are defined in this sub-clause (Fig. 4.76).

- Step 3: Impact rating
  The damage scenarios are assessed regarding the possible impact on road users based on safety, financial, operational, and privacy (S, F, O, P) (RQ-15-04). The rating is classified as:
  - severe
  - major
  - moderate
  - negligible (RQ-15-05)

Regarding safety-related impacts, the rating is derived from ISO 26262-3:2018, 6.4.3 (RQ-15-06. For this purpose, the standard provides an example table based on ISO 26262. The ratings from the HARA are taken over (see Sect. 4.4.2 "Functional safety") (Fig. 4.77).

Work products
The following work products are defined in this sub-clause (Fig. 4.78).

- Step 4: Attack path analysis
  The possible attack paths for the threat scenarios are determined (RQ-15-08) and linked to the attack path (RQ-15-09).

Work product	Requirements/ Recommendations
WP-15-03 Threat scenarios	RQ-15-03

Fig. 4.76 Overview of work products/requirements for identifying threat scenarios

Impact rating	Description
Severe	S3: Life-threatening injuries (survival uncertain), fatal injuries
Major	S2: Severe and life-threatening injuries (survival probable)
Moderate	S1: Light and moderate injuries
Negligible	S0: No injuries[1]
[1]: Rating for S0 can be based on ISO 26262-3-2018, Table B.1	

Fig. 4.77 Example of criteria for assessing the safety impact from ISO/SAE 21434

Work product	Requirements/ Recommendations
WP-15-04 Impact ratings with associated impact categories	RQ-15-04 RQ-15-06

Fig. 4.78 Overview of work products/requirements for impact assessment

Work product	Requirements/ Recommendations
WP-15-05 Attack paths	RQ-15-08 RQ-15-09

Fig. 4.79 Overview of work products/requirements for attack paths

Attack feasibility rating	Description
High	The attack path can be accomplished utilizing low effort.
Medium	The attack path can be accomplished utilizing medium effort.
Low	The attack path can be accomplished utilizing high effort.
Very low	The attack path can be accomplished utilizing very high effort.

Fig. 4.80 Assessment of attack feasibility with corresponding descriptions

Work products
The following work products are defined in this sub-clause (Fig. 4.79).

- Step 5: Attack feasibility rating
  Once the attack path is known, the attack feasibility can be rated (RQ-15-10) based on the following table (Fig. 4.80).
  Different approaches can be used for the rating method of the attack feasibility (RC-15-11):
  - Attack potential-based approach
  - CVSS-based approach
  - Attack vector-based approach

If an attack potential-based approach is used, the attack feasibility is based on core factors (RC-15-12). The following table, including the core factors, is proposed in the standard (Table G.6) (Fig. 4.81).
It corresponds to the calculation of the attack potential (Table B.2) from ISO/ IEC 18045 [BF-037], but in a simplified form (e.g., for the elapsed time). ISO/ IEC 18045 states that the attack potential corresponds to adding all parameters. The attack potential is illustrated using the following table, which is based on an adaptation of ISO/IEC 18045 (Table B.3) (Fig. 4.82).

Elapsed time		Specialist expertise		Knowledge of the item or component		Window or opportunity		Equipment	
Enumerate	Value	Enumerate	Value	Enumerate	Value	Enumerate	Value	Enumerate	Value
≤ 1 day	0	Layman	0	Public	0	Unlimited	0	Standard	0
≤ 1 week	1	Proficient	3	Restricted	3	Easy	1	Specialized	4
≤ 1 month	4	Expert	6	Confidential	7	Moderate	4	Bespoke	7
≤ 6 months	17	Multiple experts	8	Strictly confidential	11	Difficult/ none	10	Multiple bespoke	9
> 6 months	10								

Fig. 4.81 Example of the aggregation of the attack potential from ISO/SAE 21434 (Table G.6)

Attack feasibility rating	Values
High	0–9
	10–13
Medium	14–19
Low	20–24
Very low	≥ 25

Fig. 4.82 Example of the assignment of attack potentials from ISO/SAE 21434 (Table G.7)

If a CVSS-based approach is used, the attack feasibility rating is determined based on the exploitability metrics of the base metric group (see CVSS [BF-042] 2. Basic Metrics; RC-15-13), including the following factors:

– attack vector (AV)
– attack complexity (AC)
– privileges required (PR)
– user interaction (UI).

If an attack vector-based approach is used, the attack feasibility rating is determined based on evaluating the predominant attack vector (cf. CVSS [BF-042] 2.1.1 Attack Vector (AV); RC-15-14) of the attack path.

Work products
The following work products are defined in this sub-clause (Fig. 4.83).

– Step 6: Risk value determination

The risk value is determined for each threat scenario based on the impact of the associated damage scenarios and the attack feasibility of the associated attack paths (RQ-15-15). The risk value is between 1 and 5, where 1 represents minimal risk (RQ-15-16).

Work product	Requirements/ Recommendations
WP-15-06 Attack feasibility rating	RQ-15-10

Fig. 4.83 Overview of work products/requirements for assessing the feasibility of an attack

Work product	Requirements/ Recommendations
WP-15-07 Risk values	RQ-15-15 RQ-15-16

Fig. 4.84 Overview of work products/requirements for determining the risk value

Work product	Requirements/ Recommendations
WP-15-08 Risk treatment decisions	RQ-15-17

Fig. 4.85 Overview of work products/requirements for the decision on risk treatment

Work products
The following work products are defined in this sub-clause (Fig. 4.84):

- Step 7: Risk treatment decision
  Once the risks have been rated, one of the following risk treatment options is selected (RQ-15-17):
  - Reducing the risk
  - Sharing the risk
    This can be achieved through contracts or by transferring the risk to an insurance company.
  - Retaining the risk
    Risks that are shared or retained are recorded as cybersecurity claims. They are subject to cybersecurity monitoring and vulnerability management in accordance with the chapter "Continuous cybersecurity activities."

Work products
The following work products are defined in this clause (Fig. 4.85).

4.4.1.5 Cybersecurity: Conclusion
Proof of an established and effective cyber security management system is required, among other things, to obtain road approval. In Germany, the authority (KBA) conducts an audit based on the UN Regulation No. 155 Uniform provisions concerning the approval of vehicles regarding cyber security and cyber

security management system [BF-058]. The Requirements catalog for auditing cybersecurity/SU management systems [BF-060] is used for this purpose.

ISO/SAE 21434 forms a reasonable basis for introducing and maintaining such a management system.

Furthermore, UN Regulation No. 155 requires an appropriate program for internal CSMS audits covering all relevant areas and locations (including service). This means the organization must undergo regular audits, which it can only do based on ISO/PAS 5112.

Regarding suppliers, the ENX association [BF-075] has introduced an audit scheme based on ISO/PAS 5112 (see Fig. 4.86).

The goal is to provide suppliers with an auditing standard based on the audit proof of the implemented cybersecurity management system. For this purpose, ENX supplied an audit questionnaire.

Continual cybersecurity activities also mean that these activities in an organization do not end until the project is decommissioned. This means the organization must have a dedicated unit with the necessary skills to ensure this.

All in all, this requires an additional effort for both OEMs and suppliers. Most of them have adequately prepared and introduced corresponding processes.

The growing functional complexity, the increasing number and interaction of systems, and the more distributed nature of development make this task more challenging. It is undoubtedly beneficial here for OEMs and the concerned suppliers

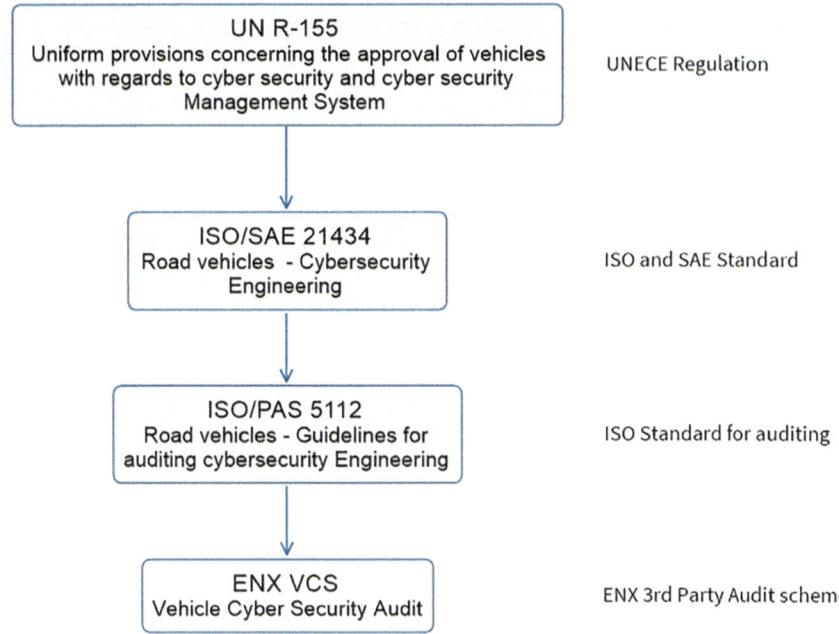

Fig. 4.86 ENX 3rd party audit scheme based on ISO/PAS 5112

to develop concepts openly and based on the overall concept of the vehicle's E/E architecture and coordinate them with each other. Otherwise, there is always a residual risk due to the restricted view (for example, the supplier looking only at its system) and possibly over-designed solutions (i.e., more measures are implemented than needed). The question is if the OEMs will share their concepts. The current answer is usually no.

4.4.1.6 Who Owns the Data?

In our example of unintended accelerations at Toyota [BF-015], it was mentioned that a data recorder was used in the vehicles at the time, meaning essential driving data was recorded and stored.

Of course, this becomes even more important with autonomous vehicles, as it is the only way to analyze the causes of incidents and to learn and improve over time. In this respect, the legislator has already formulated requirements for recording data, for example:

– In the Autonomous Vehicles Authorization and Operation Ordinance (AFBGV) in Germany for the approval and operation of motor vehicles with autonomous driving functions in defined operating areas [BF-043]. It specifies data transmission to the German Federal Motor Transport Authority (KBA).
– In the *Commission Implementing Regulation (EU) 2022/1426* [BF-061] of August 5th, 2022, Annex 2, Sect. 4.9, which describes ADS data requirements and specific data elements for event data recorders (EDR) for fully automated vehicles. Data is specified, as well as the conditions for storage. Section 5.4.1 is also interesting in this context: "The manufacturer must enable the transport service provider to make available to the type-approval authorities, market surveillance authorities, or other authorities designated by the Member States the vehicle data referred to in Sect. 5.4, the ADS data, and the specific data elements for the event data recorder collected following Annex II, Sect. 4.9."

Specifying the data limits the possibilities here, but how can it be ensured that only this data is actually recorded?

We mentioned V2x in the chapter on cybersecurity. V2x includes the vehicle's communication with all possible interfaces, for example:

– With the infrastructure to obtain information about accidents, traffic jams, etc.
– With other vehicles to avoid traffic jams or accidents
– With infrastructure networks
– With clouds, for example, in the case of backends, to process information or commands from services or applications
– With pedestrians, for example, to increase safety
– With devices that are connected directly
– With the power grid.

The software vendors are already active with their applications for some OEMs (e.g., CarPlay from Apple).

These diverse forms of communication also allow the data used to be collected in parallel (see [BF-044, BF-045]). Considering the changes that electromobility will mean for manufacturers, suppliers, and workshops in terms of future sales (with the reduction of numerous sensors, actuators, and other spare parts, along with fewer repairs), it becomes clear that new business models will be necessary to make a profit.

Transportation service providers who rent out autonomous vehicles can unlock significant additional revenue opportunities, such as offering drinks and snacks during rides.

From today's perspective, the data seems to be mainly controlled by car manufacturers, but who knows if others are not already involved (for example, via smartphone connectivity) and are already evaluating drivers' habits and shopping behavior by combining different data?

When it comes to charging, the potential is not yet clear. It is suffice to observe the location of charging stations in obscure backyards or where the charging stations are located at gas stations. One can ask oneself why ICE drivers can comfortably fuel their cars under a roof while EV owners must stand in the rain or snow. As it takes at least 20 to 30 min to recharge, it becomes apparent that, with an intelligent strategy, more revenue streams can be unlocked. The data generated during charging could be worth gold.

4.4.2   ISO 26262—Functional Safety

In the early 1980s, digital injection systems were introduced, followed by electronic control systems for automatic transmissions and various by-wire technologies, such as electronic throttle control (e-gas), which replaced the mechanical cable between the accelerator pedal and the throttle. Further innovations included shift-by-wire systems, electronic stability control, and steer-by-wire systems, all contributing to the growing integration of electronic controls in modern vehicles.

With the spread of electronic car systems, the risk of malfunctions also increased. The first error messages and problem cases for electronic systems were reported in the 1980s, but these could not always be substantiated. In 1986/87, for example, it was claimed in the USA that unintended acceleration could occur with automatic transmission in Audi vehicles [BF-014]. At that time, it was possible to shift into drive mode without standing on the brake. Obviously, drivers would then confuse the brake and gas pedals when the vehicle started. A shift-lock mechanism was introduced to eliminate this hazard, ensuring that forward and reverse gears could only be engaged when the brake pedal was pressed.

Over two decades later, unintended acceleration cases at Toyota in the USA were reported [BF-015]. As data recorders were by then in use, the relevant data could be analyzed, leading to the conclusion that drivers likely confused the brake and gas pedals.

Over time, standards were developed to reduce the risks of such dangerous malfunctions. In the automotive sector, IEC 61508 was initially used in part for this purpose. This standard was developed primarily for mechanical engineering, so it was unsuitable for the automotive industry. The demand for a specific standard arose, and the first version of ISO 26262 was published in 2011 (Fig. 4.87).

A total of twelve volumes have been published to date:

- ISO 26262 Road vehicles—Functional safety—Part 1: Vocabulary [BF-046]
- ISO 26262 Road vehicles—Functional safety—Part 2: Management of functional safety [BF-047]
- ISO 26262 Road vehicles—Functional safety—Part 3: Concept phase [BF-048]
- ISO 26262 Road vehicles—Functional safety—Part 4: Product development at the system level [BF-040]
- ISO 26262 Road vehicles—Functional safety—Part 5: Product development at the hardware level [BF-050]
- ISO 26262 Road vehicles—Functional safety—Part 6: Product development at the software level [BF-051]
- ISO 26262 Road vehicles—Functional safety—Part 7: Production, operation, service and decommissioning [BF-052]
- ISO 26262 Road vehicles—Functional safety—Part 8: Supporting processes [BF-053]
- ISO 26262 Road vehicles—Functional safety—Part 9: Automotive safety integrity level (ASIL)-oriented and safety-oriented analyses [BF-054]
- ISO 26262 Road vehicles—Functional safety—Part 10: Guidelines on ISO 26262 [BF-055]
- ISO 26262 Road vehicles—Functional safety—Part 11: Guidelines on application of ISO 26262 to semiconductors [BF-056]
- ISO 26262 Road vehicles—Functional safety—Part 12: Adaptation of ISO 26262 for motorcycles [BF-057].

Figure 4.88 shows an overview of the individual volumes, which also shows some of the different phases:

- the organizational and project-specific management of functional safety with the planning of project-specific activities
- the concept phase
- Product development with the differentiation between system, hardware, and software and a clear reference to the V-model
- production, operation, service, and decommissioning.

The supporting processes, such as automotive safety integrity level (ASIL)-oriented and safety-oriented analyses and guidelines, are presented as an additonal layer (Fig. 4.88).

As previously mentioned, the standard deals with possible hazards caused by the malfunction of safety-related E/E systems and the interaction of these systems.

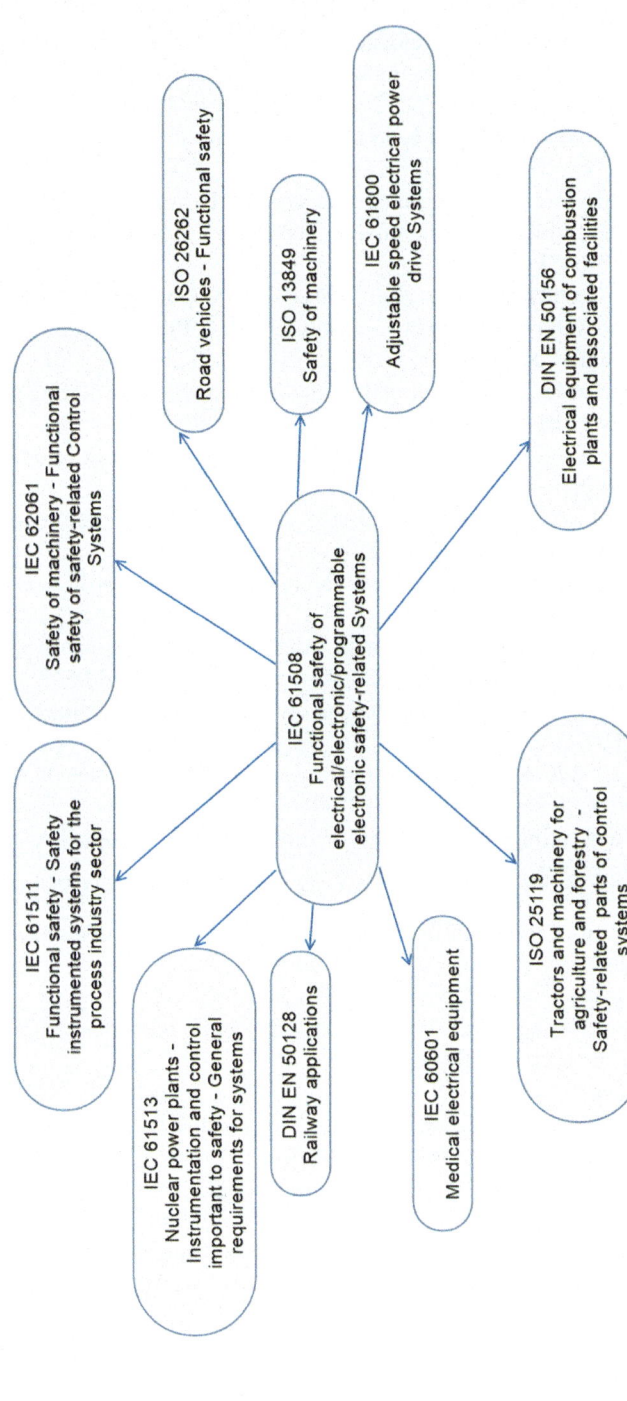

Fig. 4.87 Functional safety standards

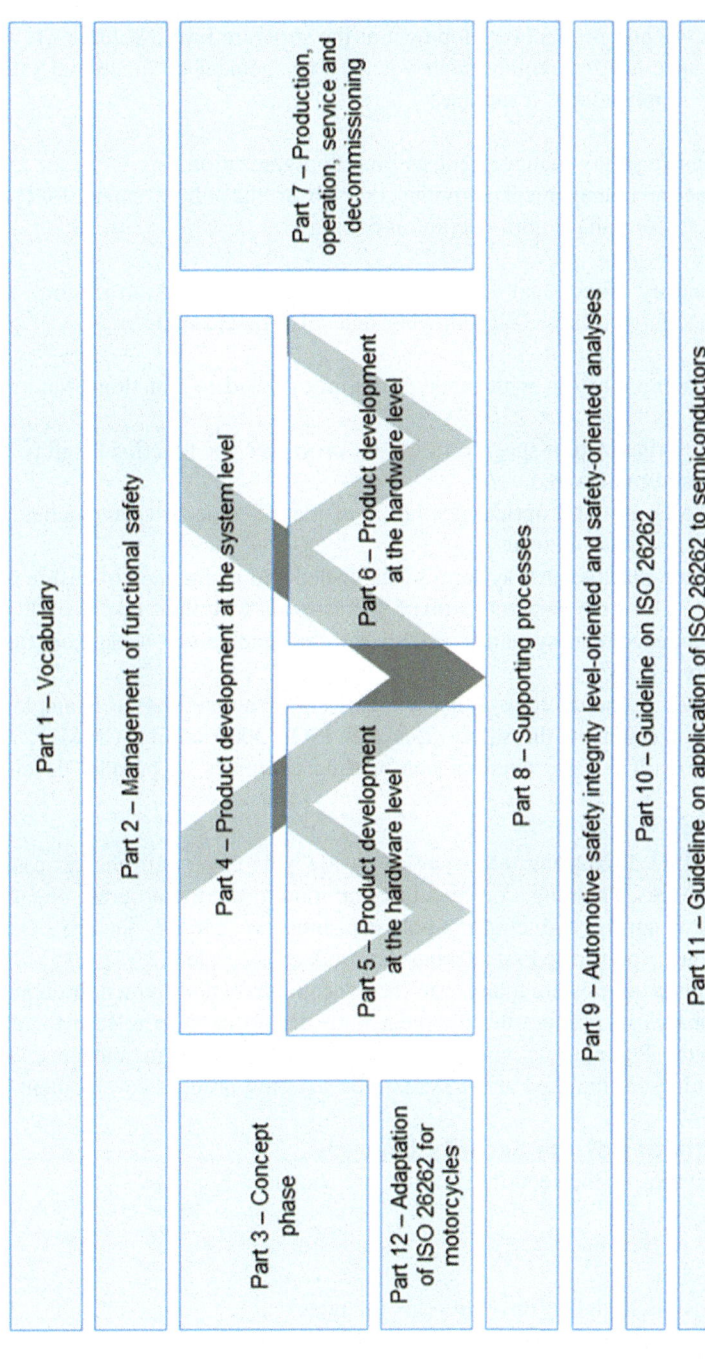

Fig. 4.88 Overview ISO 26262 standards

In this book, we will focus primarily on the management of functional safety (Volume 2), the concept phase (Volume 3), product development at the system level (Volume 4), and product development at the software level (Volume 6).

Volume 2 specifies the requirements for the management of functional safety for automotive applications. It includes:

- Comprehensive safety management within the organization
- Project-specific management activities that cover the whole safety lifecycle (from the concept phase until decommissioning).

Like cybersecurity, functional safety management stipulates introducing and maintaining a safety culture. The following must also be ensured:

- Effective communication with other disciplines related to functional safety is promoted.
- Appropriate organization-specific rules and processes for functional safety are introduced and maintained.
- Processes to ensure appropriate resolution of identified security anomalies are implemented and maintained.
- A competency management system is established and maintained to ensure that the competence of the persons involved corresponds with their responsibilities.
- A quality management system is established and maintained to support functional safety
  IATF 16949, in conjunction with ISO 9001, is mentioned as an example of quality management in the safety lifecycle. ISO 33000 family, CMMI®, and Automotive SPICE® are cited as non-binding examples of product development.

According to ISO 26262, the reference safety lifecycle covers all phases' essential safety activities. Planning, coordinating, and monitoring the progress of safety activities and ensuring that confirmation measures are carried out are central management tasks performed throughout the product lifecycle.

Development activities are planned in the concept phase and refined throughout the various phases of product development until the decision to release for production is made. Planning activities for production, operation, maintenance, and decommissioning are initiated at the system level during product development.

4.4.2.1 Activities of the Security Lifecycle

The main activities of the security lifecycle are shown in Fig. 4.89.

- Item definition
  The first task is to define the *item*. This includes:
  - Clarifying the requirements in this regard, i.e.
    - Legal requirements, standards (national and international)

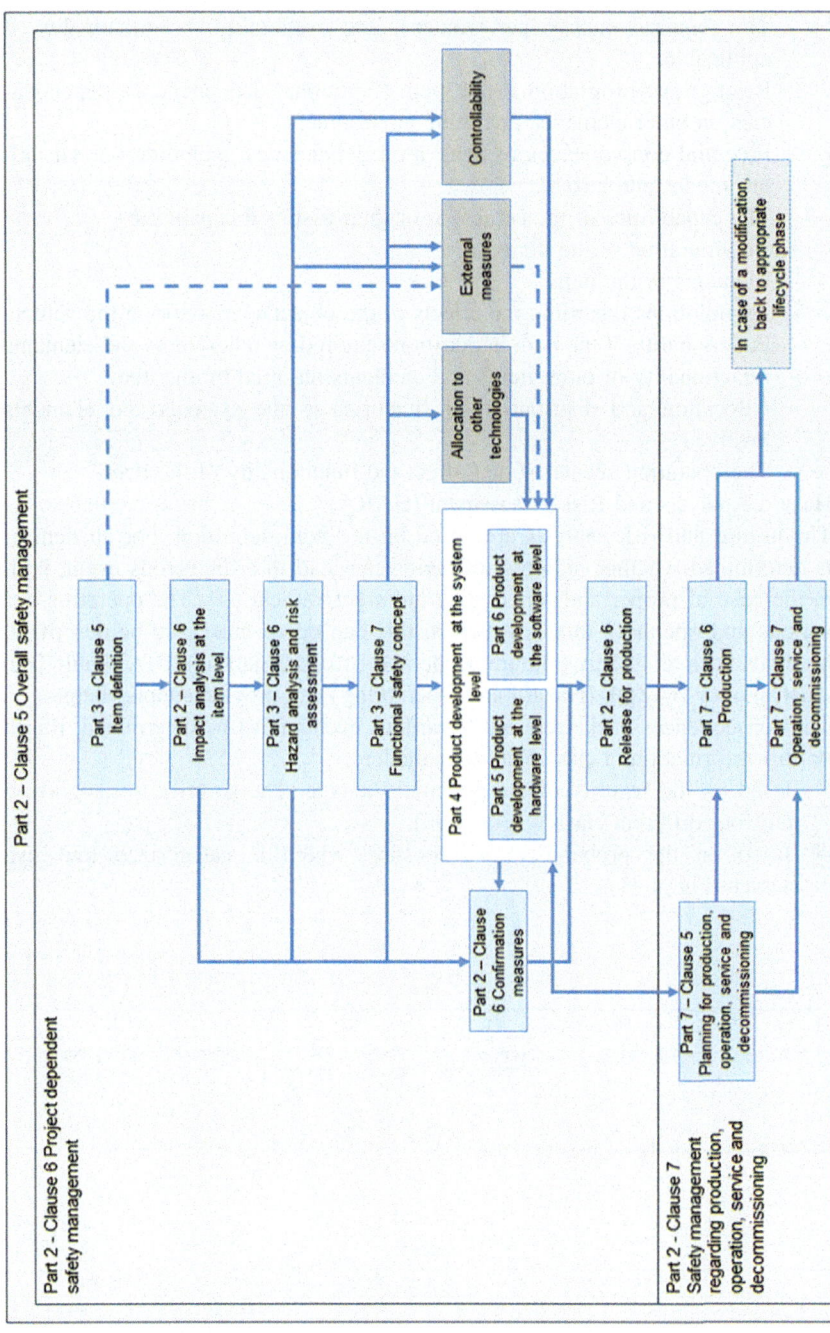

Fig. 4.89 Management activities in the security life cycle (ISO 26262–2:2018)

– The functional behavior at the vehicle level, including the operating modes or states
– The required quality, performance, and availability of functionality, if applicable
– Restrictions in relation to the item (functional dependencies, dependencies on other elements, operating environment)
– Potential consequences of behavioral deficiencies, including known failure modes and hazards
– The capabilities of the actuators, or their assumed capabilities
– The delimitation of the item, i.e.
  – Elements of the item
  – Assumptions regarding the effects of the object's behavior on the vehicle
  – Functionality of the item in question required by other items and elements
  – Functionality of other items and elements required by the item
  – Allocation and distribution of functions to the systems and elements involved
  – The operating scenarios that affect the functionality of the item.
– Hazard Analysis and Risk Assessment (HARA)
  The hazard and risk analysis are based on the item definition. For an item, it is determined whether incorrect behavior can lead to a dangerous event, both in the case of proper and foreseeable incorrect vehicle use. The operating situations and operating modes in which this can occur must also be described. Methods such as FMEA (Failure Mode and Effects Analysis), FTA (Fault Tree Analysis), or HAZOP (Hazard and Operability) can provide support here.
  The consequences of the relevant hazardous events must be determined. Based on this information, a classification is made:
  – Based on the severity of the potential damage. The severity is categorized into four different classes (Fig. 4.90).
  – Based on the probability of exposure, which is categorized into five classes (Fig. 4.91).

Class			
	S0	**S1**	**S2**	**S3**
Description	No injuries	Light and moderate injuries	Severe and life-threatening injuries (survival probable)	Life-threatening injuries (survival uncertain), fatal injuries

Fig. 4.90 Classes for the degree of severity

Class				
	E0	**E1**	**E2**	**E3**	**E4**
Description	incredible	Very low probability	Low probability	Medium probability	High probability

Fig. 4.91 Classes for the probability of exposure

Class				
	C0	**C1**	**C2**	**C3**
Description	Controllable in general	Simply controllable	Normally controllable	Difficult to control or uncontrollable

Fig. 4.92 Classes for the controllability

- Based on the controllability of each hazardous event by the driver or other persons involved in the operating situation. This must be estimated based on a justification for each dangerous event and is divided into four classes (Fig. 4.92).

Based on these three classifications, each hazardous event is assigned an ASIL classification divided into four different categories. The "QM" classification is also used for events that do not require compliance with ISO standard 26262 (Fig. 4.93).

A safety goal must be defined for each hazardous event with an ASIL rating. The determined ASIL is assigned to the safety target. The safety objectives represent the top-level safety requirements for the item. Subsequent phases derive

ASIL determination				
Severity class	Exposure class	Controllability class		
		C1	C2	C3
S1	E1	QM	QM	QM
	E2	QM	QM	QM
	E3	QM	QM	A
	E4	QM	A	B
S2	E1	QM	QM	QM
	E2	QM	QM	A
	E3	QM	A	B
	E4	A	B	C
S3	E1	QM	QM	A*
	E2	QM	A	B
	E3	A	B	C
	E4	B	C	D

• If several unlikely situations are combined that lead to a lower probability of exposure than E1, QM can be argued for S3, C3 on the basis of this combination.

Fig. 4.93 ASIL classification based on severity, probability of exposure, and controllability

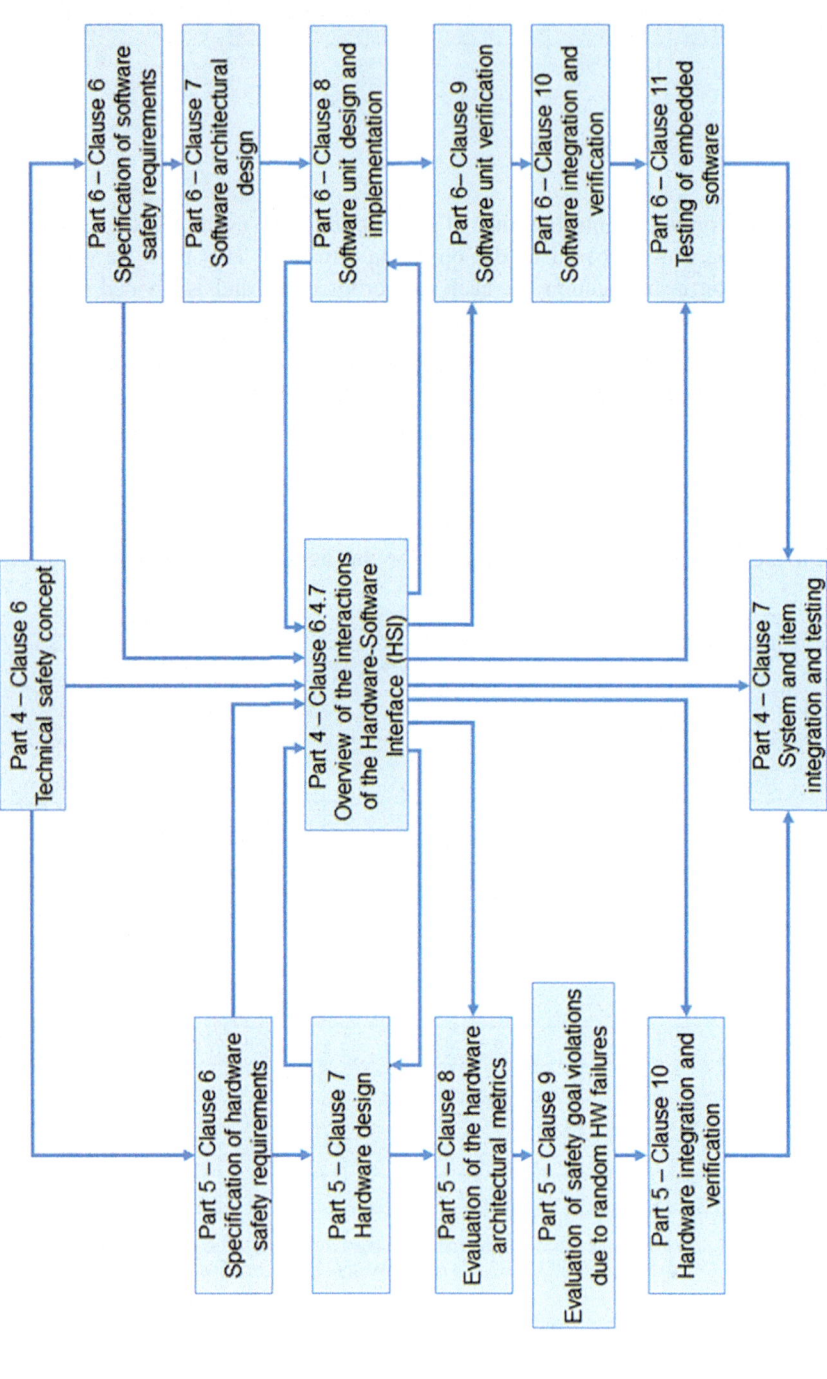

Fig. 4.94 Overview of hardware–software interface interactions (HSI) (ISO 26262—Part 4)

detailed safety requirements from the safety objectives. Each safety requirement inherits the ASIL of the corresponding safety goal.

It is possible to break down a safety requirement into several requirements ("decomposition"). The following decompositions are permitted:

- An ASIL D requirement shall be decomposed as one of the following:
  - one ASIL C(D) requirement and one ASIL A(D) requirement, or
  - one ASIL B(D) requirement and one ASIL B(D) requirement, or
  - one ASIL D(D) requirement and one QM(D) requirement.
- An ASIL C requirement shall be decomposed as one of the following:
  - one ASIL B(C) requirement and one ASIL A(C) requirement, or
  - one ASIL C(C) requirement and one QM(C) requirement.
- An ASIL B requirement shall be decomposed as one of the following:
  - one ASIL A(B) requirement and one ASIL A(B) requirement, or
  - one ASIL B(B) requirement and one QM(B) requirement.
- An ASIL A shall only be decomposed, if needed, as one ASIL A(A) requirement and one QM(A) requirement.

The hazard analysis and risk assessment, including the safety goals, shall be verified (by Part 8, Clause 9) [BF-053].

- Functional Safety Concept

  A functional safety concept is developed based on the preliminary architecture assumptions and safety goals. It derives functional safety requirements from the safety goals.
- Product Development on the System Level

  The technical safety concept is developed once the functional safety concept has been specified. This comprises the technical safety requirements and the design of the system architecture. The technical safety requirements are assigned to the elements of the system architecture and further refined. If requirements arise from the system architecture, they are added, including the hardware-software interface (HSI) (Fig. 4.94).

Following the V-model, the hardware and software elements are integrated and tested after they develop into a product (see Volume 4, Clause 7). This is then incorporated into a vehicle, and a safety validation is performed (see Part 4, Clause 8) to provide proof of functional safety regarding the safety goals.

- Software Product Development

  The software development process is based on the V-model. On the left, the software requirements specification, the software architecture design, and the implementation take place. Software integration, verification, and testing occur on V's right side.

When comparing Fig. 4.95 with the process overview in Automotive SPICE®, one might get the impression that at least the traceability requirements in ISO 26262 are not as complex. However, a closer look at the safety requirements quickly changes this impression (see Sect. 4.4.2.2)

The standard also recommends software architecture design, implementation, integration, verification, and testing. We have just picked out some examples.

The first one is related to software unit design and implementation (Fig. 4.96).

++ means the method is highly recommended, + means the method is recommended, and o means the method has no recommendation for or against its usage.

The listed principles are generally described in coding guidelines.

ASPICE does not cover functional safety, but these principles could, for example, be checked as a specific requirement for an ASIL-classified product (e.g., avoidance of global variables). Regarding the evaluation of resource usage, ASPICE requires the analysis of software architecture in terms of relevant technical design aspects. One technical aspect is resource consumption (e.g., ROM, RAM, external/internal EEPROM, or Data Flash or CPU load; SWE.2.BP3 in PAM 4.0).

The following example is related to software unit verification (Fig. 4.97).

Regarding reviews, walk-throughs or inspections are recommended depending on the ASIL classification. Static code analysis is also highly recommended, allowing us to check the principles described in Fig. 4.96. In principle, MISRA C [BF-034] covers many of these principles for software units coded in C language and is available in static code analysis tools. In the same way, tool solutions are available for modeling tools.

The last example is related to metrics on the software unit level (Fig. 4.98).

These metrics can often be found in general requirements from the car manufacturers (e.g., as part of a software dashboard). This is usually also true for the methods and principles shown in the two previous examples.

– Production, Operation, Service, and Decommissioning

The planning of this phase (see Part 7, Clause 5) begins during product development at the system level. Here, exchanging information and requirements (e.g., safety-relevant special features) is essential to improve the product's ability to manufacture.

This also involves defining responsibilities within the organization to achieve and maintain functional safety during production, operation, service, and decommissioning.

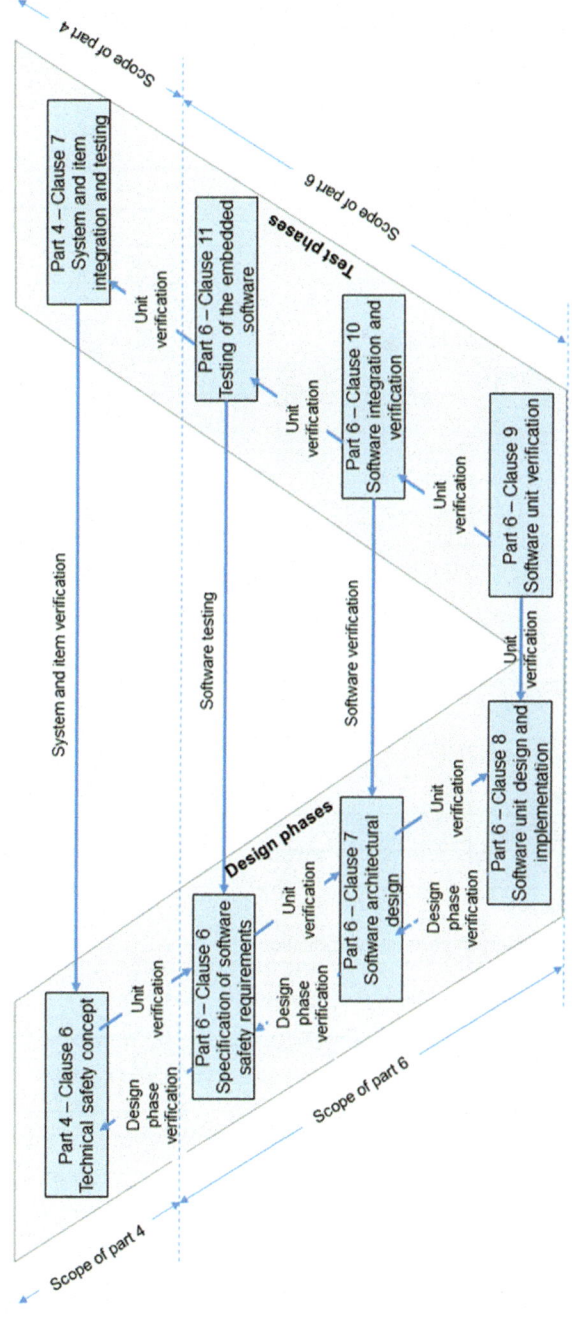

Fig. 4.95 Reference phase model for product development at software level

Principle		ASIL			
		A	B	C	D
1a	One entry and one exit point in subprograms and functions	++	++	++	++
1b	No dynamic objects or variables, or else online test during their creation	+	++	++	++
1c	Initialization of variables	++	++	++	++
1d	No multiple use of variable names	++	++	++	++
1e	Avoid global variables or else justify their usage	+	+	++	++
1f	Restricted use of pointers	+	++	++	++
1g	No implicit type conversions	+	++	++	++
1h	No hidden data flow or control flow	+	++	++	++
1i	No unconditional jumps	++	++	++	++
1j	No recursions	+	+	++	++

Fig. 4.96 Design principles for software unit design and implementation

Methods		ASIL			
		A	B	C	D
1a	Walk-through	++	+	o	o
1b	Pair-programming	+	+	+	+
1c	Inspection	+	++	++	++
1d	Semi-formal verification	+	+	++	++
1e	Formal verification	o	o	+	+
1f	Control flow analysis	+	+	++	++
1g	Data flow analysis	+	+	++	++
1h	Static code analysis	++	++	++	++
1i	Static analyses based on abstract interpretation	+	+	+	+
1j	Requirements-based test	++	++	++	++
1k	Interface test	++	++	++	++
1l	Fault injection test	+	+	+	++
1m	Resource usage evaluation	+	+	+	++
1n	Back-to-back comparison test between model and code, if applicable	+	+	++	++

Fig. 4.97 Methods for software unit verification

Methods		ASIL			
		A	B	C	D
1a	Statement coverage	++	++	+	+
1b	Branch coverage	+	++	++	++
1c	MC/DC (Modified Condition/Decision Coverage)	+	+	+	++

Fig. 4.98 Structural coverage metrics at the software unit level

4.4.2.2 Expectations Concerning Safety Requirements

Safety requirements must be:

- comprehensible
- atomic
- internally consistent
- feasible and achievable
- verifiable
- necessary
- implementation free
- complete
- conforming to standards (for example, legal requirements, automotive standards, etc.).

Furthermore, they must have an ASIL rating and a unique identifier that remains unchanged throughout the safety lifecycle and status.

Note: The requirements' characteristics correspond to those expected by Automotive SPICE®.

The management of safety requirements also ensures that:

- Requirements are hierarchically organized. This means they are structured according to the security activities.
- The safety requirements are grouped according to this structure.
- Completeness is ensured, i.e., one level's requirements cover entirely the previous level's requirements.
- External consistency is ensured, i.e., several security requirements do not contradict each other.
- There is no duplication of information within one level of the hierarchical structure.
- Requirements are maintainable, i.e., they can be changed or expanded (Fig. 4.99).

Traceability is another essential FuSa requirement (similar to Automotive SPICE). It must be ensured that:

- Safety requirements refer to a security requirement of the next higher hierarchy level.
- Safety requirements refer to a derived safety requirement at the lower hierarchy level or its realization in the design.
- A safety requirement has a reference to a verification specification, i.e.
  - Review or analysis checklists, or
  - Simulation scenarios, or
  - Test cases, test data, and test objects.

A complete peer review of the safety requirements is indispensable. Simple reviews are allowed for QM requirements. Systematic *walkthroughs* are permitted up to ASIL A, while formal _inspections_ are strongly recommended for ASIL B and above (Fig. 4.100).

Similar to Automotive SPICE, the safety requirements are relevant regarding configuration management, i.e., they must be managed and frozen according to corresponding baselines. The relevant configuration management strategy must be documented in the configuration management plan.

4.4.2.3 Conclusion Regarding Functional Safety

In principle, ISO 26262 is not a requirement for the homologation of vehicles. However, FuSa is now regarded as "state-of-the-art"; thus, it has become a de facto regulatory requirement.

The increasing complexity of systems and components and more distributed development processes lead to increased liability risks.

Another issue arises when immature or separate engineering processes are explicitly developed for functional safety, creating a parallel workflow rather than integrating it with the rest of the project. Whether this approach represents an integrated safety culture is still discussed in our industry.

A notorious lack of well-trained and qualified Functional Safety (FuSa) experts presents another challenge. These experts must maintain a comprehensive overview, which is crucial given various systems' increasing complexity and interdependencies. This is especially important in areas such as developing comprehensive safety concepts.

Another aspect is the introduction of new Electrical/Electronic (E/E) architectures, such as the division into domains with corresponding domain control units that handle overarching tasks, simplifying the overall architecture and reducing the number of control units. However, this can also result in a more complex safety concept if the safety requirements are distributed across multiple control units.

The ISO 26262 standard is an exceedingly complex and comprehensive regulatory requirement with the ultimate goal of managing precisely stipulated risks of safety malfunctions. However, while this standard is widely recognized and adopted, some critical industry and safety experts question its scope and limitations. They argue that the complexity and scope of ISO 26262 can lead to a focus on compliance rather than improving safety. This can lead to a "checklist mentality", where ticking off requirements becomes more important than understanding

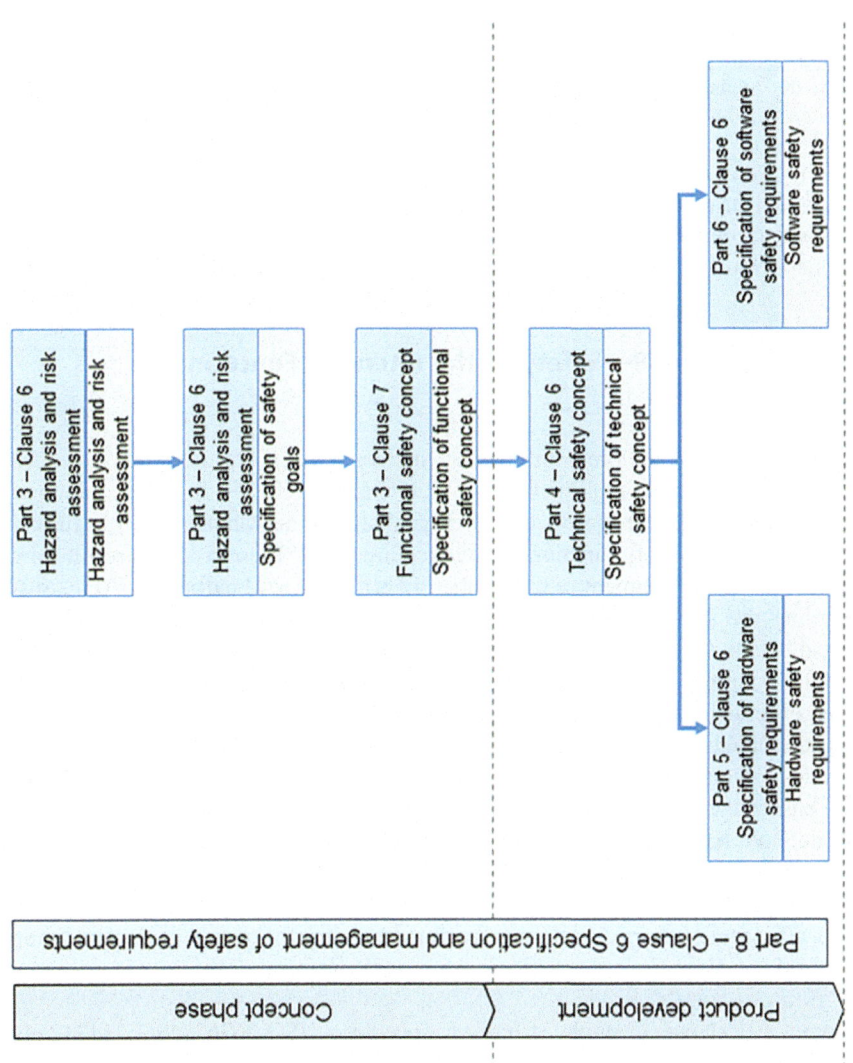

Fig. 4.99 Structure of the security requirements

Methods		ASIL			
		A	B	C	D
1a	Verification by walk-through	++	+	o	o
1b	Verification by inspection	+	++	++	++
1c	Semi-formal verification[a]	+	+	++	++
1d	Formal verification[a]	o	+	+	+
[a] Verification can be supported by executable models					

Fig. 4.100 Methods for checking the safety requirements (ISO 26262-8:2018)

and effectively implementing safety principles. Others point out that the high complexity and strict requirements of ISO 26262 can be a significant hurdle, especially for smaller companies and start-ups. This can slow innovation or exclude smaller suppliers from developing safety-relevant systems altogether.

4.4.3   ISO 21448—Safety of the Intended Functionality (SOTIF)

As we have seen in the previous chapter, functional safety was designed to reduce the risks of dangerous malfunctions. In autonomous vehicles, however, the complex system of multiple sensors and intelligent algorithms can be subject to further risks even if they satisfy functional safety requirements. These risks can result from changes in the environment (e.g., weather, light), other road participants (e.g., disregarding the traffic rules), misuse (e.g., not using the system as specified), or critical events (e.g., a child suddenly crossing the road). Indeed, some older and random examples of accidents where the function worked as designed but still posed excessive risks are demonstrable (see [BF-081–BF-083]).

These challenges led to the development of ISO 21448 (SOTIF—Safety of the Intended Functionality) [BF-079], first published in 2019. The scope of ISO 21448 is to ensure the safety of the intended functionality. The absence of unreasonable risk due to a hazard caused by functional insufficiencies, i.e.:

– Insufficiencies of specification of the intended functionality at the vehicle level
– Insufficiencies of specification or performance insufficiencies in implementing electric and/or electronic (E/E) elements in the system

Figure 4.101 shows a sample of hazards covered by ISO 26262, ISO 21434, and ISO 21448.

Systems for automated driving are usually structured in the key elements Sense–Plan–Act, as shown in Fig. 4.102.

The hazards can be classified into four areas based on known or unknown scenarios resulting from given use cases, as shown in Fig. 4.103. Unknown scenarios can, for example, result from triggering conditions that have been identified (e.g.,

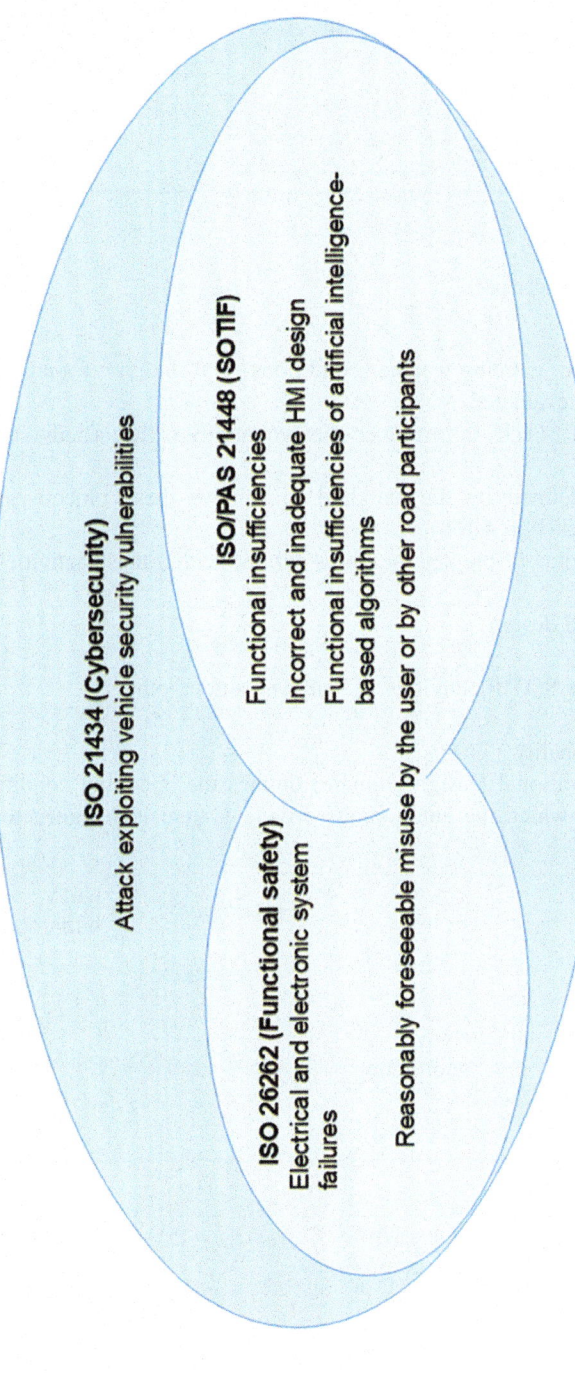

Fig. 4.101 Sample hazards covered by ISO 26262, ISO 21434 and ISO 21448

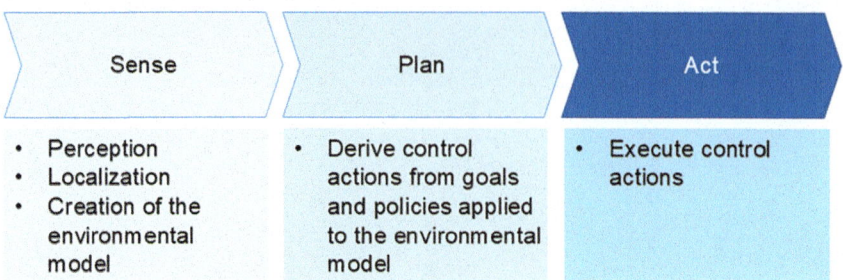

Fig. 4.102 Sense-Plan-Act model

extreme temperature, extreme weather conditions), but the corresponding system response cannot be evaluated.

The objective of SOTIF is to reduce the probability of hazardous known and unknown scenarios.

The activities defined by the standard to achieve these objectives can be described as follows (Fig. 4.104).

Here is an overview of the key measures defined within these activities:

– Specification and design

This is the basis for SOTIF activities. It includes, among others:

– Intended functionality
– The ODD (Operational Design Domain) defines the operating conditions and scenarios under which the autonomous driving system is designed to operate

Fig. 4.103 Scenario categories

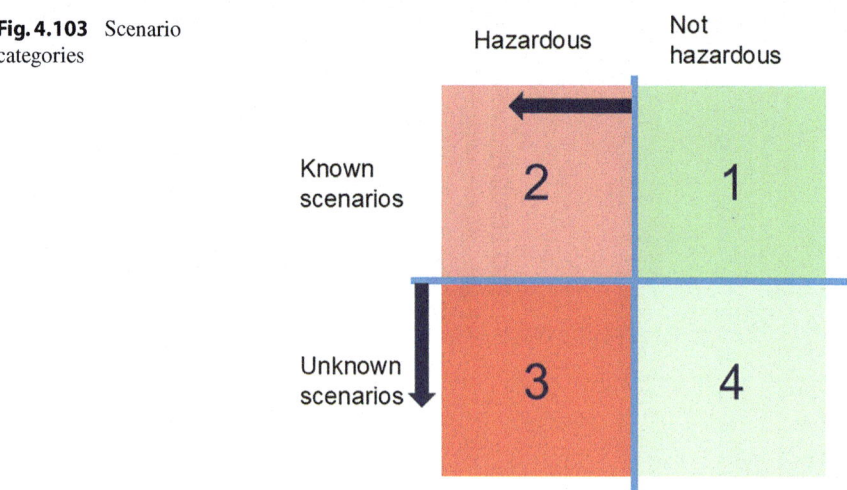

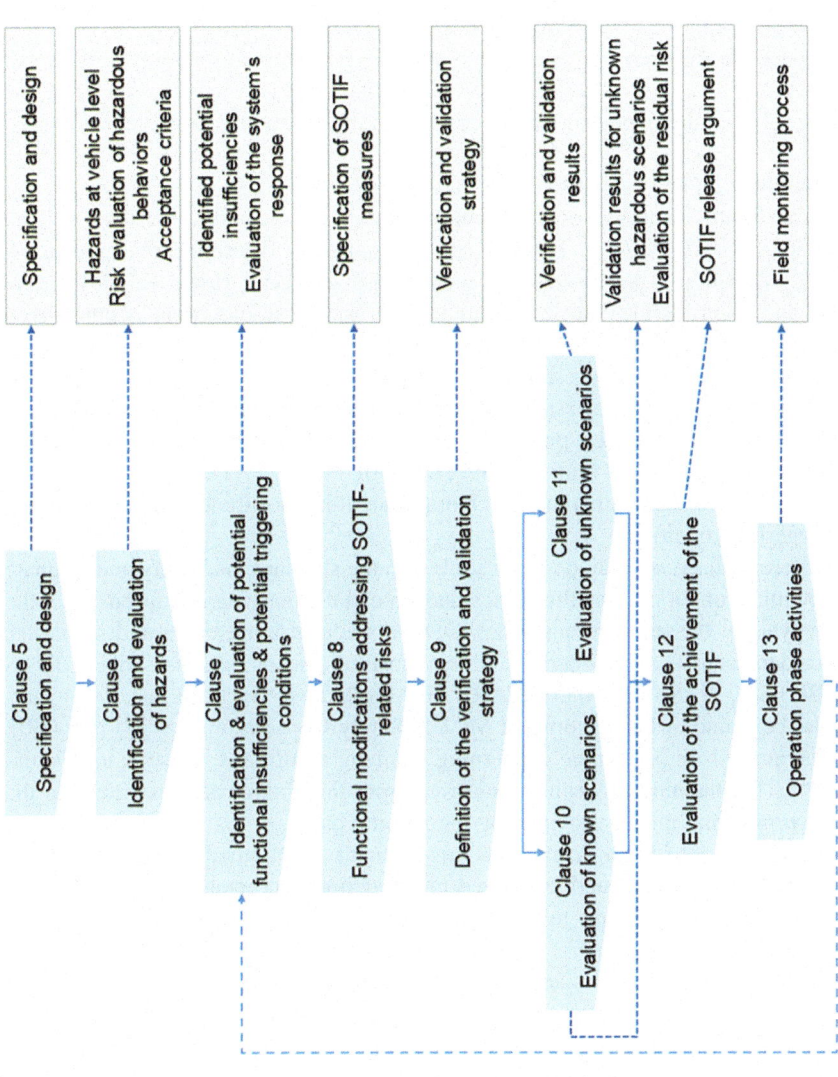

Fig. 4.104 Simplified activity flow

safely (see, for example, Fig. 9 "ODD Classification Framework With Top-Level Categories and Immediate Subcategories" in *A Framework for Automated Driving System Testable Cases and Scenarios* from NHTSA [BF-111])

- System architecture
- Warning strategies
- Degradation concept
- Interface with the back-office operator
- Expectable misuse

The SOTIF requirements can be integrated into the functional safety requirements.

- Identification and evaluation of hazards
  Hazards are identified on the vehicle level. The methods described in ISO 26262-3 [BF-048] (e.g., HARA) can be applied to determine the hazards. A risk evaluation is performed for each hazardous event. However, only severity and controllability are considered here. The hazardous event is considered SOTIF-related if controllability is not rated as "controllable in general" or severity is not rated as "no resulting harm." In this case, acceptance criteria must be specified. These can result from regulations, existing functions, and standard driver performance. Best practices can, for example, be found at the Automated Vehicle Safety Consortium (AVSC) [BF-080].
- Identification and evaluation of potential functional insufficiencies and potential triggering conditions
  Different analysis methods are described in the standard to identify and evaluate potential functional insufficiencies and potential triggering conditions, like the analysis of the requirements (incl. ODD boundaries), use cases and scenarios, accident statistics provided by official institutions, or databases like GIDAS [BF-084], NASS [BF-085], CARE [BF-086]. Best practices can, for example, be found at the Automated Vehicle Safety Consortium (AVSC) [BF-080]. The analysis is performed concerning planning algorithms, sensors, and actuators. The planning algorithm is derived from the control actions based on the environmental model provided by the sensing part.
  An analysis of (direct or indirect) misuse is also performed. Finally, the residual risk is estimated and compared to the defined acceptance criteria. Further functional modifications must be implemented if the residual risk exceeds the acceptance criteria.
- Functional modifications addressing SOTIF-related risks
  Measures are defined to reduce the residual risk under the acceptance criteria limit. These measures can induce specification and design updates, which must be re-evaluated through Clauses 6 and 7 activities.
- Definition of the verification and validation strategy
  The SOTIF verification and validation strategy is defined. It must include:
  - Validation goals
  - Evaluation of potentially hazardous scenarios
  - Sufficient coverage of the relevant scenario

- Documentation of the results
- Methods to derive verification and validation activities
- A documented rationale for the selected verification and validation activities
- Evaluation of known scenarios
  The objective is for verification results to show that the validation goals have been fulfilled. Verification is performed for:
  - Sensing

    Done by proving the functional performance, timing, accuracy, and robustness.
  - The planning algorithm

    Based on the environmental model, proving the ability to react as required and showing no unwanted reactions.
  - The actuation

    Reacting as intended based on the planning algorithm.
  - The integrated system

    Verify the robustness and controllability of the system on the vehicle level.

    If the validation goals for known hazardous scenarios are met.
- Evaluation of unknown scenarios

  The objective is that the verification results show that the acceptance criteria are met with sufficient confidence. As unknown scenarios happen in real life, one evaluation method is to perform tests on open roads.
- Evaluation of the achievement of the SOTIF

  The work products of the SOTIF activities are reviewed for completeness, correctness, and consistency. Based on these work products and the fulfillment of the objectives of each clause, an argument for achieving the safety of the intended functionality is documented and evaluated. This argument allows us to recommend the approval or rejection of the SOTIF release.
- Operation phase activities

  A field monitoring process must be defined before release and implemented during operation. Depending on field monitoring results (e.g., incidents in the field, incidents recorded in databases, the evolution of regulations) and the detection of new hazards, development activities starting with 'identification and potential functional insufficiencies' (Clause 7) must be planned and implemented.

ISO 21448 (SOTIF) closes a critical gap in the safety of autonomous cars, where traditional functional safety may prove insufficient. By focusing on the risks that emerge from the intended functionality, especially in unknown or dynamic scenarios, SOTIF plays a crucial role in overall safety management. SOTIF aims to ensure that autonomous systems can meet the highest safety standards in real-world environments. It will play a critical role in higher levels of autonomous driving.

4.5   Conclusion: The Role of Automotive Standards

In this chapter, we have focused only on the most critical quality standards from the perspective of automotive development. The progressively developed functions for autonomous driving will undoubtedly lead to increased legal regulations and standards while simultaneously posing a challenge to innovation in our industry.

At the same time, this will also present a challenge for integrated process management. However, this is just one of many challenges organizations face. As demonstrated by cybersecurity and SOTIF requirements, organizations must continue to allocate development resources during production and throughout the entire lifecycle, from the start of production until the systems' obsolescence.

While the development activities on the left side of V are crucial, we should not underestimate the challenge posed by validating and verifying automotive systems, emphasizing the proper usage of modern test tools. The complexity of testing is steadily growing, and for some, a massive amount of data must be stored safely for a long time. In particular, test scenarios are often country-specific, which may imply that thousands of additional test cases are needed to meet all regulatory requirements. This is why quality will never be "free," regardless of what some experts claim. On the contrary, automotive quality will continue to be more expensive and complex, posing a challenge to the related process quality.

The final challenge will be to integrate all required information to ensure vehicle safety (e.g., from functional safety, cybersecurity, and SOTIF) to evaluate the safe release of a system. To achieve this, safety assurance cases can be created and structured into claims backed by well-founded argumentation and supported by concrete evidence. This approach provides a clear framework for assessing and demonstrating system safety. Further information can be found in ISO/IEC/IEEE 15026-2 Systems and software engineering—System and software assurance—Part 2: Assurance case [BF-089] or UL 4600 Standard for safety—Evaluation of autonomous products [BF-087].

To conclude, here is an overview of existing and planned standards required for automated driving:

- ISO/PAS 8800 Road vehicles—Safety and artificial intelligence [BF-088]
- ISO/PAS 11585:2023 Road vehicles—Partial driving automation—Technical characteristics of conditional hands-free driving systems [BF-091]
- ISO/AWI PAS 11585-2 Road vehicles—Partial driving automation—Part 2: Test method to evaluate the performance of partial driving automation conditional hands-free driving systems [BF-092]

- ISO/AWI 23792-1:2023 Intelligent transport systems—Motorway chauffeur systems (MCS)—Part 1: Framework and general requirements, ISO, ISO—International Organization for Standardization [BF-093]
- ISO/AWI 23792-2 Intelligent transport systems—Motorway chauffeur systems (MCS)—Part 2: Requirements and test procedures for discretionary lane change, ISO, ISO—International Organization for Standardization [BF-094]
- ISO 23793-1:2024 Intelligent transport systems—Minimal risk manoeuvre (MRM) for automated driving—Part 1: Framework, straight-stop and in-lane stop [BF-095]
- ISO/AWI 23793-2 Intelligent transport systems—Minimal risk manoeuvre (MRM) for automated driving—Part 2: Road shoulder stop—Minimum requirements and test procedures [BF-096]
- ISO/AWI TR 17720 Intelligent transport systems—Operational Design Domain Boundary and Attribute Awareness for an Automated Driving System [BF-090]
- ISO 23374-1:2023 Intelligent transport systems—Automated valet parking systems (AVPS) Part 1: System framework, requirements for automated driving and for communications interface [BF-097]
- ISO/TS 23374-2:2023 Intelligent transport systems—Automated valet parking systems (AVPS) Part 2: Security integration for type 3 AVP [BF-098]
- ISO 20900:2023 Intelligent transport systems—Partially-automated parking systems (PAPS)—Performance requirements and test procedures [BF-099]
- ISO/CD 12768-1 Intelligent transport systems—Automated Valet Driving Systems (AVDS)—Part 1: Requirements, System Framework, Communication Interfaces and Test Procedures [BF-100]
- ISO/AWI 12768–2 Intelligent transport systems—Automated Valet Driving Systems (AVDS)—Part 2: System framework, security procedures and requirements [BF-101]
- ISO/DIS 7856 Intelligent transport systems—Remote support for low speed automated driving systems (RS-LSADS)—Performance requirements, system requirements and performance test procedures [BF-102]
- ISO 4273:2024 Intelligent transport systems—Automated braking during low-speed manoeuvring (ABLS)—Requirements and test procedures [BF-103]
- ISO/CD TS 5083 Road vehicles—Safety for automated driving systems—Design, verification and validation [BF-069]
- ISO 23150-1 Road vehicles—Logical interface between sensors and data fusion unit for automated driving functions—Part 1: General information and principles [BF-070]
- ISO 23150-11 Road vehicles—Logical interface between sensors and data fusion unit for automated driving functions—Part 11: Radar detection interface [BF-071]
- ISO 23150-12 Road vehicles—Logical interface between sensors and data fusion unit for automated driving functions—Part 12: Lidar detection interface [BF-072]

- ISO 23150-13 Road vehicles—Logical interface between sensors and data fusion unit for automated driving functions—Part 13: Camera detection and feature interfaces [BF-073]
- ISO 23150-14 Road vehicles—Logical interface between sensors and data fusion unit for automated driving functions—Part 14: Ultrasonic detection and feature interfaces [BF-108]
- ISO 23150-15 Road vehicles—Logical interface between sensors and data fusion unit for automated driving functions—Part 15: Microphone detection interface [BF-109]
- ISO 23150-20 Road vehicles—Logical interface between sensors and data fusion unit for automated driving functions—Part 20: Supportive and sensor input interface [BF-110]
- ISO 34501:2022 Road vehicles—Test scenarios for automated driving systems—Vocabulary [BF-104]
- ISO 34502:2022 Road vehicles—Test scenarios for automated driving systems—Scenario based safety evaluation framework [BF-105]
- ISO 34503:2023 Road Vehicles—Test scenarios for automated driving systems—Specification for operational design domain [BF-106]
- ISO 34504:2024 Road vehicles—Test scenarios for automated driving systems—Scenario categorization [BF-107]
- ISO/TS 22133:2023 Road vehicles—Test object monitoring and control for active safety and automated/autonomous vehicle testing—Functional requirements, specifications, and communication protocol [BF-062]
- ISO 22733-1:2022 Road vehicles—Test method to evaluate the performance of autonomous emergency braking systems—Part 1: Car-to-car [BF-063]
- ISO 22733-2:2023 Road vehicles—Test method to evaluate the performance of autonomous emergency braking systems—Part 2: Car to pedestrian [BF-064]
- ISO 37181:2022 Smart community infrastructures—Smart transportation by autonomous vehicles on public roads [BF-065]
- ISO 39003:2023 Road traffic safety (RTS)—Guidance on ethical considerations relating to safety for autonomous vehicles [BF-066]
- ISO/TR 9241-810:2020 Ergonomics of human-system interaction—Part 810: Robotic, intelligent and autonomous systems [BF-067]
- ISO 37168:2022 Smart community infrastructures—Guidance on smart transportation by Electric, Connected and Autonomous Vehicles (eCAVs) and its application to on-demand responsive passenger services with shared vehicles [BF-068].

It is necessary to continue working on the legal framework for autonomous driving, addressing issues like respfonsibility and insurance while guiding development in a regulated and structured direction.

The growing avalanche of legal regulations and standards will add a significant workload to all involved parties, ultimately driving up costs. This makes it increasingly necessary to maintain integrated management systems, streamline and structure processes efficiently, and continuously invest in employees' skills to manage these challenges effectively.

References

[BF-001] VDA QMC: Automotive SPICE Process Assessment/Reference Model V3.1, https://automotivespice.com

[BF-002] ISO/IEC/IEEE 12207, ISO, https://www.iso.org

[BF-003] VDA QMC Band Automotive SPICE Guidelines—Process assessment using the Automotive SPICE PAM 4.0, 2nd revised edition, November 2023

[BF-004] Gallup State of the Global Workplace: 2023 Report

[BF-005] International Organization for Standardization

[BF-006] ISO/IEC 33000 Family, ISO, https://www.iso.org

[BF-007] ISO/IEC 33001 Information technology—Process assessment—Concepts and terminology, ISO, https://www.iso.org

[BF-008] ISO/IEC 33002 Information technology—Process assessment—Requirements for performing process assessment, ISO, https://www.iso.org

[BF-009] IATF 16949 Quality system requirements for automotive production and relevant service parts organizations

[BF-010] DIN EN ISO 9001 Qualitätsmanagementsysteme—Anforderungen (ISO 9001:2015)

[BF-011] Leading Change, John P. Kotter, ISBN: 978-1-4221-8643-5

[BF-012] ISO/IEC TR 15504-10:2011 Information technology—Process assessment—Part 10: Safety Extension

[BF-013] VDI 5702 Blatt 1: Medizinprodukte-Software—Medical SPICE Prozessassessmentmodell

[BF-014] The Center for Auto Safety, Audi Sudden Acceleration, Audi Sudden Acceleration

[BF-015] U.S. Department of Transportation, U.S. Department Of Transportation Releases Results From NHTSA-NASA Study Of Unintended Acceleration In Toyota Vehicles

[BF-016] CVE-2015-5611 https://www.cve.org/CVERecord?id=CVE-2015-5611

[BF-017] NIST Vulnerability Database (NVD) https://nvd.nist.gov

[BF-018] Common Attack Pattern Enumeration and Classification (CAPEC) https://capec.mitre.org/index.html

[BF-019] Common Vulnerabilities and Exposures (CVE) Home | CVE

[BF-020] Common Weakness Enumeration (CWE) CWE–Common Weakness Enumeration (mitre.org)

[BF-021] MITRE Solving Problems for a Safer World | MITRE

[BF-022] MITRE ATT&CK® MITRE ATT&CK®

[BF-023] NIST IR 8473—Cybersecurity Framework Profile for Electric Vehicle Extreme Fast Charging Infrastructure https://nvlpubs.nist.gov/nistpubs/ir/2023/NIST.IR.8473.pdf

[BF-024] Cybersecurity Framework Profile for Connected Vehicle Environments https://www.its.dot.gov/research_areas/cybersecurity/docs/4_CSF_for_CVE_Dot_Chart.xlsx

[BF-025] SAE J3061—Cybersecurity for Cyber-Physical Vehicle Systems

[BF-026] ISO/SAE 21434—Road vehicles—Cybersecurity engineering

[BF-027] ENISA https://www.enisa.europa.eu/t

[BF-028] BSI (Bundesamt für Sicherheit in der Informationstechnik) https://www.bsi.bund.de/DE/Home/home_node.html

[BF-029] ASRG (Automotive Security Research Group) https://asrg.io/

[BF-030] AUTO-ISAC: Automotive Information Sharing and Analysis Center https://automotiveisac.com/

[BF-031] Upstream Security Global Automotive Cybersecurity Report 2022 https://upstream.auto/

[BF-032] Upstream H1'2023 Automotive Cyber Trend Report https://upstream.auto/

[BF-033] Upstream Security-Global_Automotive_Cybersecurity_Report_2020

[BF-034] MISRA C:2023 Guidelines for the use of the C language in critical systems, The MISRA Consortium Limited, https://misra.org.uk/

[BF-035] SEI CERT C Coding Standard: Rules for Developing Safe, Reliable, and Secure Systems (2016 Edition), SEI, SEI CERT C Coding Standard: Rules for Developing Safe, Reliable, and Secure Systems (2016 Edition) (cmu.edu)

[BF-036] Ansys medini analyze for Cybersecurity, Ansys, https://www.ansys.com/

[BF-037] ISO/IEC 18045 Information security, cybersecurity and privacy protection—Evaluation criteria for IT security—Methodology for IT security evaluation, ISO

[BF-038] E-safety vehicle intrusion protected applications (EVITA), EVITA (evita-project.org)

[BF-039] Security requirements for automotive on-board networks based on dark-side scenarios (EVITA Deliverable 2.3), Security requirements for automotive on-board networks based on dark-side scenarios (EVITA Deliverable 2.3) (zenodo.org)

[BF-040] Method and pro forma for Threat, Vulnerability, Risk Analysis (TVRA), ETSI TS 102 165-1 V5.2.3 (2017–10), TS 102 165-1—V5.2.3—CYBER; Methods and protocols; Part 1: Method and pro forma for Threat, Vulnerability, Risk Analysis (TVRA) (etsi.org)

[BF-041] UcedaVélez, Tony and Morana, Marco M. Risk Centric Threat Modeling: Process for Attack Simulation and Threat Analysis. Hoboken, New Jersey: Wiley, May 2015. ISBN: 978-1-118-98835-0

[BF-042] FORUM OF INCIDENT RESPONSE AND SECURITY TEAMS (FIRST). Common Vulnerability Scoring System (CVSS), Common Vulnerability Scoring System v3.1: Specification Document, CVSS v3.1 Specification Document (first.org)

[BF-043] Verordnung (Autonome-Fahrzeuge-Genehmigungs-und-Betriebs-Verordnung—AFGBV)—Anlage 1 Anforderungen an Kraftfahrzeuge mit autonomer Fahrfunktion, AFGBV.pdf (gesetze-im-internet.de)

[BF-044] Your New Car Is Watching You And Collecting Your Data, José Rodríguez Jr., Your New Car Is Watching You And Collecting, Selling Your Data (jalopnik.com)

[BF-045] Das Auto als Spion, Melanie Böff und Antje Erhard und Constantin Röse, ARD-Finanzredaktion, https://www.tagesschau.de/wirtschaft/verbraucher/auto-fahrerdaten-besitz-100.html?xing_share=news

[BF-046] ISO 26262-1 Road vehicles—Functional safety—Part 1: Vocabulary, ISO, https://www.iso.org

[BF-047] ISO 26262-2 Road vehicles—Functional safety—Part 2: Management of functional safety, ISO, https://www.iso.org

[BF-048] ISO 26262-3 Road vehicles—Functional safety—Part 3: Concept phase, ISO, https://www.iso.org

[BF-049] ISO 26262-4 Road vehicles—Functional safety—Part 4: Product development at the system level, ISO, https://www.iso.org

[BF-050] ISO 26262-5 Road vehicles—Functional safety—Part 5: Product development at the hardware level, ISO, https://www.iso.org

[BF-051] ISO 26262-6 Road vehicles—Functional safety—Part 6: Product development at the software level, ISO, https://www.iso.org

[BF-052] ISO 26262-7 Road vehicles—Functional safety—Part 7: Production, operation, service and decommissioning, ISO, https://www.iso.org

[BF-053] ISO 26262-8 Road vehicles—Functional safety—Part 8: Supporting processes, ISO, https://www.iso.org

[BF-054] ISO 26262-9 Road vehicles—Functional safety—Part 9: Automotive safety integrity level (ASIL)-oriented and safety-oriented analyses, ISO, https://www.iso.org

[BF-055] ISO 26262-10 Road vehicles—Functional safety—Part 10: Guidelines on ISO 26262, ISO, https://www.iso.org

[BF-056] ISO 26262-11 Road vehicles—Functional safety—Part 11: Guidelines on application of ISO 26262 to semiconductors, ISO, https://www.iso.org

[BF-057] ISO 26262-12 Road vehicles—Functional safety—Part 12: Adaptation of ISO 26262 for motorcycles, ISO, https://www.iso.org

[BF-058] UN Regulation No. 155 Uniform provisions concerning the approval of vehicles with regards to cyber security and cyber security management system, UNECE, UN Regulation No. 155—Cyber security and cyber security management system|UNECE

[BF-059] UN Regulation No. 156 Software update and software update management system, UNECE, UN-Regulation No. 156—Software Update and Software Update Management System

[BF-060] Anforderungskatalog Auditierung von Cybersecurity/SU-Managementsystemen, KBA, Anforderungskatalog, Auditierung von Cybersecurity/SU-Managementsystemen, Stand: 01/2021 (kba.de)

[BF-061] Commission implementing regulation (EU) 2022/1426, EUR-Lex, EUR-Lex–32022R1426–EN-EUR-Lex (europa.eu)

[BF-062] ISO/TS 22133:2023 Road vehicles—Test object monitoring and control for active safety and automated/autonomous vehicle testing—Functional requirements, specifications, and communication protocol, ISO, ISO–International Organization for Standardization

[BF-063] ISO 22733-1:2022 Road vehicles—Test method to evaluate the performance of autonomous emergency braking systems—Part 1: Car-to-car, ISO, ISO–International Organization for Standardization

[BF-064] ISO 22733-2:2023 Road vehicles—Test method to evaluate the performance of autonomous emergency braking systems—Part 2: Car to pedestrian, ISO, ISO - International Organization for Standardization

[BF-065] ISO 37181:2022 Smart community infrastructures—Smart transportation by autonomous vehicles on public roads, ISO, ISO–International Organization for Standardization

[BF-066] ISO 39003:2023 Road traffic safety (RTS)—Guidance on ethical considerations relating to safety for autonomous vehicles, ISO, ISO–International Organization for Standardization

[BF-067] ISO/TR 9241-810:2020 Ergonomics of human-system interaction—Part 810: Robotic, intelligent and autonomous systems, ISO, ISO–International Organization for Standardization

[BF-068] ISO 37168:2022 Smart community infrastructures—Guidance on smart transportation by Electric, Connected, and Autonomous Vehicles (eCAVs) and its application to on-demand responsive passenger services with shared vehicles, ISO, ISO–International Organization for Standardization

[BF-069] ISO/CD TS 5083 Road vehicles—Safety for automated driving systems—Design, verification and validation, ISO, ISO–International Organization for Standardization

[BF-070] ISO 23150-1 Road vehicles—Logical interface between sensors and data fusion unit for automated driving functions—Part 1: General information and principles, ISO, ISO–International Organization for Standardization

[BF-071] ISO 23150-11 Road vehicles—Logical interface between sensors and data fusion unit for automated driving functions—Part 11: Radar detection interface, ISO, ISO–International Organization for Standardization

[BF-072] ISO 23150-12 Road vehicles—Logical interface between sensors and data fusion unit for automated driving functions—Part 12: Lidar detection interface, ISO, ISO–International Organization for Standardization

[BF-073] ISO 23150–13 Road vehicles—Logical interface between sensors and data fusion unit for automated driving functions—Part 13: Camera detection and feature interfaces, ISO, ISO–International Organization for Standardization

[BF-074] Automotive SPICE® Guidelines Process assessment using the Automotive SPICE PAM 4.0 (2nd revised edition, November 2023), VDA QMC, VDA QMC—Qualitätsmanagement Center im VDA (vda-qmc.de)

[BF-075] ENX Association, ENX Association

[BF-076] Ross R, McEvilley M, Winstead M (2022) Engineering Trustworthy Secure Systems. (National Institute of Standards and Technology, Gaithersburg, MD), NIST Special Publication (SP) NIST SP 800–160v1r1, Appendix D. Trustworthy Secure Design, https://doi.org/10.6028/NIST.SP.800-160v1r1.

[BF-077] ISO 31000:2018, Risk management—Guidelines, ISO, ISO–International Organization for Standardization

[BF-078] Organization SPICE PRM/PAM v3.00, VDA-QMC, VDA QMC—Qualitätsmanagement Center im VDA (vda-qmc.de)

[BF-079] ISO 21448 Road vehicles—Safety of the intended functionality, ISO, ISO–International Organization for Standardization

[BF-080] Automated Vehicle Safety Consortium, AVSC, Home Page (sae-itc.com)

[BF-081] Collision Between Vehicle Controlled by Developmental Automated Driving System and Pedestrian, NTSB, Accident Report NTSB/HAR-19/03 PB2019–101402, Collision Between Vehicle Controlled by Developmental Automated Driving System and Pedestrian, Tempe, Arizona, March 18, 2018 (ntsb.gov)

[BF-082] Collision Between a Sport Utility Vehicle Operating With Partial Driving Automation and a Crash Attenuator, NTSB, Accident Report NTSB/HAR-20/01 PB2020–100112, Collision Between a Sport Utility Vehicle Operating With Partial Driving Automation and a Crash Attenuator, Mountain View, California, March 23, 2018 (ntsb.gov)

[BF-083] Collision Between a Sport Utility Vehicle Operating With Partial Driving Automation and a Crash Attenuator, NTSB, Accident Report NTSB/HAR-17/02 PB2017–102600, Collision Between a Car Operating With Automated Vehicle Control Systems and a Tractor-Semitrailer Truck Near Williston, Florida, May 7, 2016. (ntsb.gov)

[BF-084] German In-Depth Accident Study, GIDAS, https://www.gidas.org/

[BF-085] NASS General Estimates System, NHTSA, https://www.nhtsa.gov/national-automotive-sampling-system/nass-general-estimates-system

[BF-086] CARE database, European Union, https://road-safety.transport.ec.europa.eu/european-road-safety-observatory/methodology-and-research/care-database_en

[BF-087] UL 4600 Standard for safety—Evaluation of autonomous products, third edition March 17, 2023, UL, UL Solutions

[BF-088] ISO/PAS 8800 Road vehicles—Safety and artificial intelligence, ISO, ISO–International Organization for Standardization

[BF-089] ISO/IEC/IEEE 15026–2 Systems and software engineering—System and software assurance—Part 2: Assurance case, ISO, ISO–International Organization for Standardization

[BF-090] ISO/AWI TR 17720 Intelligent transport systems—Operational Design Domain Boundary and Attribute Awareness for an Automated Driving System, ISO, ISO–International Organization for Standardization

[BF-091] ISO/PAS 11585:2023 Road vehicles—Partial driving automation—Technical characteristics of conditional hands-free driving systems, ISO, ISO–International Organization for Standardization

[BF-092] ISO/AWI TR 17720 Intelligent transport systems—Operational Design Domain Boundary and Attribute Awareness for an Automated Driving System, ISO, ISO–International Organization for Standardization

[BF-093] ISO/AWI 23792-1:2023 Intelligent transport systems—Motorway chauffeur systems (MCS)—Part 1: Framework and general requirements, ISO, ISO–International Organization for Standardization

[BF-094] ISO/AWI 23792-2 Intelligent transport systems—Motorway chauffeur systems (MCS)—Part 2: Requirements and test procedures for discretionary lane change, ISO, ISO–International Organization for Standardization

[BF-095] ISO 23793-1:2024 Intelligent transport systems—Minimal risk manoeuvre (MRM) for automated driving—Part 1: Framework, straight-stop and in-lane stop, ISO, ISO–International Organization for Standardization

[BF-096] ISO/AWI 23793-2 Intelligent transport systems—Minimal risk manoeuvre (MRM) for automated driving—Part 2: Road shoulder stop—Minimum requirements and test procedures, ISO, ISO–International Organization for Standardization

[BF-097] ISO 23374-1:2023 Intelligent transport systems—Automated valet parking systems (AVPS)—Part 1: System framework, requirements for automated driving and for communications interface, ISO, ISO–International Organization for Standardization

[BF-098] ISO/TS 23374-2:2023 Intelligent transport systems—Automated valet parking systems (AVPS)—Part 2: Security integration for type 3 AVP, ISO, ISO–International Organization for Standardization

[BF-099] ISO 20900:2023 Intelligent transport systems—Partially-automated parking systems (PAPS)—Performance requirements and test procedures, ISO, ISO–International Organization for Standardization

[BF-100] ISO/CD 12768-1 Intelligent transport systems—Automated Valet Driving Systems (AVDS)—Part 1: Requirements, System Framework, Communication Interfaces and Test Procedures, ISO, ISO–International Organization for Standardization

[BF-101] ISO/AWI 12768-2 Intelligent transport systems—Automated Valet Driving Systems (AVDS)—Part 2: System framework, security procedures and requirements, ISO, ISO–International Organization for Standardization

[BF-102] ISO/DIS 7856 Intelligent transport systems—Remote support for low speed automated driving systems (RS-LSADS)—Performance requirements, system requirements and performance test procedures, ISO, ISO–International Organization for Standardization

[BF-103] ISO 4273:2024 Intelligent transport systems—Automated braking during low-speed manoeuvring (ABLS)—Requirements and test procedures, ISO, ISO–International Organization for Standardization

[BF-104] ISO 34501:2022 Road vehicles—Test scenarios for automated driving systems—Vocabulary, ISO, ISO–International Organization for Standardization

[BF-105] ISO 34502:2022 Road vehicles—Test scenarios for automated driving systems—Scenario based safety evaluation framework, ISO, ISO–International Organization for Standardization

[BF-106] ISO 34503:2023 Road Vehicles—Test scenarios for automated driving systems—Specification for operational design domain, ISO, ISO–International Organization for Standardization

[BF-107] ISO 34504:2024 Road vehicles—Test scenarios for automated driving systems—Scenario categorization, ISO, ISO–International Organization for Standardization

[BF-108] ISO 23150-14 Road vehicles—Logical interface between sensors and data fusion unit for automated driving functions—Part 14: Ultrasonic detection and feature interfaces, ISO, ISO–International Organization for Standardization

[BF-109] ISO 23150-15 Road vehicles—Logical interface between sensors and data fusion unit for automated driving functions—Part 15: Microphone detection interface, ISO, ISO–International Organization for Standardization

[BF-110] ISO 23150-20 Road vehicles—Logical interface between sensors and data fusion unit for automated driving functions—Part 20: Supportive and sensor input interface, ISO, ISO–International Organization for Standardization

[BF-111] A Framework for Automated Driving System Testable Cases and Scenarios, NHTSA, A Framework for Automated Driving System Testable Cases and Scenarios (nhtsa.gov), Accessed on 09.24.2024

Know the Risks: Mastering Automotive Project Management

5

Abstract

This chapter explores the complex landscape of automotive project management, examining various methodologies from traditional to agile approaches. It delves into risk management strategies, project orientation versus matrix organization, and the standard lifecycle in automotive development. The chapter discusses how project managers must navigate between quality demands, cost constraints, and time pressures while dealing with increasingly complex regulatory requirements. It analyzes different project management frameworks, including PMI, PRINCE2, and agile methodologies, such as Scrum, SAFe, and Kanban, and compares their effectiveness in the automotive context. The chapter concludes by examining future trends in automotive project management and the challenges project managers face in balancing technological innovation with economic constraints.

With its long-standing tradition and unique dynamics, automotive engineering has evolved into an extraordinarily diverse industry spanning mechanics, electronics, software, and more. Business aspects include mass production, logistics, environmental elements, socio-political considerations, economic challenges, and basic research and development.

System development in our industry follows a rigorous process. This process is divided into standard phases: project preparation, pre-development, actual product development, design validation, industrialization, and start of production (SOP). As explicitly assumed by standards such as ISO 26262, the development process follows the ASPICE V-model, with the exact details left to the project supplier.

In the following chapter, we highlight various development concepts relevant to us and used in developing automotive systems, such as ADAS systems, and explain some of their strengths and weaknesses.

© The Author(s), under exclusive license to Springer Fachmedien Wiesbaden GmbH, part of Springer Nature 2025
R. Mildner et al., *Car IT Reloaded*, https://doi.org/10.1007/978-3-658-47691-5_5

5.1   Project Methodologies

Different project management methods and processes can be used to manage a development project. As there is no standard project management methodology, it is up to the client to decide which one to use. The spectrum ranges from traditional project management to agile approaches.

5.1.1   Traditional Project Management

Various project management standards, such as GPM, PMI, or PRINCE2, are used. But there is also a whole range of standards in this area:

– ISO 15188:2001 Project management guidelines for terminology standardization [BF-006]
– ISO 21500:2021 Project, programme and portfolio management – Context and concepts [BF-007]
– ISO 21502:2020 Project, programme and portfolio management – Guidance on project management [BF-008]
– ISO 21503:2022 Project, programme and portfolio management – Guidance on programme management [BF-009]
– ISO 21504:2022 Project, programme and portfolio management – Guidance on portfolio management [BF-010]
– ISO 21505:2017 Project, programme and portfolio management – Guidance on governance management [BF-011]
– ISO/IEC/IEEE 16326:2019 Systems and software engineering – Life cycle processes – Project management [BF-012].

Below is a brief overview of the aforementioned project management standards.

GPM
GPM stands for the German Association for Project Management, the German professional association for project management. GPM defines standards and best practices for project management in Germany and promotes the further development of this specialist area. According to the GPM guidelines, project management comprises the targeted planning, control, and implementation of projects to achieve project objectives.

According to GPM, project management is based on internationally recognized standards such as those of the International Project Management Association (IPMA) and the USA's Project Management Institute (PMI). GPM also develops its standards to consider the specific requirements of German project management.

Typical characteristics of project management according to GPM are:

– Project definition: Monitoring the progress of the project, identifying deviations, and taking measures to keep the project on schedule and within budget

- Risk management: Identification and assessment of risks and implementation of risk reduction measures
- Quality management: Ensuring the quality of project work and results
- Communication and stakeholder management: Effective communication with the project participants and stakeholders
- Completion and project reporting: Documentation of project success, handover of results, and collection of experience for future projects

In contrast to other approaches to project management, the focus here is on competence. As a project manager in projects ranging from small to large, "soft skills" are crucial to properly lead, guide, and motivate the employees in the project. Naturally, the same applies to the project's key stakeholders, who can influence the project in both positive and negative ways.

The principles behind this are outlined in the IPMA Individual Competence Baseline [BF-013] and are categorized into three key areas:

- Perspective
- People
- Practice

These areas highlight GPM's comprehensive approach, which focuses on technical skills and interpersonal competence in achieving successful project outcomes.

Prince2

PRINCE2 [BF-014] stands for "**PR**ojects **IN** Controlled **E**nvironments" and is a widely used project management method developed by the British government in 1989, primarily focusing on IT projects.

PRINCE2 Version 7 is based on seven principles, seven practices, seven processes, the project context, and the people element. The basic principles are:

- Ensure continued business justification
  According to the business case, a project fulfills a specific business benefit and must achieve a defined profitability. These aspects must be monitored and reported continuously to ensure the intended added value for the organization concerned. If this is not the case, the project should be terminated.
- Experience-based learning ("learn from experience")
  Project teams must be aware of the experiences of previous projects ("lessons learned") and make their own experiences and mistakes available to other projects.
- Definition of roles, responsibilities, and relationships ("defines roles, responsibilities, and relationships")
  The roles are defined in the processes. Depending on the use case, additional roles must be defined. All stakeholders in a project must be described as roles. This includes customers, suppliers, and service providers. Relationships must be established with all stakeholders.

– Management based on phases ("manage by stages")
A project is divided into manageable subsections or phases. Detailed planning takes place before the start of each phase. The progress of the project is managed by controlling and monitoring the phases.
– Manage by exception
The main advantage of PRINCE2 lies in this principle; project planning does not occur within fixed limits, but deviation tolerances are defined. This gives the project manager a certain amount of leeway and prevents too frequent escalations.
– Focus on the product ("focus on products")
The result of a project is usually a product (an exception is, for example, an organizational improvement project). The product and its characteristics, therefore, determine the work. This is based on the quality expectations and requirements of the users.
– Tailored to suit the project
Every project is different. PRINCE2 can be scaled and adapted to the project's needs ("tailoring").

The following practices are defined:

– Business case
This is about how the original idea, which has added value for the organization, is developed into a viable investment proposal and how success is defined and monitored throughout the project.
– Organizing
Project work must be managed, delegated, and assigned by those responsible to those who will deliver the required products. To this end, all roles, responsibilities, and relationships must be defined. Projects are cross-functional, so the usual line management structures do not fit (see Abschn. 5.6).
– Plans
Project work must be planned. The design and development of plans must be defined; they must also be tailored to the people's needs and at the center of all communication and management.
– Quality
The aim is to ensure that everyone involved understands the quality characteristics of the products to be delivered and that project management ensures the relevant requirements are met.
– Risk
Uncertainties and risks are managed.
– Issues
Open issues can be problems, change requests, or non-specification-compliant implementations. This describes how the project manages these issues and their impact (influence on objectives).

– Progress
  Progress is continuously monitored to determine whether the plans can be adhered to and what the escalation process looks like if the plans are not being adhered to.

The following processes are defined:

– Starting a project
– Directing a project
– Initiating a project
– Controlling a stage
– Managing product delivery
– Managing a stage boundary
– Closing a project

The people element consists of three sections:

– Leading successful change
– Leading successful teams
– Communication

In contrast to the Magic Triangle (see Sect. 4.3.1), PRINCE2 considers seven variables for project performance:

– Benefits
– Costs
– Time
– Quality
– Scope
– Sustainability
– Risk

PRINCE2 provides a flexible framework that ensures project alignment with business objectives and emphasizes adaptability, continuous learning, and effective team leadership to drive successful project outcomes. It also perfectly aligns with the IT service operations standard ITIL (Information Technology Infrastructure Library).

PMBOK (PMI)
PMBOK, provided by PMI (Project Management Institute) [BF-002], is a leading standard for project management worldwide. It is currently described in two documents:

– PMBOK Guide [BF-003]
  The project management principles and the intended results are described here.

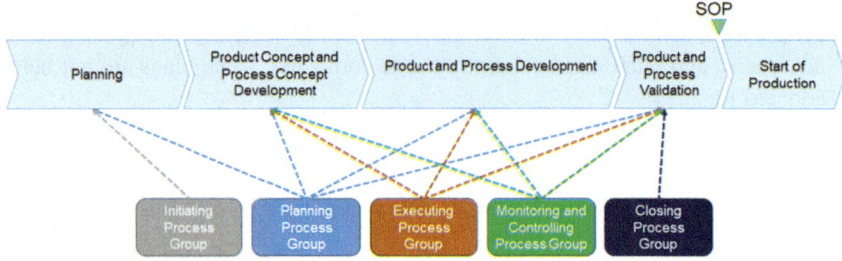

Fig. 5.1 Example assignment of process groups to product development phases

- Process Groups Practice Guide [BF-004]
  The project management principles are further detailed in the form of process groups and processes that each organization must adapt or tailor for itself.

The process groups are divided as follows:

- Initiating
- Planning
- Executing
- Monitoring and Controlling
- Closing

PMI provides a foundation for project execution based on the process groups and the assigned 49 processes. Each project must determine the required processes for each phase of the product development process/lifecycle.

Figure 5.1 shows an example of such an assignment.

The initialization processes are only used at the start of the project, the planning, implementation, monitoring, and control processes in all phases up to the SOP (Start of Production), and the closing processes at the end of the project. However, using the closing processes at the end of each phase makes sense. Since aspects such as assumptions, estimates, changes, quality, problems, risks, and the business case are considered here, applying these processes at the end of the phase would be appropriate. In many organizations, such topics are examined in so-called quality reviews (quality gates) at the end of each phase. The business case is usually not considered because these reviews are carried out on the development side. However, how often have you realized at the end of a development phase that the final price has exceeded the original target price several times over? And how often have you almost finished developing the product and then realized during customer acquisition that there is no interest in it? These issues could be avoided using suitable methods (e.g., QFD – Quality Function Deployment) [BF-015].

5.1.1.1 PMI (PMBOK) versus ASPICE: A Comparison

In Sect. 4.3.5, we discussed the ASPICE aspects of project management. At this point, we will examine the differences between ASPICE and PMI in terms of project management. PMI is more detailed than ASPICE, but there are some similarities. Another aspect to consider is that PMI processes have a phase dependency that does not exist in ASPICE, as ASPICE is a model without reference to a phase model. Inevitably, this results in differences in the naming of PMI processes and the basic practices in ASPICE.

The primary purpose of this comparison is to demonstrate that ASPICE does not impose extraordinary demands on project management. Project managers often claim that ASPICE requirements are challenging to implement. This suggests that, until now, some projects may have been "flying blind." Of course, this could also be attributed to the project manager's resource constraints.

In this comparison, we limit ourselves to the PMI processes and the basic practices from ASPICE.

Initiating

Initialization includes the commissioning of the project manager and the required information: the Project Charter, with the most important key data on the project, and a rough definition of the product to be developed. At this point, the project manager can find out about the project stakeholders critical to the project's success and document them in the stakeholder list.

In this early phase, there is no significant difference between PMI and ASPICE (see Fig. 5.2). ASPICE is more generic in describing the basic practices, but the outcome reflects similar content, if not as detailed as PMI.

PMI clearly describes the expectations of the Project Charter (see Fig. 5.3).

In ASPICE, the expectations for the scope of work are outlined in *VDA QMC Volume Automotive SPICE Guidelines—Process Assessment Using the Automotive SPICE PAM 4.0, 2nd Revised Edition, November 2023* [BF-001], and these expectations are more abstract. The scope of work must include the motivation (objectives), the boundaries, both the project and product scope and the limitations of the project. It is not sufficient to describe only the product to be developed.

PMI	Project Management ASPICE (MAN.3)	
Initiating Process Group	Base Practices	Description
4.1 Develop Project Charter	BP1 Define the scope of work	• Identify the project's goals, motivation and boundaries[1]
4.2 Identify Stakeholders	BP7 Define and monitor project interfaces and agreed commitments	• Identify and agree interfaces of the project with affected stakeholders and monitor agreed commitments. • Define an escalation mechanism for commitments that are not fulfilled. • Note 9: Affected stakeholders may include other projects, organizational units, sub-contractors, and service providers.[2]

[1] Im ersten Schritt im **Project Charter** definiert

[2] Erste Version der **Stakeholder-Liste**

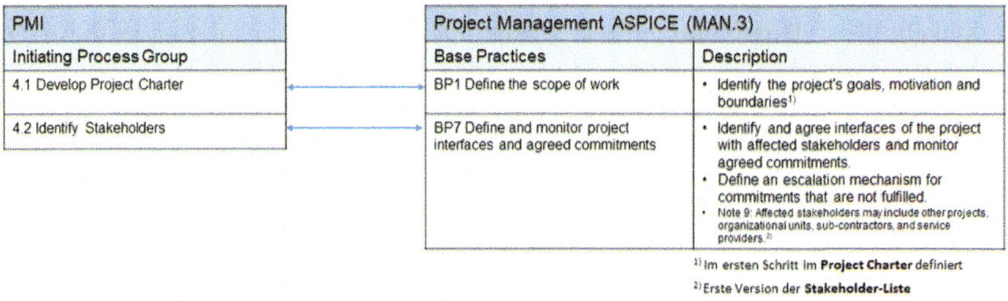

Fig. 5.2 Mapping PMI to ASPICE: project initiating

PMI - Project Charter
Project purpose
Measurable project objectives and related success criteria
High-level requirements
High-level project description, boundaries and key deliverables
Overall project risk
Summary milestone schedule
Preapproved financial resources
Key stakeholder list
Project approval requirements (i.e. what constitutes success, who decides the project is successful, who signs off on the project)
Project exit criteria (i.e. what are the conditions to be met in order to close or to cancel the project or phase)
Assigned project manager, responsibility and authority level
Name and authority of the sponsor or other person(s) authorizing the project charter

Fig. 5.3 PMI: requirements to the project charter

("The scope of work has to cover the motivation (goals), the boundaries including project and product scope, and the constraints of the project. Describing only the product to be developed is not sufficient.").

In Sect. 4.3.5, we discussed key aspects and shortcomings of project management. The following can be summarized about the Project Charter:

– Measurable project goals: These are the standard *on time*, *on spec*, *on cost*, and *on quality*.
– Boundaries: Both internal (e.g., technology) and external boundaries must be determined, and the customer must share a common understanding of these.
– The project/product risks must be identified (e.g., qualifications/competencies, new technologies, time-critical deliveries, trials, etc.)
– Milestone plan: This is the basis for further and detailed planning.
– Financial resources: Never start a project without sufficient financial resources!
– Stakeholders: Direct and indirect stakeholders must be considered, as these can influence or sabotage the project objectives.
– The project manager must also have authority in the collaborating domains for projects spanning multiple product categories. Without this authority, organizational silos can influence or undermine project objectives in a similar manner.

Another essential aspect to consider is the assumptions made during the project decision-making process. These can be, for example, technical issues such as EE architecture, the computer structure used, network architecture, etc. However, this can also be market-based assumptions, such as being the first to offer a new solution and thus gaining significant market share.

Planning

In the first step, project planning can be roughly based on the project order (charter) and the framework conditions known up to this point.

The planning itself can be purely milestone planning in the first step. Only then is detailed planning carried out on a milestone-by-milestone basis:

- The expected content to a milestone
- The expenses required for this
- The necessary resources (human and material)
- The sequence to be observed
- The acceptance criteria of the single content.

Of course, this also depends on the type of product development that is to take place:

- Is it a completely new product?
- Is it a known product with new features?
- Is it a known product with slight adaptations (for example, to specific customer requirements)?

In the latter case, we speak of a "variant project." If the customer is known, the adjustments can be estimated, and the entire project can be planned from the outset, as the content is largely known.

This is the simplest type of project, but even here, customers can cause chaos by constantly reprioritizing projects.

The following is a rough comparison of the planning process group in PMI and project management in ASPICE (Fig. 5.4).

"Collect requirements" is covered in ASPICE via the "SYS.1 Requirements elicitation" process Fig. 5.5).

The first process in PMI during planning is developing the Project Management Plan (PMP). This plan defines the approach to the project, i.e., how it will be implemented, monitored, controlled, and completed.

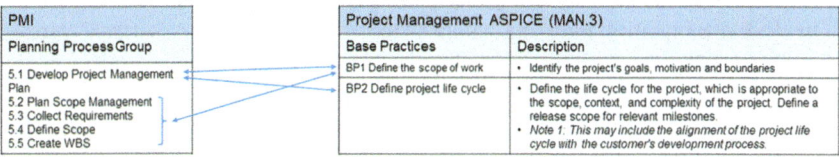

Fig. 5.4 Mapping PMI to ASPICE: project planning (part 1)

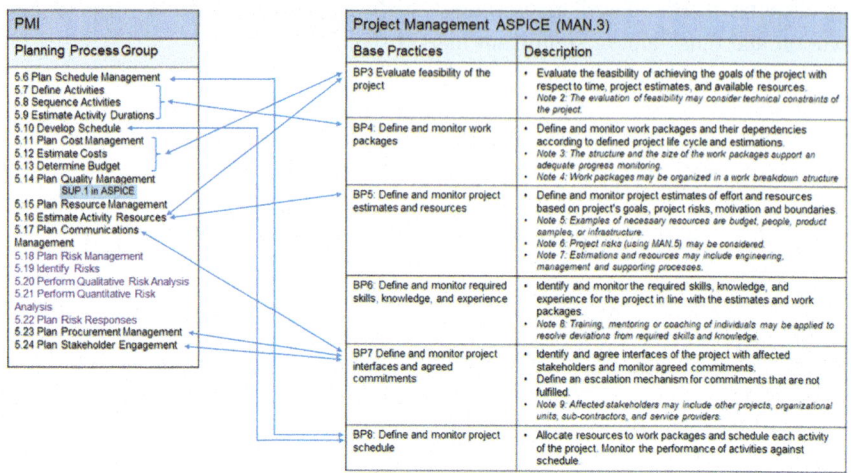

Fig. 5.5 Mapping PMI to ASPICE: project planning (part 2)

The Project Management Plan contains all Subsidiary Management Plans, such as the Scope Management Plan, the Requirements Management Plan, and the Schedule Management Plan. The baselines for scope, scheduling, and costs are also included, as well as additional components (e.g., Change Management Plan, Configuration Management Plan, Project Lifecycle, and Management Reviews).

ASPICE has provided the following generic plan description in PAM v3.1 (Fig. 5.6).

This generic definition overlaps well with the planning process in PMI. Unfortunately, it is no longer included in PAM v4.0.

However, it must also be added that keeping a project management plan or a project manual is now standard practice in projects.

PMI details this in the following processes: Plan Scope Management, Collect Requirements, and Define Scope.

Risk management (MAN.5) is included in ASPICE as an option in the VDA scope. Even if it is not in the scope of a project, a risk assessment is still expected to underline essential risk-related decisions and measures. Furthermore, risk management in engineering is required at the system level anyway, for example, to comply with standards such as ISO 26262 (e.g., through FMEA, FTA, etc.). We will discuss risk management in more detail later in Abschn. 5.2.

In ASPICE, quality planning takes place in the Quality Assurance process (SUP.1), whereby the majority of quality assurance measures are naturally ensured by the engineering processes (e.g., through reviews and tests).

PMI considers the competence, experience, and knowledge of the resources used in project implementation.

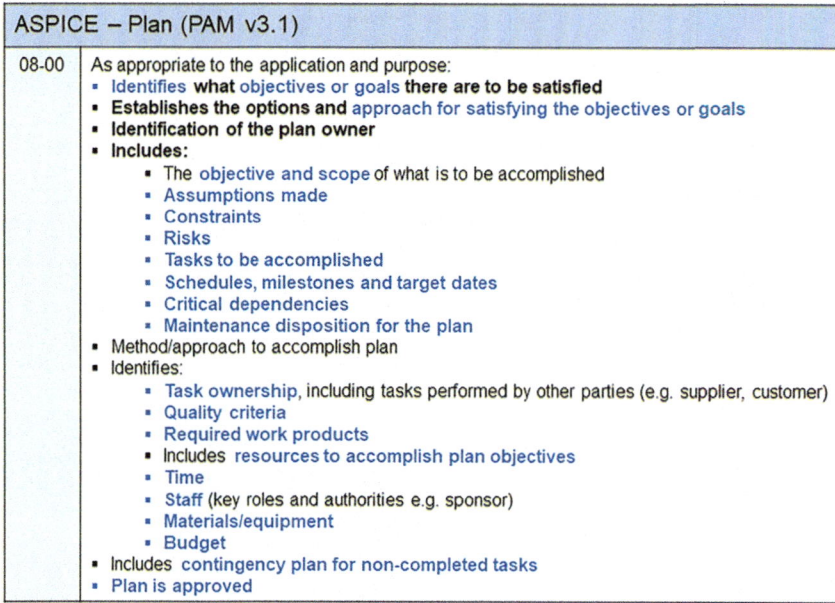

ASPICE – Plan (PAM v3.1)	

08-00	As appropriate to the application and purpose:
	• Identifies **what** objectives or goals **there are to be satisfied**
	• **Establishes the options and** approach for satisfying the objectives or goals
	• **Identification of the plan owner**
	• **Includes:**
	    • The objective and scope of what is to be accomplished
	    • Assumptions made
	    • Constraints
	    • Risks
	    • Tasks to be accomplished
	    • Schedules, milestones and target dates
	    • Critical dependencies
	    • Maintenance disposition for the plan
	• Method/approach to accomplish plan
	• Identifies:
	    • Task ownership, including tasks performed by other parties (e.g. supplier, customer)
	    • Quality criteria
	    • Required work products
	    • Includes resources to accomplish plan objectives
	    • Time
	    • Staff (key roles and authorities e.g. sponsor)
	    • Materials/equipment
	    • Budget
	• Includes contingency plan for non-completed tasks
	• Plan is approved

Fig. 5.6   ASPICE: generic definition of a plan in PAM v3.1

Executing

According to PMBOK, resources are acquired during implementation, so competencies and skills are also considered in this context.

One key aspect is the team's development ("6.5 Develop Team"). This point is not directly addressed in ASPICE. Note 8 of BP6 only mentions that additional training needs to be provided in case of discrepancies between expected and actual competencies.

As skills are reviewed annually in larger companies, this should not be an issue (even considering the wide range of available training opportunities). Nevertheless, it is not uncommon to find surprising weaknesses in this area during assessments or consultations. Of course, this point can only be checked in exceptional cases in Europe due to the General Data Protection Regulation (GDPR). However, this can certainly be presented for a group or department as a general overview without going into detail about individuals (i.e., anonymously) (Fig. 5.7).

ASPICE addresses several fundamental aspects of project management found in PMI processes, though its coverage is less comprehensive. Notably, risk management in ASPICE can be included as an optional component, whereas PMI provides more detailed and structured guidance on this and other project management topics.

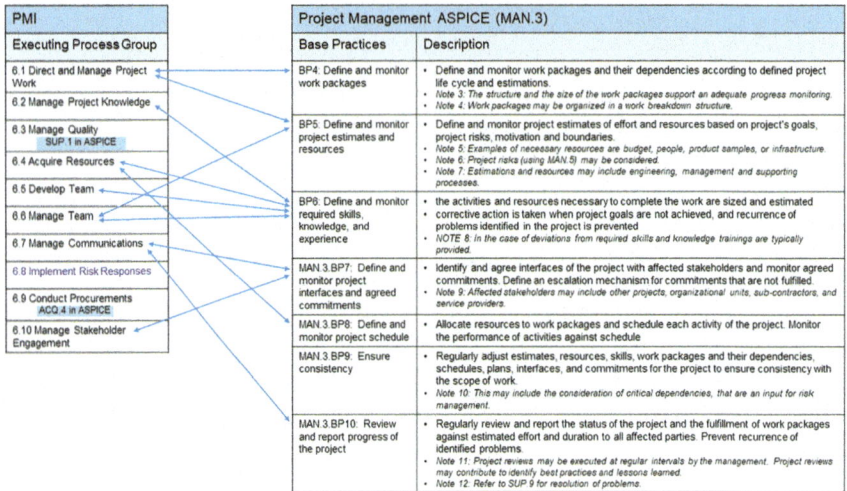

Fig. 5.7 Mapping PMI to ASPICE: project executing

Similarly, an integral part of the project management concept in PMBOK deals with supplier management and monitoring in more detail ("6.9 Conduct Procurements"), while ASPICE in the VDA scope only focuses on ACQ.4 "Supplier Monitoring."

Monitoring and Controlling
The following topics are monitored and controlled in PMI:

- Project work
- Integrated change management
- Scope
- Planning
- Costs
- Quality
- Resources
- Communication
- Risks
- Supplies
- Stakeholder engagements (see Fig. 5.8).

These topics are also considered in project management in ASPICE, with the following exceptions:

- Integrated change management is considered in "SUP.10 Change management"
- Quality is considered in "SUP.1 Quality Assurance"
- Supplies are considered in "ACQ.4 Supplier monitoring"

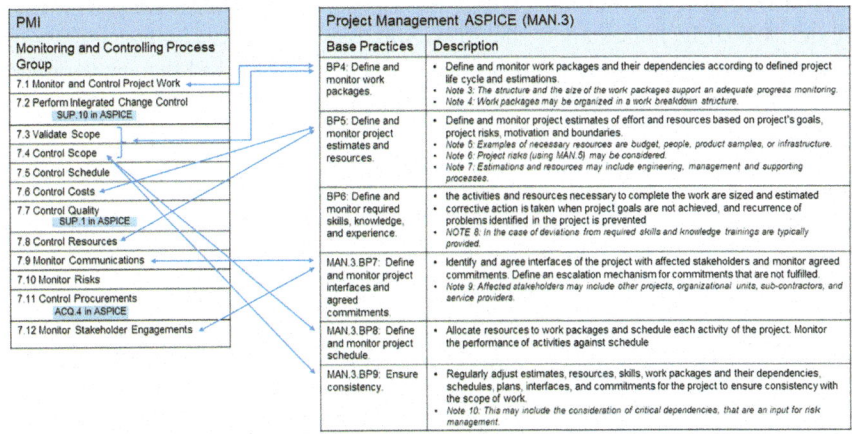

Fig. 5.8 Mapping PMI to ASPICE: project monitoring and controlling

PMI	Project Management ASPICE (MAN.3)	
Closing Process Group	**Base Practices**	**Description**
8.1 Close Project or Phase	MAN.3.BP10: Review and report progress of the project.	• Regularly review and report the status of the project and the fulfillment of work packages against estimated effort and duration to all affected parties. Prevent recurrence of identified problems. • Note 11: Project reviews may be executed at regular intervals by the management. Project reviews may contribute to identify best practices and lessons learned. • Note 12: Refer to SUP.9 for resolution of problems

Fig. 5.9 Mapping PMI to ASPICE: Project closure

Closing

According to PMI, documentation is updated accordingly at the end of the project, "lessons learned" from the project are determined, handover to production and service is ensured, the final report is prepared, and, if necessary, organizational processes are updated.

The closure occurs at the end of the project but can also occur at the end of each phase and, if necessary, when the project is canceled (see Fig. 5.9).

In ASPICE, this topic is addressed in MAN.3.BP.10 with regard to "Best Practices" and "Lessons Learned" in Note 11. In PAM v3.1, the topic of archiving was also addressed in SUP.8.BP9. This is no longer mentioned in PAM v4.0. As ASPICE assessments take place before the SOP, the project has not yet been completed, and nothing else can be checked apart from the archiving requirements.

5.1.1.2 Conclusion

As mentioned at the beginning of this chapter, ASPICE provides good coverage of project management topics, although it cannot achieve the level of detail of a project management standard with a single process.

However, there are additional interesting aspects in the 'Process Groups Practice Guide' [BF-015]. For example, the "Develop Team" process is about improving competencies and the interaction between project members and the entire project environment to improve project performance. The following topics are listed here:

− Maintain open and effective communication
− Create team-building opportunities
− Develop trust between team members
− Manage conflicts constructively
− Promote collaborative decision-making

Another interesting aspect is described in the section "Project Success Measures." Here, in addition to the project objectives mentioned in Sect. 4.3.5.5, further objectives are considered, which are derived, for example, from the organization's business objectives (financial and non-financial objectives from the business case, stakeholder satisfaction, fulfillment of contractual conditions, customer acceptance, etc.).

5.1.2   Agile and Lean Approaches

Agile has become a popular approach in fast-paced software development and project management. Below, we summarize some leading agile concepts to broaden the context of the general topic of automotive project management.

5.1.2.1 Agile Principles

The concept of agility emerged in response to the challenges associated with the waterfall model. Traditional methods, including those embedded in process assessment models like CMM, CMMI, and Automotive SPICE, often prove rigid and cumbersome. The waterfall stages in these models can lead to significant issues, especially when not all requirements are clearly defined from the outset.

The *Agile Manifesto*, published in 2001, was a direct reaction to these rigid processes. The core principles contained therein were:

− Individuals and interactions via processes and tools
− Functioning software via comprehensive documentation
− Cooperation with the customer via contract negotiation
− Responding to change by following a plan

The fundamental idea is the flexibility of direct communication and quick reactions to project changes.

Building on this, the *Agile Manifesto* formulates twelve agile principles:

1. Our highest priority is to satisfy the customer through the early and continuous delivery of valuable software.
2. Welcome changing requirements, even late in development. Agile processes harness change for the customer's competitive advantage.
3. Deliver working software frequently, from a couple of weeks to a couple of months, with a preference for a shorter timescale.
4. Business people and developers must work together daily throughout the project.
5. Build projects around motivated individuals. Give them the environment and support they need, and trust them to get the job done.
6. Face-to-face conversation is the most efficient and effective method of conveying information to and within a development team.
7. Working software is the primary measure of progress.
8. Agile processes promote sustainable development. The sponsors, developers, and users should be able to maintain a constant pace indefinitely.
9. Continuous attention to technical excellence and good design enhances agility.
10. Simplicity—the art of maximizing the amount of work not done—is essential.
11. The best architecture, requirements, and designs emerge from self-organizing teams. At regular intervals, the team reflects on how to become more effective, then tunes and adjusts its behavior accordingly.

These principles are designed to enable teams to react quickly to change. Changes to project requirements are possible at any time and should be implemented in direct cooperation with the customer. In addition, functioning software is consistently delivered.

Although agility offers many advantages, there are also disadvantages, especially in our highly regulated and safety–critical industry. Possible disadvantages of agile methods in the automotive industry include:

- Many agile methodologies, which promote rapid iteration and flexibility, may fail to meet the strict requirements of quality standards such as Automotive SPICE or ISO 26262 due to the rigid documentation required.
- Agile methods are often geared towards smaller teams. However, larger teams are usually required to create comprehensive, formal documentation in the automotive industry. Practical experience shows that purely agile projects often don't scale well.
- Agility requires a cultural shift. Prioritizing informal work and direct collaboration between teams and customer stakeholders over strict process conformity can be challenging to reconcile with traditional work methods.
- Many companies in our industry have already invested heavily in assessment models, such as Automotive SPICE. Integrating agile methods into these existing structures often provokes resistance.
- The automotive industry requires long-term planning horizons and schedules (often from the start of the project to SOP), which is usually incompatible with agile methods' short-term, iterative methodologies.

– The numerous dependencies within the supply chain (e.g., vehicle platform development) require a firm commitment for specific deliveries. This expectation is generally not compatible with agility.
– Traditional budgeting is generally not possible with agile principles.

Adapting the corporate culture to agile ideas is usually carried out through an "agile transformation." However, this is only possible if the entire supply chain consistently follows a specific agile methodology. A shift towards agile working by a single company, such as an automotive supplier, often clashes with the expectations of OEMs that depend on deliveries from numerous suppliers and must, therefore, insist on the "Bermuda Triangle," as described in Chap. 4.

There are numerous agile methodologies, including SCRUM, Kanban, Extreme Programming (XP), Lead Software Development, Feature-Driven Development (FDD), Cristal, SAFe, Scrum of Scrums, etc. Below, we will briefly outline some of these methodologies (including SCRUM and SAFe).

5.1.2.2 SCRUM

SCRUM is an agile project management methodology used in software development. It was developed by the US consultants Jeff Sutherland and Ken Schwaber at the end of the 1990s. The idea was to develop software with small teams in close-knit, small steps.

The SCRUM process enjoys wide popularity in software development circles. Schwaber later founded Scrum.org, which is an organization dedicated to spreading SCRUM and monetizing the practice of SCRUM consultancy. Like traditional project management methodologies such as PMBOK or GPM, Scrum.org offers training and various certifications.

SCRUM is based on the principle of rigorous delivery of work results, corrective review, and continuous adjustment of the project scope. SCRUM is iterative and incremental. Iterations (called "sprints") usually last two to four weeks. Within these sprints, the development team works on a selected set of functions (increments), which should result in a potentially deliverable product increment at the end of the sprint.

Only three roles are defined in SCRUM: the Product Owner, the Scrum Master, and the team. The Product Owner is responsible for the product backlog, in which the sprint details are put together. The Scrum Master has the task of adhering to the SCRUM process defined by Scrum.org. The development team organizes itself.

SCRUM is a reactive process where the team quickly responds to objective deviations and scope changes without focusing on detailed long-term planning. Sprints function as feedback and learning loops, with retrospectives playing a key role. Stand-up meetings, or daily scrums, are mandatory for all team members. SCRUM emphasizes the importance of these daily meetings, regular retrospectives, and reviews to ensure continuous improvement and adaptation.

The advantage of SCRUM is the extreme simplicity of the process. SCRUM is emphasized as a methodology for small teams of experts (typically up to around eight developers). However, it is precisely the simplicity of the approach that sometimes poses a challenge. Developing safety-relevant systems, as required in the automotive industry under ISO 26262 for functional safety and ISO/SAE 21,434 for cybersecurity, typically accompanied by ASPICE compliancy requirements, implies a high degree of short- and long-term planning (usually up to SOP). In addition, such an environment requires comprehensive and strict documentation, formal results testing, and extensive validation of the results. These processes must be backed up by comprehensive risk assessments, extensive testing, and frequent formal reviews (e.g., inspections). They must undergo strict, predefined tests and formal, external certifications. There is also a requirement for strict change management, which must be meticulously followed and documented.

At the same time, SCRUM promotes a flexible and adaptable way of working in a small team. However, it is challenging to implement a non-trivial project with a small team of around eight people with all these formalisms, in which systematic, long-term planning is also carried out periodically. The two approaches usually collide, limiting the use of SCRUM in the context of safety-relevant projects.

5.1.2.3 SAFe

SAFe, or Scaled Agile Framework, is an agile project methodology developed by Dean Leffingwell. SAFe adds roles and methods to SCRUM to help implement scalable agility. SAFe complements agile with "lean management elements."

In contrast to SCRUM, SAFe defines several roles, including the Release Train Engineer (RTE, also known as Chief Scrum Master), Product Owner, Scrum Master, and System Architect. The Scrum Master now has the task of supporting SAFe instead of SCRUM. The Product Owner has the same role as in SCRUM. The System Architect has the task of ensuring that the technical implementation is carried out following the requirements. Other roles are Customer, Business Owner, Agile Team, Domain Expert, Specialist, Technical Expert, Independent Tester, Integration etc.

The development teams are organized in the Agile Release Train. The results of the small iterations (sprints) and the larger work packages (epics) are implemented in PIs (program increments), whereby the PIs also have a fixed release rhythm, for example, quarterly. In this way, several teams are synchronized throughout the project.

SAFe defines ten principles:

1. Take an economic view: Prioritize the most economically advantageous measures by understanding the trade-offs between options to maximize value and minimize costs.
2. Apply systems thinking: Recognize and optimize the overall system instead of individual components to better understand dependencies and improve the overall process and performance.
3. Assume variability preserves options: Maintain decision-making flexibility by keeping several options open to adapt to uncertainties and environmental changes.
4. Build incrementally with fast, integrated learning cycles: Develop solutions iteratively, which enables quick feedback and adjustments, reduces risks, and delivers value faster.
5. Base milestones on an objective evaluation of working systems: Measure progress through tangible, working products rather than activities or milestones to ensure value is delivered.
6. Make value flow without interruptions: Streamline the process to continuously deliver value by removing bottlenecks, reducing batch sizes, and eliminating waste.
7. Apply cadence synchronize with cross-domain planning: Use regular, predictable work intervals to establish a rhythm for development. Iterations must be globally synchronized across all teams to integrate activities seamlessly.
8. Unlock the intrinsic motivation of knowledge workers: Create an environment in which employees are motivated by their interest in the work, which promotes commitment, innovation, and productivity.
9. Decentralize decision-making: Empower individuals and teams to make decisions at the local level, where they have the best information available and are closest to the problem, to improve responsiveness and outcomes.
10. Organize around value: Structure teams and activities around delivering value to the customer to ensure your efforts are directly linked to value creation and customer satisfaction.

SAFe includes various ideas and concepts, such as Value Stream Management, Contract, Portfolio and Portfolio Flow, Solution, Roadmap, the OKR principle (Objects and Key Results), Vision, System Team (but no Software Team), and so on. Lean (lean and lean portfolio management or lean budget) is generally emphasized. Many topics are addressed, including the use of artificial intelligence.

SAFe is a framework combining numerous buzzwords, making the methodology seem extensive or even confusing. While the concepts outlined in the *Agile Manifesto* emphasize flexibility, SAFe often contradicts these principles. For instance, the rigid scope defined by SAFe is fundamentally unagile. Similarly, the "roadmap" required by SAFe shares this strict nature. SAFe mandates a series of documents (artifacts) that are essentially traditional documentation. The outcome of PI planning is also a comprehensive document, and the Architectural Runway resembles a conventional architecture specification.

Developers familiar with SCRUM and agile principles are often frustrated by the complexities of SAFe, finding it overly bureaucratic and contradictory to the core agile values they are accustomed to. They complain about SAFe's demand to produce extensive documentation for new projects, which feels more like a return to bureaucratic, heavy processes than embracing true agility. This perceived "un-agile" nature of SAFe can be discouraging for those who prioritize flexibility and iterative development.

SAFe is a complex methodology marketed as "agile"; arguably, it does not align closely with the principles of the *Agile Manifesto*. Because SAFe attempts to incorporate all trending ideas, its implementation often leads to unnecessary misunderstandings and ideological conflicts, frequently resolved during ongoing projects. Additionally, safety and security-relevant documentation is optional, which can result in development teams failing to address these critical requirements on time.

SAFe has much potential, but it is pretty extensive and difficult to interpret and implement efficiently. In addition, V-model compliance, indispensable in our industry, is not sufficiently emphasized, making the compatibility of SAFe with ISO 26262 and ISO/SAE 21434 more difficult.

5.1.2.4 Kanban

Kanban (Japanese for "drawing board") was derived from the Toyota Production System (TPS), in which "waste" was minimized and manufacturing processes were optimized for just-in-time production. "Waste" refers to negative aspects such as unnecessary transportation, redundant work activities, waiting times, overproduction, defects, etc. David J. Anderson [BF-005] formulated the Kanban method on this basis. It was later used for activities other than production, such as software development.

In our context, Kanban is a simple process for managing project tasks. These are displayed on a Kanban board that visualizes the progress, with columns for each phase of the workflow (for example, "to do", "in progress", and "done"). Limits are set on the board for the number of tasks in a particular development phase to avoid bottlenecks. By regularly checking the Kanban board, teams can measure the work speed and make necessary corrections and improvements. Processes are constantly optimized and improved based on Kanban activities.

Unlike other methods, Kanban does not define specific roles. Responsibilities are defined as part of the team organization.

Kanban can be integrated into various project methodologies, promoting and demanding continuous improvement and adaptability of project processes. Unlike SCRUM, Kanban requires no specific planning methodology or timing, such as sprints. It is not a self-contained development process but rather a universal, methodologically-agnostic building block. Kanban is also available in numerous popular project management tools, such as Atlassian Jira, and is often favored by development teams.

5.1.2.5 Summary

Many agile methodologies have been developed in the 25 years since agility found its way into development offices. It's not possible to list all the agile methodologies as comprehensively as possible; here is a non-exhaustive attempt.

- Adaptive Test-Driven Development (ATDD)
- Adaptive Software Development (ASD)
- Agile Business Analysis (AgileBA)
- Agile Contract Management (ACM)
- Agile Digital Services (AgileDS)
- Agile Fluency
- Agile Modeling (AM)
- Agile Portfolio Management (AgilePfM)
- Agile Program Management (AgilePgM)
- Agile Project Management (AgilePM)
- Agile Quality Management (AQM)
- AgileSHIFT
- Agile Testing
- Agility Scales
- Behavior-Driven Development (BDD)
- Bimodal Portfolio Management (Bimodal PfM)
- Continuous Integration/Continuous Deployment (CI/CD)
- Crystal Methods
- Design Thinking
- DevOps/Business DevOps (BusDevOps)
- Disciplined Agile (DA)
- Dynamic Systems Development Method (DSDM)
- eXtreme Programming (XP)
- Evidence-Based Portfolio Management (E-B PfM)
- Experiment-Driven Development (EDD)
- Feature-Driven Development (FDD)
- Holacracy
- Kanban
- Kanban Maturity Model (KMM)
- Large-Scale Scrum (LeSS)
- Lean IT
- Lean Kanban University (LKU)
- Lean Startup
- Lean Software Development
- Lean UX
- Management of Portfolios (MoP)
- Mob Programming
- Nexus
- Open Space Agility (OSA)
- Pair Programming

- PMI Agile Certified Practitioner (PMI-ACP)
- PRINCE2 Agile
- Project Half Double
- Rapid Application Development (RAD)
- SAFe
- Scaled Agile Framework (SAFe)
- Scaled Agile Lean Development (ScALeD)
- SCRUM
- Scrum@Scale (S@S)
- Scrumban
- Sociocracy
- Specification by Example (SbE)
- Spotify Model
- Standard for Portfolio Management (SfPfM)
- Test-Driven Development (TDD)
- User Experience Design (UX Design)

Agility is a popular idea, and agile elements have even been integrated into many traditional frameworks, such as PMI/PMBOK, PMI-Agile, or PRINCE2 Agile.

Many agile methodologies are challenging because they are usually incompatible with strict quality standards such as ISO 26262. The best results are achieved when agile and traditional approaches are combined (hybrid approaches). However, no dominant process methodology currently implements hybrid methods. While pure agility alone won't suffice in our industry, it is clear that we must continue to adapt, finding the right balance between agility and rigorous standards – or risk falling behind.

5.2   Risk Management

The increasing complexity and rapid changes in the automotive industry (for example, the transition to EV, further digitalization within and around modern cars, new mobility services, and autonomous driving) bring uncertainty and new potential risks, forcing legacy car manufacturers to rethink and reduce risk in order to survive in the market in the long run. The answer to these challenges lies in adopting new technologies, introducing new functionalities, and making organizational adjustments, such as restructuring and establishing new specialized roles to manage change effectively.

How can project risks be effectively and quickly identified in the early stages of a project? One approach could be, for example, using the SWOT analysis (Strengths, Weaknesses, Opportunities, Threats). This method is often used in business consulting (see Chap. 7 for more information) (see Fig. 5.10).

Based on the strengths, opportunities, weaknesses, and threats identified, strategies are developed to:

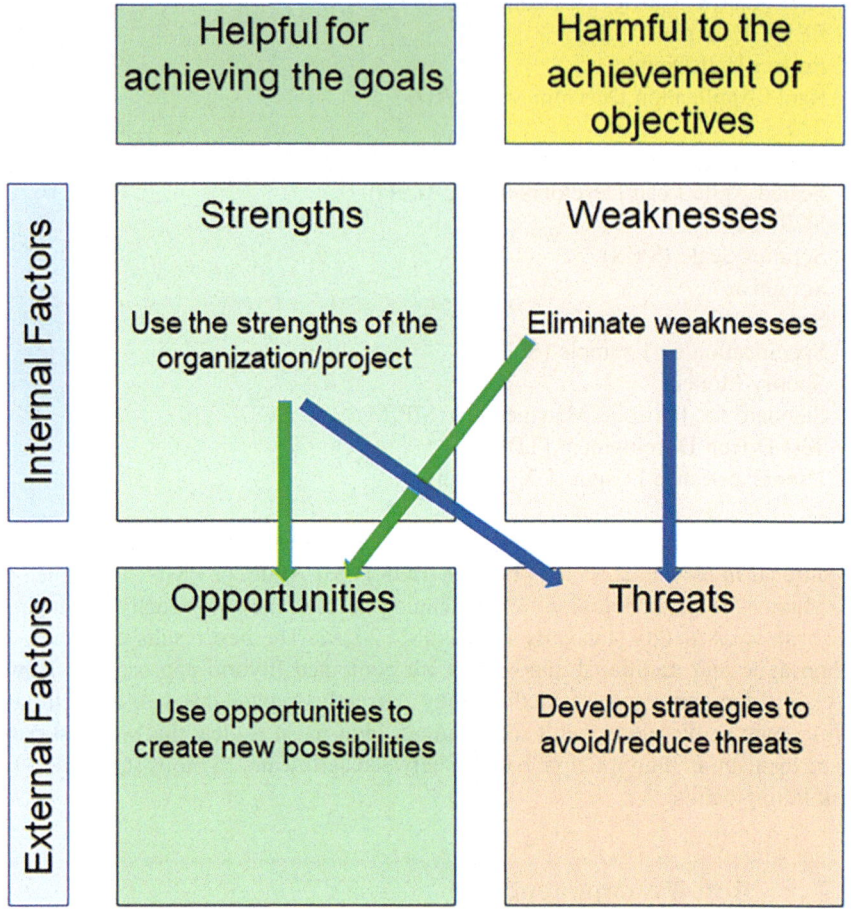

Fig. 5.10 SWOT Analysis

- Leverage strengths to capitalize on opportunities
- Address weaknesses to exploit opportunities
- Use strengths to mitigate threats
- Prevent weaknesses from turning into threats

The weaknesses and threats in Fig. 5.10 can result in risks.

In the project environment, typical weaknesses can include limited resources, inadequate team skills, project schedules that are too short, tight budgets, and a technology backlog.

Probability					
5	10	15	20	25
4	8	12	16	20
3	6	9	12	15
2	4	6	8	10
1	2	3	4	5

Impact

Fig. 5.11   Example of a risk matrix

Typical threats can be changes in the competitive landscape, new technologies from competitors (for example, new batteries with significantly higher performance and shorter charging times), changes in legislation, a declining market, or pandemics (for example, the COVID-19 pandemic).

Throughout the project, new risks may arise that were not identified in the initial phase. Technical risks can arise from methods such as FMEA, FTA, HARA (see Sect. 4.3.7), or TARA (see Sect. 4.3.6).

Risk management generally involves the following steps:

– Identify risks
– Analyze and assess risks, i.e., determine probability and impact (see Fig. 5.11).

A matrix as shown in Fig. 5.11 can be used for this purpose. The risks can then be classified (see Fig. 5.12).

These steps are carried out periodically in the project:

– Determine measures based on the risk classification to:
    – Mitigate the impacts
    – Avoid the risks
    – Share or transfer the risks
– Plan and implement risk reduction measures
– Monitor the status of risks, i.e.:
    – Update risk assessment

Fig. 5.12   Example of risk classification

25	Critical
15 – 20	High
6 – 12	Medium
1 – 5	Low

- Evaluate the impact of the measures and adjust the risk assessment if necessary

When risks materialize, they become problems. Therefore, it is crucial to identify potential risk indicators and prepare appropriate contingency plans (see, for example, Tom DeMarco, *Waltzing With Bears: Managing Risk on Software Projects*) [BF-016]. These indicators should help identify risk at an early stage.

Once the risk indicators are identified, they must also be continuously monitored. It is therefore important to point out that risks must not only be *managed* but also actively *reduced* so that project management can be carried out effectively.

- Organizations often consciously or unconsciously accept risks without taking any action; for example, taking on a critical project despite insufficient resources, hoping to achieve favorable economic results by the SOP.
- The management is responsible for promoting and maintaining a risk management culture. Active risk reduction is essential for effective automotive project leadership.
- Risk reduction efforts should start during project startup and be applied consistently throughout the project.
- Risk *management* cannot solely focus on risk reduction; however, risk *minimization* is a substantial, indispensable component of successful project management. It is irresponsible merely to manage risks. Identified risks must not only be detected but actively reduced.
- Product quality can only be achieved through effective risk management with a strong focus on risk reduction.

Risk management and reduction are pursued on several levels: product quality, process quality, employee competence, supply chain management, technology innovation, and continuous improvement. Quality systems like Automotive SPICE and ISO 26262 exist primarily to facilitate active, systematic risk management. In Chap. 6, we will return to risk management in the context of process quality.

5.3   Project Orientation

Traditionally, the automotive industry has been dominated by a line-oriented product management structure driven by a strong organizational matrix. These type of organizations emphasize risk avoidance and stability, meaning slower but more reliable product development dominates car manufacturing. A matrix organization is inherently risk-averse and resistant to change because the primary goal is the company's stability. A strong matrix is especially beneficial for companies offering consistently repeatable products or services that don't change often or quickly.

An alternative to the strong matrix is a project-oriented company organization ("projectized" organization); however, this operates quite differently from the matrix organization.

Comparing a Diplodocus to a T-Rex could help explain the difference between a traditional matrix organization and a project-based organization [RM-143]. Diplodocuses lived around 150 million years ago, while the T-Rex came around 84 million years later, but this time lag has little effect on the significance of the comparison. A Diplodocus was a giant herbivore. It was up to 35 m long and weighed an estimated 30 or, according to some estimates, even 80 tons. A significant advantage of the Diplodocus was its enormous body size, which protected it from most predators. Slow and ponderous, it had few natural enemies, as T-Rex did not yet exist and other tetrapods rampant at the time had little desire to attack the massive animal. It was safest in a crowd as a herd animal and, usually, only sick and younger animals could become prey. On the other hand, T-Rex, about 13 m long and weighing 6 tons, was a predator. Its advantages lay in its power, speed, and powerful teeth.

In an organizational ecosystem, matrix organizations—the Diplodocuses—and project organizations—the T-Rexes—have different advantages and disadvantages. Diplodocuses did not have to hunt; they simply picked their food. Such organizations include government agencies, insurance companies, energy providers, financial organizations such as some banks and insurance companies, electricity, gas, and water companies, and most educational institutions. In this metaphorical example, the T-Rexes are technology companies that deal with fast-moving products and services such as software, biotechnology, and FinTech—or EV makers.

Traditional car manufacturers went through their *Sturm und Drang* period—their T-Rex phase—over a century ago. Chapter 1 of this book shows how thousands of automotive startups went bankrupt or were "swallowed up" by competitors. The winners of this natural selection eventually became diplodocuses—the usual course of events, the typical "corporate lifecycle." Incidentally, this development is similar for OEMs and leading automotive suppliers. However, the recent EV wave is forcing these companies to rejuvenate: they must become T-Rexes again or become extinct. Traditional line management structures, like departments and head offices—many rooted in Prussian bureaucratic thinking—are inherently inflexible. They cannot easily cope with faster technological changes. The growing role of software, especially in the ongoing EV revolution, focuses on the ability to drive new product and architecture concepts in vehicle construction faster. Our industry is no longer operating "as usual"; we are now embracing new and different ways of doing things.

As traditional matrix and line organizations are neither agile nor fast enough to implement disruptive platform concepts in such an environment, the "projectized" organization offers a solution. Following our evolutionary example, projects are T-Rexes: they are change agents—drivers of change—that do not shy away from the new and take the implementation risk while making it manageable. This approach offers flexibility, speed, and adaptability, precisely the attributes that have become indispensable in our industry. A consistent project orientation makes it possible to break away from often cumbersome, rigid, hierarchical structures and instead form teams that can focus on rapidly evolving goals and react quickly to new challenges. In a project-oriented organization, risk minimization (rather than just risk management) is not seen as an obstacle but as a positive mindset.

However, implementing a project-oriented corporate structure also requires a cultural transformation. Traditional methods and ways of thinking must be replaced, and a culture of openness, experimentation, and continuous learning, which is natural for an agile project culture, must be relentlessly promoted. It is crucial that company leadership actively supports and drives this change.

To successfully navigate this transformation, certain foundational principles must be embraced. The following premises are at the heart of the T-Rex model:

- Project orientation (or project-centric mindset)
- Technical career paths
- The pooling principle

Project and program orientation means elevating a conscious decision to prioritize projects to the strategic core business. T-Rex organizations *consistently* focus on projects. They recognize that real progress and innovation are best achieved through project-oriented structures. As a result, risk-taking and skillful risk reduction become routine. A set of related projects is referred to as a program, meaning that product variants are coordinated and managed through overarching program management.

Technical career paths complement the project orientation concept. This involves the *creation* of technical career paths. Instead of employees having to move into management positions to earn higher salaries, they are offered career paths within their specialist technical areas (e.g., established as Subject Matter Experts). For example, a software expert could progress from Software Developer to Senior Software Developer to Software Architect to Chief Software Architect, with the salary at each level comparable to that of a corresponding line manager level. This technical career ladder allows experts to focus on their technical skills while being appropriately compensated.

The pool principle in a T-Rex organization refers to the centralization of experts in specialist groups in which they are involved in training and developing new ideas and technologies during "pool times"—i.e., phases in which they are not engaged in a project. These experts are integrated into projects as needed, whereby they are fully assigned to the respective project team during their project work and do not report to their pool superiors (see Fig. 5.13).

An especially effective approach involves how the project team is composed. Project managers (or project leads) decide on the team's composition and monitor team members' performance. This creates a stronger team bond and encourages team members to always act in the best interests of the project's success. In this model, project leads actively select the most suitable experts from the pool for their projects, which promotes performance-oriented competition. This approach ensures that the most sought-after and competent employees are assigned to projects, while less sought-after experts, who may receive fewer project assignments, remain in the pool. This process serves as an efficient market mechanism

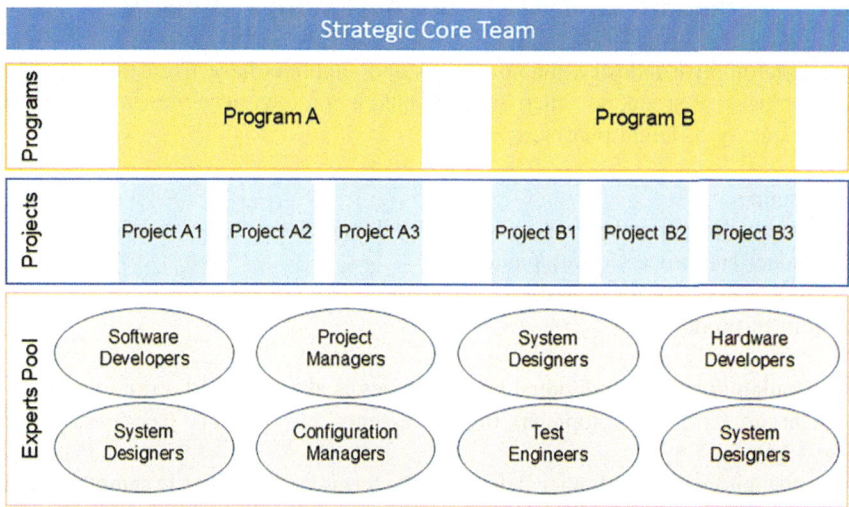

Fig. 5.13 The pooling principle in a project-oriented organization

that prioritizes performance and the demand for expertise. If there are not enough top performers in the pool, younger or less experienced employees, for example, can work on a project and thus move up in the pool organization through their successes. The bracket for the pool principle is formed by a pool manager, which could be compared to a leading HR role, for example.

In addition to the project management concept, in which the individual project plays a central role, overarching program management is essential so that platforms and variant management, for example, remain consistent. Program managers act as a link between platforms and individual projects.

Of course, a project-oriented T-Rex organization set up in this way also has some disadvantages. For example, some tasks cannot be performed on a project-specific basis, such as maintenance of test equipment, provision of drinks and food for the teams, or maintenance and equipping of workstations—incredibly remote workstations. This also includes maintaining production facilities and shop floors, managing logistics and supply chains, supporting the IT infrastructure and networks, ensuring building management and corporate security, maintaining the corporate culture and employee training, and monitoring compliance with legal regulations.

Consistent project orientation enables transformation to an agile, "projectized" organization—a T-Rex company that functions effectively and is relentlessly goal-oriented. Teams can act flexibly and creatively thanks to technical career paths and the pooling principle. By positioning ourselves as agile, dynamic hunters in the automotive ecosystem, we are in the best starting position to master the challenge of electromobility.

5.4   Standard Lifecycle in the Automotive Industry

In the automotive industry, manufacturers and suppliers have fixed lifecycles for developing new products, which are presented as product development processes. These specify different phases, such as:

– Planning
– Product concept and process concept development
– Product and process development
– Product and process validation
– Start of production.

The availability of the individual sample stages is also specified during the development or further development of components and systems (see example in Fig. 5.14).

In the automotive and other industries, the terms A, B, C, and D samples (often referred to as "prototypes") are used to describe different stages in the development and testing of products. Each sample stage represents a specific point in the development process, with the requirements and characteristics of the samples becoming increasingly complex and detailed.

– **A-sample** (prototype/concept sample)
   A-samples are prototypes or concepts of a new product. They are sometimes handled as a "showcase" and used to visualize initial ideas and concepts and test essential functions. They are sometimes also used as a "proof of concept." These may be models or simulations that validate the concept, test essential technical solutions, and gain initial insights into feasibility.
– **B-sample** (functional sample/development status)
   B-samples are more advanced prototypes that are produced from auxiliary tools and already have many features of the final product. They are used for detailed tests and to check the system design.

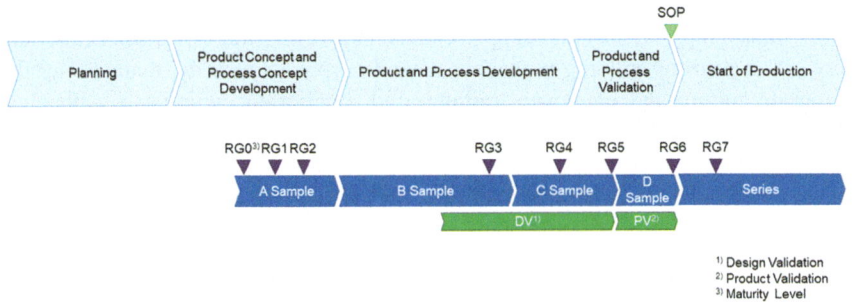

Fig. 5.14 Example product development process (PDP) with sample phases

These samples often contain the actual materials and components and are used to perform design verification and ensure that the product meets technical requirements. B-samples are usually used in the DV (**d**esign **v**alidation) test.

– **C-sample** (pre-series sample/near-to-series production)
C-samples are system versions suitable for series production. They correspond as closely as possible to the planned series production and are used for Product Validation (PV, sometimes referred to as Production Validation) and final tests. These samples are produced using a series of tools to check whether the product can be manufactured reproducibly and to the required quality in the production environment. Process validation by the supplier and the customer also starts at this stage.

– **D-sample** (series sample/pilot series)
D-samples are the final products manufactured under regular production conditions. They are configured using the actual production ("calibration") data and installed directly in the end-customer's vehicle. They represent the final quality and specifications of the product.
They are used to confirm production readiness, perform final quality checks, and are often part of the first production batch (e.g., initial samples) delivered to customers.

– **Design Validation (DV)**
Design Validation (DV) is a process that ensures that the design of a product fulfills all specified requirements. This phase checks the design is correct and that all essential functions and features meet customer requirements. This is typically done with the B-sample version of the system. Sometimes, this is called a "design freeze" at the system or software level. DV can be realized through a series of tests and analyses to confirm the design's integrity and functionality. The aim is to verify that the product has been designed to fulfill its intended purpose. DV activities include laboratory tests, simulations, prototyping, EMC tests, mechanical robustness tests (e.g., drop tests), chemical compatibility tests, and basic functional tests. After a successful DV, further features, particularly software-related features, are often completed, so the 'software design freeze' only occurs later, often as late as the C-model phase.

– **Product Validation (PV)**
Product Validation (PV) refers to the production process. The aim is to ensure the product can be manufactured reproducibly and consistently in the manufacturing environment. This can involve checking the relevant production facilities. PV verifies that the production processes and equipment can produce the product according to the defined design specifications and to the required quality. PV activities often involve creating a certain number of system units, which are then subjected to extensive testing to ensure that the production processes are reliable and that the end product meets manufacturing quality standards. As a rule, PV is carried out with the C- or D-sample.

All OEMs have their specifications for testing components (e.g., VW 80000). However, there are also many standards for the validation of new parts:

- IEC 60,529 Degrees of protection provided by enclosures (IP Code) [BF-017]
  - Protection against water
- ISO 16750–2:2023 Road vehicles–Environmental conditions and testing for electrical and electronic equipment – Part 2: Electrical loads [BF-018]
  - Functionality at minimum and maximum supply voltage
  - Protection against overvoltage
  - Protection against short circuit
  - Protection against polarity inversion
- ISO 16750–4:2023 Road vehicles – Environmental conditions and testing for electrical and electronic equipment–Part 4: Climatic loads [BF-019]
  - Thermal shock
  - Temperature cycle
- IEC 60,068-2-64:2008 Environmental testing–Part 2–64: Tests–Test Fh: Vibration, broadband random and guidance [BF-020]
  - Protection against vibrations
- CISPR 25:2021 Vehicles, boats, and internal combustion engines–Radio disturbance characteristics–Limits and methods of measurement for the protection of on-board receivers [BF-021]
  - Protection of on-board receivers
- ISO 11452–2:2019 Road vehicles–Component test methods for electrical disturbances from narrowband radiated electromagnetic energy–Part 2: Absorber-lined shielded enclosure [BF-022]
  - Protection against electromagnetic interference
- ISO 10605:2023 Road vehicles–Test methods for electrical disturbances from electrostatic discharge [BF-023]
  - Protection against electrostatic discharge (ESD)
- ISO 7637–2:2011 Road vehicles–Electrical disturbances from conduction and coupling–Part 2: Electrical transient conduction along supply lines only [BF-024]
  - Protection against electrical interference based on capacitive or inductive coupling.
- SOP
  SOP stands for "Start of Production"–the "moment of truth": formally completing system development. All components are installed in the customer's vehicle.

Several releases cover the phases. In the case of complex systems such as "system of systems", for example, these phases and the deliveries or releases must be thoroughly coordinated so the overall system can represent the expected platform maturity. This makes the planning and coordination of individual deliveries or releases and their corresponding documentation in the form of baselines all the more critical.

Understanding the automotive industry's standard lifecycle and product development processes is essential for understanding the challenges in project management. Embedded in this overall structure, the project management discipline is critical in ensuring that each phase is executed efficiently, risks are managed effectively, and the final product meets quality standards and customer expectations.

5.5   Project Management: Between a Rock and a Hard Place

Automotive project management is a delicate balance of technological innovation, economic constraints, and stringent regulations. Despite its critical importance to our industry, project management is often undervalued and sometimes dismissed merely as a "nice-to-have." Partly, this is due to the matrix mentality described in the previous chapter; on the other hand, expectations of "the project" are increasing due to ever-shorter deadlines. If, as we sometimes hear in informal conversations, it takes a Chinese OEM just two years to develop a new vehicle instead of four, the question of how this is supposed to work, especially if we have to meet all these regulations and occasionally abstract quality requirements, becomes ever more pressing.

The answer to this question remains open.

Our industry is characterized by relentless competitiveness, deeply rooted in the historical developments and supplier problems outlined in Chap. 1. Added to this are the "Bermuda Triangle" challenges—quality, cost, and time—plus scope, as described in Chap. 4. OEMs insist fully on all these aspects—or else.

Car part suppliers, in particular, are feeling the effects of this situation. Since the Ignacio López era in the 1990s, automotive companies have put increasing pressure on suppliers. This long-term OEM strategy has led to a culture of constant cost-cutting ("cost optimization" is a popular euphemism), often at the expense of innovation and quality. Suppliers are under continuous pressure to reduce their prices, while, at the same time, the demands on quality and technology are increasing. This balancing act between efficiency and costs is prevalent in our industry.

Project management in the automotive industry must navigate this unrelenting pressure while constantly adapting to rapidly evolving technologies, market conditions, and a flood of new regulations. The tricky "Bermuda Triangle" is elevated to a categorical imperative and often deters new suppliers, especially in software development, where these conditions can be perplexing.

We urgently need ways out of this dilemma. One option would be to question the regulations, of which there are more and more at an ever-increasing pace. After all, having too many rules leads to the opposite of the intended improvement in the quality and reliability of our products. However, it is easy for legislators to impose new regulatory constraints on the industry—after all, it costs them nothing; the industry bears the consequences—and it is nearly impossible to question the meaningfulness of new rules and regulations. As a result, it is rare for rules to be relaxed. That would require civil courage—a feat that could jeopardize the regulator's career.

Consequently, cost-cutting will probably remain the only option for OEMs and suppliers in their constant struggle for meaningful profit.

5.6   Future Trends in Automotive Project Management

Given the exponential growth of functions and regulations, the fundamental question is whether traditional project management or its newer, more progressive approaches (such as agile or hybrid methodologies) can succeed. From our perspective, the answer is "yes," provided we focus on pragmatic approaches that internalize "best of breed" as a principle and put it into practice.

The following aspects should be considered when weathering the electrification storm.

- **Embrace Structural Change**
  Ever more ambitious projects are set up in response to change and pressure from competitors, and more or less elusive deadlines are set. In most cases, the existing structures are often retained, i.e., a project manager is appointed to manage hundreds of project members, with sub-project managers appointed to support the project manager, who usually has the corresponding line responsibility. This means the project manager must not only master problems in the project but also constantly make compromises to consider the interests of the relevant stakeholders. In addition, project responsibilities are often not clearly defined or are interpreted following individual career goals. The result can be a lack of project progress, a weakening of the innovative strength, and more project risks. These only become apparent at a late stage.
  Therefore, lighter structures are needed to avoid mixing project and line responsibilities. The matrix must be weakened. Product leads must be *effectively* established to plan and manage their products' progress.
  In our view, a clear project orientation, as described in section 5.3 is a suitable way to improve product development capabilities.
- **Extension/adaptation of the "Bermuda Triangle"**
  Insisting on completeness within a "Bermuda Triangle" is no longer a worthwhile goal. Other vital factors must be considered. Aspects listed in the PRINCE2 standard, such as customer benefit, risk, and sustainability, could be included instead.
- **Implement Project-Oriented Funding Models and Career Paths**
  More diverse career paths must be considered:
  - Management career
  - Project career
  - Specialist career

Unfortunately, the management career path often provides the most compensation and influence. To motivate and retain talented employees long-term, it's crucial to equally value other career paths and reward competence, expertise, and experience. With the increase in systems complexity, experienced specialists and architects are in high demand. These experts are likely to leave if companies don't financially support them and promote them via a specialist career path.

– **Build/promote Skills**

Competence and experience always go hand in hand. The theoretical basis can be acquired through training courses and seminars, but practical experience in successfully implementing this knowledge is more important. Training on-the-job and formal training options are manifold and must be carefully planned with today's segmentation and "segregation" of specialist areas.

– **Keep a Hands-on Perspective**

Central roles like quality assurance or process ownership should not be held too long, or at least should not be their sole role. A rotating system between theory and practice is essential; for example, after three years, a quality assurance employee should return to a development role for a defined period. This maintains necessary practical relevance and helps employees be rounded in project reality.

– **Cultivate Motivation and a Passion for Professional Work**

In the current development environment, developers often lose their motivation and enjoyment of their work due to factors such as:

– Incomprehensible management decisions
– An uncoordinated tool landscape ("tool fragmentation") that makes work slower and more difficult.
– Continuously mushrooming processes that pointlessly restrict developers and limit their creativity.
– Suppressing meaningful discussions. This prevents a vital feedback loop between developers and management, contributing to frustration.
– Reporting becomes increasingly superficial as it moves up the management hierarchy, with facts being increasingly distorted or downplayed. This culture undermines the integrity of the management structure at all levels of the company.

– **Taskforce as a Universal Remedy**

In troubled projects, so-called "task forces" are frequently installed with the aim of getting the project back on track. However, taskforces only make sense if sufficient competence and talent are available. It is also essential to thoroughly analyze the causes of the failure and make appropriate organizational improvements. Thus, similar mistakes can be avoided in the future.

Corporate culture is sometimes structured so that without a task force, it is almost impossible to deliver a project within its original scope. In such a case, perhaps the task force process should become the standard development process.

When a project encounters difficulties, solid project management skills are required. Chapter 6 examines "project restructuring" as a possible remedy.

– **Harmonize Processes, Standards, and Models**

Processes, standards, and models play an essential role in our industry. However, the increasing speed of development should go hand in hand with the complexity of required standards. Therefore, the focus should not be exclusively on promoting ideas such as "agility" but on achieving genuine processual "leanness." This means simplifying processes and designing them effectively to avoid unnecessary complexity and formal "waste." As a top priority, processes should comply with functional safety (ISO 26262) as fundamental. This ensures that safety aspects are integrated and systematically implemented into the development process right from the start.

In this context, CORE SPICE (see Chap. 6) is essential. It provides a project coaching framework for the pragmatic improvement and evaluation of system and software processes, emphasizing the functional safety standard.

As a forward-thinking observation, integrating automation, artificial intelligence, and advanced tools is the key to increasing project management efficiency and informed decision-making in the age of the software-defined car and autonomous driving functions. These technologies can play a vital role in automating repetitive and manual project tasks, accelerating data analysis and interpretation, and thus increasing the productivity of project teams. By combining these approaches, organizations can work faster, smarter, and safer.

The future of automotive project management doesn't rest in traditional product development organizations, where formal power enforces a rigid matrix structure. Instead, intelligent project management skills are required. The task of a project manager should not be limited to mastering spreadsheets and PowerPoint presentation "engineering" and a disjointed tool landscape under formal pressure, excessive overtime, and near burnout. Instead, the role of a modern automotive project manager is to inspire and to be able to avoid demotivating and often seemingly pointless tasks. Organizations that structure project management this way will not just survive rapid changes–they will lead the industry and set the standard for others to follow.

References

[BF-001] VDA QMC Band Automotive SPICE Guidelines—Process assessment using the Automotive SPICE PAM 4.0, 2nd revised edition, November 2023

[BF-002] PMI—Project Management Institute https://www.pmi.org/

[BF-003] PMI—Project Management Institute, PMBOK Guide, https://www.pmi.org/

[BF-004] PMI—Project Management Institute, Process Groups Practice Guide, https://www.pmi.org/

[BF-005] History of Kanban, KanbanTool, https://kanbantool.com/kanban-guide/kanban-history, Zugriff am 09-Nov-2023

[BF-006] ISO 15188:2001 Project management guidelines for terminology standardization, ISO, ISO—International Organization for Standardization

[BF-007] ISO 21500:2021 Project, programme and portfolio management—Context and concepts, ISO, ISO—International Organization for Standardization

[BF-008] ISO 21502:2020 Project, programme and portfolio management—Guidance on project management, ISO, ISO—International Organization for Standardization

[BF-009] ISO 21503:2022 Project, programme and portfolio management—Guidance on programme management, ISO, ISO—International Organization for Standardization

[BF-010] ISO 21504:2022 Project, programme and portfolio management—Guidance on portfolio management, ISO, ISO—International Organization for Standardization

[BF-011] ISO 21505:2017 Project, programme and portfolio management—Guidance on governance management, ISO, ISO—International Organization for Standardization

[BF-012] ISO/IEC/IEEE 16326:2019 Systems and software engineering—Life cycle processes—Project management, ISO, ISO—International Organization for Standardization

[BF-013] IPMA Individual Competence Baseline ICB 4.0, IPMA, https://ipma.world

[BF-014] Prince2®7, Prince2, https://www.prince2.com

[BF-015] Yoji Akao, Quality Function Deployment (QFD): Integrating Customer Requirements into Product Design, Productivity Press, ISBN 978–1563273131

[BF-016] Tom Demarco, Waltzing With Bears: Managing Risk on Software Projects, Dorset House, ISBN 0932633609

[BF-017] IEC 60529 Degrees of protection provided by enclosures (IP Code), International Electrotechnical Commission, https://www.iec.ch/homepage

[BF-018] ISO 16750–2:2023 Road vehicles—Environmental conditions and testing for electrical and electronic equipment—Part 2: Electrical loads, ISO, ISO—International Organization for Standardization

[BF-019] ISO 16750–4:2023 Road vehicles—Environmental conditions and testing for electrical and electronic equipment—Part 4: Climatic loads, ISO, ISO—International Organization for Standardization

[BF-020] IEC 60068–2–64:2008 Environmental testing - Part 2–64: Tests—Test Fh: Vibration, broadband random and guidance, International Electrotechnical Commission, https://www.iec.ch/homepage

[BF-021] CISPR 25:2021 Vehicles, boats, and internal combustion engines—Radio disturbance characteristics—Limits and methods of measurement for the protection of on-board receivers, International Electrotechnical Commission, https://www.iec.ch/homepage

[BF-022] ISO 11452–2:2019 Road vehicles—Component test methods for electrical disturbances from narrowband radiated electromagnetic energy—Part 2: Absorber-lined shielded enclosure, ISO, ISO—International Organization for Standardization

[BF-023] ISO 10605:2023 Road vehicles—Test methods for electrical disturbances from electrostatic discharge, ISO, ISO—International Organization for Standardization

[BF-024] ISO 7637–2:2011 Road vehicles—Electrical disturbances from conduction and coupling—Part 2: Electrical transient conduction along supply lines only, ISO, ISO—International Organization for Standardization

CORE SPICE

6

Abstract

This chapter introduces CORE SPICE, a lean project management coaching concept that supports high-quality automotive systems development. It explains how CORE SPICE addresses the limitations of traditional assessment models while maintaining compliance with safety standards like ISO 26262. The chapter outlines CORE SPICE's basic principles, approaches, and role models, including the innovative Team Capability Coach (TCC) role. It compares CORE SPICE with traditional Automotive SPICE, demonstrating how it can improve project efficiency while maintaining quality standards. The chapter also introduces Effective Critical Systems Thinking (ECST) as a broader perspective for automotive development and explains how CORE SPICE can be used to turn around distressed projects.

The ability to react to changes in the automotive industry—or, even better, to act proactively—determines the success or failure of a new vehicle platform. Gone are the days when you could take four to five years—the latest competition— whether Tesla or Chinese manufacturers such as BYD—is setting an enormous pace. At the same time, there is a weak correlation between conventional development process quality and product quality in the traditional car design business. In our industry, complex processes are often designed and introduced throughout development teams, and there is a perception that meticulously detailed development processes should be the source of creativity and product quality. Traditional models such as ASPICE are frequently developed as extensions of the ASPICE specification, leading to complex and bloated development processes. A closer look at how assessment models are misused as naïve process blueprints requires professional scrutiny.

However, our practical experience has shown that we cannot completely abandon the process concept and the entire Automotive SPICE principle in favor of

© The Author(s), under exclusive license to Springer Fachmedien Wiesbaden GmbH, part of Springer Nature 2025
R. Mildner et al., *Car IT Reloaded*, https://doi.org/10.1007/978-3-658-47691-5_6

purely agile methods. That would be impossible in a safety-relevant industry such as automotive engineering. The idea introduced in this book, CORE SPICE, represents our approach, which leverages decades of process improvement knowledge and experience while avoiding the complexity of ASPICE. This approach concentrates on essential process quality elements, the engineering mindset, and effective project management. CORE SPICE emphasizes a quality and professional collaboration culture among system and software development teams. It also emphasizes customer involvement at an early stage of a development project.

CORE SPICE is a project quality management improvement that can be used as a project coaching concept to increase the pace of systems development (time-to-market), maintain development quality, and minimize bureaucracy, which is often criticized as "process satisfaction syndrome," that is, a self-serving bureaucratic activity that usually offers little benefit to the customer.

6.1   Why ASPICE Exists

Our consulting practice has witnessed countless attempts to improve processes in software and system development organizations. The rate of failed CMMI and ASPICE process improvement projects remains unknown. However, in informal conversations with consultants from leading process consulting firms and groups in this industry, the rumor is that the rate of failed process improvement projects may be as high as 80 percent. This is not an attempt to discredit the industry; we have helped many automotive businesses improve their processes using assessment models such as Automotive SPICE. However, our community has observed its fair share of failed projects.

Process improvements are costly and fraught with numerous risks. They can be difficult for project managers and their superiors, potentially jeopardizing their careers. Therefore, changes in a company typically require a great deal of courage, resourcefulness, and determination to undertake such transformations.

The reason ASPICE is still routinely used in car development projects is not solely due to the necessity of demonstrating an ASPICE certificate to the OEM. The truth is that, following the demise of the CMMI model, ASPICE is the only process assessment model usually required by OEMs. 'There is no other alternative' is the popular justification. This is also why safety-relevant standards such as ISO 26262 (functional safety) reference ASPICE as a viable option to fulfill safety compliance. In other words, when ISO 26262 is required, it implicitly means demonstrating ASPICE compliance.

Development processes are a complex requirement. In Chap. 4, we discussed the criticality of software complexity in the context of the CMM model, which was the first attempt to address the process quality issue. ASPICE is simply another incarnation of these efforts. The "software crisis" mentioned in Chap. 4 illustrates the ongoing frustration with software quality.

Therefore, the primary purpose of ASPICE is not the (supposed) harassment of car manufacturers to increase the burden on them. The ultimate reason why

Automotive SPICE is required is the effort to **minimize project and product risks** in car systems design. Such risks include, for example, the following:

- Defects in safety-relevant components (particularly critical in self-driving cars).
- Systems security (a.k.a. cybersecurity) risks, such as data theft, data manipulation.
- Incorrect integration of software and hardware (e.g., software-based sensor technology).
- Inadequate software product quality.
- Regulatory risks in software design (see "Diesel gate").
- HMI (Human–Machine Interface) problems (e.g., confusing user guidance in some infotainment systems).

The often-cited "software complexity" is also a common concern. The rationale for the increased number of such complaints is the growing tendency to use software instead of hardware to implement car functions. In car systems design, the role of software has long been underestimated by automotive managers, who usually have little software expertise. That also explains the emphasis on software quality as a "problem." While software enables shorter innovation cycles, product quality must keep pace. Otherwise, embarrassing, expensive, and even safety-related problems arise. For instance, the innovative Daimler W211 platform, which rolled off the production line from 2002 to 2009, was notorious. Using new technologies and components led to an increased complexity of the W211 series' systems. Software was perhaps too often blamed for quality issues, as the difficulties were platform-wide. The vehicle contained many electronic control units (ECUs) connected to the car network. The on-board network was overwhelmed with such a high number of ECUs, which caused timing and performance problems.

Such challenges are familiar to modern car manufacturers. While in the past, quality assurance was possible through thorough design validation (DV), it is now generally recognized that in the era of software, not everything can be "simply" tested. In software-based car systems, out of hundreds of millions of lines of code, only a subset can be verified by software test engineers. For this reason, software development processes have moved into the spotlight. When implemented throughout the V-model, traceability allows, at least in theory, a systematic selection of critical software components and aspects that need to be thoroughly scrutinized. When well-implemented, traceability allows for systematic regression and makes accurate handling of new features possible.

Despite these and other efforts, software continues to be a critical complaint, which means the pressure on suppliers and OEMs remains high and will likely continue to remain high or even increase. The expectation now is that software quality must improve. But what type of "quality" does it mean? Process quality? Product quality? Both?

6.2   About Product Quality and Process Quality

In the automotive industry, since the early era of CMM and process improvement hype, process quality" has become part of contractual obligations typically included in supplier contracts. Automotive SPICE is intended to contribute to the quality of the development process and, thus, implicitly, to the quality of resulting products.

As described in Chap. 4 on "Process Models and Maturity Levels," process assessment models such as ASPICE are supposed to provide a *quantifiable* measure of process quality in terms of the capability to manage project risks within the Bermuda Triangle." Process improvement is a secondary objective. Therefore, an interesting question is: How does process evaluation correlate with the overarching concept of quality, specifically process and product quality?

A common definition of "quality" is formulated in the ISO 9000 standard. This standard defines quality as the 'degree to which a set of inherent characteristics of an object fulfills requirements' [RM-13]. In principle, a QMS (Quality Management System) is required, whereby continuous improvement of the QMS is postulated. The term "process quality" is not explicitly defined in this context. Instead, a process (often called a "process model") is usually referred to as another work product that must be maintained.

Furthermore, the ISO 9000 standard fundamentally postulates *process orientation.*

A process is a set of interrelated or interdependent activities that use inputs to achieve an intended result. In other words, a process generates added business value in the company.

Process quality is emphasized in process evaluation models such as CMMI or ASPICE. Interestingly, however, the term "process" itself is not explicitly defined in Automotive SPICE 3.1 or 4.0. It is also not included in the list of generic work products in the ASPICE standard. Only the *process description* is listed as a work product, which is a small but subtle difference from the term "process" (see Table 6.1).

Table 6.1 Process description (a generic ASPICE work product)

10–00	Process description	Process description of a standard or defined process (e.g., after tailoring), including: – Scope and the intended use of the process – Process activities, including description and dependencies – Entry and exit criteria such as input information needed and expected outputs for activities – Roles assigned to process activities (e.g., as RASIC) or work products – Guidelines – Templates – Specific methods/work instructions

Therefore, a process is implicitly understood through its description (which strongly correlates with "process definition" but is not identical). The reluctance of consultants who have defined this standard suggests that the term "process" is ambivalent. However, this is an unfortunate state of affairs because, to put it casually, anyone with a sharpened pencil and a piece of paper can now define development processes.

The question "What is a process?" could be further dialectically analyzed to polemicize the discussion further. A cheeky question: "Why do you want to know what a process is?" could lead to a heated debate. We observe that the term "process" is not interpreted the same way by all stakeholders. The underlying issue, which may seem provocative, is that the true motivation behind process definition or improvement is often unclear. In other words, the objective of process improvement is frequently ambiguous and subject to corporate politics.

Lewis Carroll, a writer and mathematician, once said: 'If you don't know where you are going, any road will get you there.' This esoteric insight also applies to process improvement activities: where exactly you end up under such a premise is uncertain. Process improvement activities' commercial and organizational goals and their genuine intentions must be well-defined. Otherwise, the result could be "confusion and despair".

Process improvement projects carry high personal risks within an organization, which should be taken seriously. Thus, the motivation for such endeavors must be clear. If, for example, a manager is compelled to define or improve processes due to vague or external demands (e.g., from customers or assessors) while lacking a clear and thorough understanding of the science of process improvement, this is a flashing red warning sign that, whatever happens next, the responsible managers could bitterly regret their involvement.

While product improvement is intuitively clear, the "process quality" topic differs. Therefore, we want to take a closer look at the interplay between product quality and process quality. For example, CMMI-DEV (see Chap. 4) deals with the quality of processes and product quality. In CMMI, separate process areas are defined: VAL (Validation), VER (Verification), and PPQA (Process and Product Quality Assurance) (see Fig. 6.1).

One disadvantage of these differences is that testing (a specific case of verification) is not explicitly mentioned. Testing is an experimental process, whereas verification can and must extend beyond it. Process designers can decide whether certain test levels are defined, such as software integration or system qualification testing.

In ASPICE, product quality is viewed in the context of the right-hand side of the V-model, in which several test levels are defined (see Fig. 6.2).

In the ASPICE V model itself, the product quality is defined in SYS.4, SYS.5, SWE.4, SWE.5, and SWE.6. At the same time, product quality is expected to be defined in SUP.1 "Quality Assurance" support process. The objective of the SUP.1 assessment process is as follows:

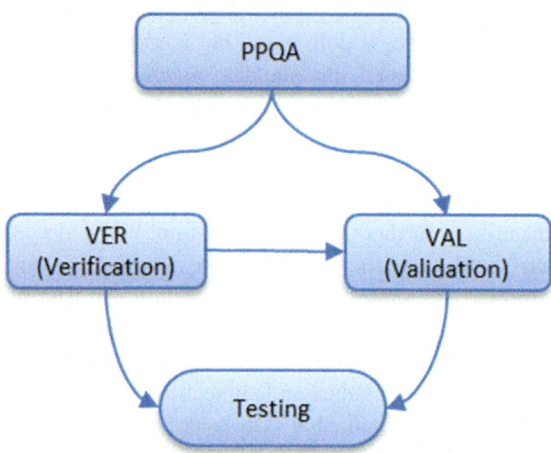

Fig. 6.1 Relationship between process quality, product quality, and testing

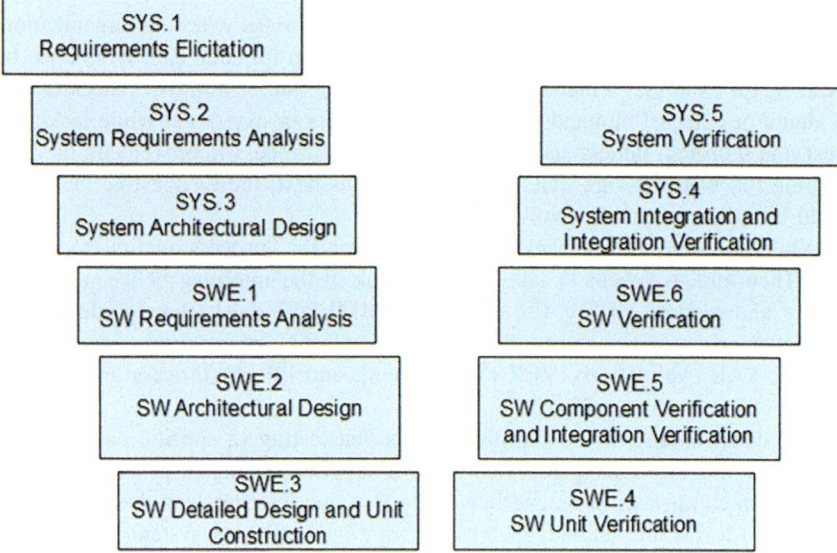

Fig. 6.2 ASPICE: V-model

> The purpose of the Quality Assurance Process is to provide independent and objective assurance that work products and processes comply with defined criteria and that non-conformances are resolved and further prevented.

The assumption that the quality assurance manager (QA manager) may be responsible for verifying work product is not entirely uncommon; there has been frequent

Fig. 6.3 Process quality
versus product quality

confusion in CMMI-related process improvement activities in the past. This was a misunderstanding. A QA manager should ensure that verification and testing occur correctly—that's it. In all areas of the right-hand side of the V-model, responsibility for testing generally lies with a test manager (e.g., Test Lead). This is where product quality is ensured through verification and testing.

Given these explanations, the question arises: Who ensures process quality and conformity with the written process definition? Should it be the QA manager? The head of the department? The BU (business unit) manager? Or "the team"? The answers vary. We have occasionally seen QA managers refusing to take responsibility for improving process quality, following the motto: 'Checklists are okay, but process design—please let others do that.' However, this seems understandable, as it is rare that the job description of the position of quality manager includes expertise in all ASPICE reference processes in the VDA scope. Sometimes, the responsibility for process quality remains open until an ASPICE assessment is suddenly due, often followed by a frantic wrangling over responsibilities. In the following sections, we will discuss this topic in more detail and offer CORE SPICE as a solution for defining project responsibilities.

We can conclude that quality can be broken down into two fundamental areas: process and product quality (see Fig. 6.3).

Product quality is a very clearly defined task. Many suitable and well-established methods and standards (such as ISTQB) can be used to ensure product quality.

Process quality, on the other hand, is not so easy to grasp. An entire consulting industry thrives on the fact that system and software developers are repeatedly confronted with many confusing process questions, such as:

– What does "verification criteria" mean for system and software requirements?
– What is meant by "correct results"?
– Do you need a CCB (Change Management Board)? Why?
– Are change management and change request management the same thing?
– What exactly is the difference between software integration and software integration testing?
– How do I ensure traceability in all V-model areas?

The list could go on for pages.

ASPICE 3.1 was already a voluminous document of about 130 pages. ASPICE 4.0 has swollen to 153 pages. Of course, the practices listed there also require explanation, while others appear deceptively simple. Therefore, another ASPICE-specific document is indispensable to understand the expectations for ASPICE-based process improvement and preparation for an ASPICE assessment: Automotive SPICE Guidelines [RM-16]. However, more nuanced knowledge of ASPICE is still needed. Thus, further specialist literature is required; a good ASPICE book helps to understand the mindset of an ASPICE consultant or assessor, for example [RM-17]. Such books are beneficial if they have been written by consultants involved in creating the ASPICE specification.

Process quality is an opaque and almost esoteric matter. It presents a similar challenge for both ASPICE and other traditional process assessment models like CMMI. Therefore, it is essential to understand what a "process" *truly* is.

Apart from the usual contractual demands imposed by OEMs, according to which Automotive SPICE is not negotiable, fundamental questions arise: Why do we need "process quality," *how much* process quality do we need, and how strongly does a high-quality process correlate with the quality of the end product delivered to the customer? If ASPICE was not a contractual "must-have," would OEMs and automotive suppliers still use it to ensure process quality? What *measurable* added value does ASPICE bring in terms of business profitability? Does introducing ASPICE in projects and business units lead to a measurable increase in quality, as is often claimed?

In other words, what exactly is the "added value" of ASPICE in quantitative terms?

An ROI (return on investment) analysis is necessary to answer this question. We will address this aspect in the next section.

6.3   Return on Investment (ROI) from ASPICE

The return on investment (ROI) of development processes is a task fraught with considerable uncertainty. In this section, we explain why it can be difficult to calculate the ROI of processes. We present some calculation approaches, discuss the value of process improvement that is not directly expressed in figures, and explain why it is so difficult to give concrete ROI figures.

6.3.1   The Challenge of Measuring ROI in Process Improvements

As discussed earlier, the primary purpose of ASPICE is for OEMs to require at least the second maturity level in the VDA scope. Without this, it is practically impossible for a supplier to receive a nomination for its product development.

Fortunately, the OEM's demands aside, many automotive systems vendors are genuinely interested in improving development processes to ensure the quality of the manufactured end products. To this end, the associated development processes are usually designed to be ASPICE-compliant.

Since process improvements often demand a seven-figure investment, analyzing such initiatives' return on investment (ROI) makes strong business sense. It is also a very challenging task for several reasons, such as:

– It is often difficult to evaluate quality in euros and dollars. Many benefits of a process improvement, such as increased team morale or low staff turnover, are not easily quantifiable.
– The process improvement procedure is usually lengthy and extends over several years, requiring a complex, holistic, long-term perspective.
– Process improvement does not scale linearly, depending on factors such as team size, the number of dependent and independent projects in the respective scope, time constraints, the complexity of different projects and industries, etc. Improvements often occur in leaps and bounds—or, occasionally, not at all.
– Standards sometimes change rapidly, as with the transition from ASPICE 2.4 (ENG) to ASPICE 3.0 (SYS and SWE).
– Team effectiveness varies greatly depending on the competence, working environment, and geographical and electronic distribution of project team members.
– Personnel fluctuations distort the ROI assessment.
– There is no industry-wide or standardized definition of KPIs (Key Process Indicators) to enable a consistent ROI assessment.

At the same time, the ROI must be assessed as objectively as possible. Even after decades in this business, it's clear that evaluating the ROI of process improvements can often feel like mission impossible. Therefore, a rational justification for why large sums should be spent on such actions is crucial. Unfortunately, detailed and sensible assessments of the added value of process improvement measures are rare.

6.3.2 ROI Calculation Methods

David F. Rico's work *ROI of Software Process Improvement* [RM-018] examines the efficiency of process improvement measures in the context of CMMI. In his book, he evaluates the overall ROI and derived metrics that support a meaningful ROI calculation. In the context of process improvement (SPI—Software Process Improvement), he defines added value as follows:

The ROI of SPI is a simple ratio of all the benefits to all the costs of a new and improved software process.

That sounds deceptively simple. However, numerous complex ROI calculation models have been attempted in the past, which we no longer hear about today. David Rico, therefore, warns against using complex ROI calculations:

The difficult methods of ROI assessment are generally not worth the time and effort. They can jeopardize a person's political career and undermine their own power and status.

David Rico, therefore, recommends using the simplest possible ROI calculation methods.

This book will not detail the academic world of software process efficiency models. Instead, we want to look at what specific metrics David Rico considered worth looking at:

- Productivity.
- Defect Density.
- Quality.
- DRE (Defect Removal Efficiency).
- DRM (Defect Removal Model).
- Software Effort.
- TLC (Total Lifecycle Cost).

The metrics in detail:

Productivity

$$Productivity = \frac{\text{Lines of Code}}{\text{Hour}}$$

This simple metric calculates how many hours a developer takes to create and quality-assure a line of code. You can already guess that this metric is fragile, as the productivity of an efficient software developer can be 5 or even 10 times higher than that of a beginner. It is also clear that software development is not just about "hardcore coding"; every line of software includes necessary further attention from developers, such as specification, software design, software verification, etc. David Rico is aware of this and warns that this metric is only used out of necessity, as other models are too complicated.

Defect Density

$$Defect\, Density = \frac{\text{Defects}}{\text{Lines of Code}}$$

This metric calculates the number of errors found per line of code by the end of a project. It represents an estimate of the software's error susceptibility.

Quality

$$Quality = 1 - \frac{\text{Defects}}{\text{Lines of Code}} \times 100\%$$

In this context, the term "quality" refers to the difference between 1 and the error ratio. In other words, if we have 0 bugs, the quality is 100 percent. We don't want to confuse this interpretation of "quality" with the one discussed earlier, but this is how David Rico defines it in this context.

This metric is derived as a function of the error density.

DRE (Defect Removal Efficiency)

$$DRE = \frac{\text{Defect Removed}}{\text{Defect Escaped} + \text{Defect Injected}} \times 100\%$$

DRE is a ratio of all fixed to overlooked defects. This metric assesses the efficiency of the test and review process.

DRM (Defect Removal Model)

$$DRM = \text{Defects Escaped} + \textit{Defects injected} - \textit{Defects Removed}$$

This metric is the generally popular "defect curve" that OEMs, for example, routinely demand.

Software Effort
This metric is simply the sum of the total development effort. The expenditure for maintenance after the completed project is not considered in this case.

Total Lifecycle Cost (TLC)
These are all the software development costs, including the subsequent maintenance phase after the development project has been completed.

Some of these metrics are not uncontroversial. In particular, using lines of code as a metric to measure development efficiency is problematic. Besides the efficiency considerations already mentioned (faster vs. slower developers), counting lines of code seems anachronistic. The use of this metric is surprising because even when the book was published (2004), it was already increasingly irrelevant. While it used to make sense to count lines of code, today, developers work with millions of lines of source code hidden in a component (for example, bought from another vendor). Therefore, counting interfaces rather than lines of code makes more sense. Added to this is the variability of the number of lines of code considered. For example, if a singular software component—a black box—is broken up, as often happens in open-source projects, the question arises of how much of it is "discrete code" that should now be reviewed in detail—or should the entire software base be examined, as potential side effects distributed across various sections of code need to be investigated? Sure, a good tool can help analyze the code semi-automatically, but the challenge remains: a black box has become a white box, inevitably distorting the ROI metric.

Table 6.2 Process ROI

Method/process	ROI (%)
Personal software process	4133
Software inspection process	3272
Team software process	2826
Software capability maturity model®	871
ISO 9001	229
Capability maturity model integration®	173

However, given the rarity of publications like David Rico's book, let's set aside these shortcomings and focus on the bigger picture. Of particular interest is David Rico's ROI calculation of process improvement, which varies depending on the process improvement model used.

Rico analyzed the following process improvement models (SPIs):

- Software Inspection Process.
- Personal Software ProcessSM.
- Team Software ProcessSM.
- Software Capability Maturity Model®
- ISO 9001.
- Capability Maturity Model Integration®.

He did not consider ASPICE, as it was not yet established in the year his book was published (2004).

Using the metrics above (LoC—Lines of Code), the following assessments resulted from Table 6.2.

The calculation was based on a hypothetical project with 10,000 lines. The interesting thing here, however, is the correlation between process complexity and ROI. Methodologies such as the Software Inspection Process or ISO 9001 are much clearer than CMMI. CMMI is the most expensive process model in this list.

For CMMI conformity alone, 51 policies and procedures requiring approximately 530 working hours are estimated. Based on this, 640 working hours are needed for each project. In other words, just under 1,200 h (which, with an average workload, means approx. 1,500 h) must be estimated for "process complacency" (as it is sometimes pejoratively called). Additional expenses, such as organizing the process documentation, must be considered for operational work.

For ASPICE at the second level, which, in practice, is somewhere between CMMI levels 2 and 3, even stricter traceability requirements are to be expected, so it will not be "cheaper".

While Rico's assessment might be viewed differently today, his considerations remain consistent regarding a relative evaluation. According to his calculations, CMMI is over 40 times more expensive than the simplest process that, in Rico's view, delivers a comparable quality result. However, it should be noted that these projects were generally not explicitly considered for safety-relevant product development projects, which relativizes the result.

Rico's ROI calculation, while not without its flaws, still provides valuable insight into the world of process improvement. Despite the challenges in quantifying the exact value of a process methodology, his analysis implicitly emphasizes the importance of considering the software error frequency in a project as an indicator of process quality. This approach offers a more practical means of assessing the impact of process improvements rather than relying solely on ROI models.

6.3.3 Anecdotal Added Value of ASPICE

In 1994, SEI published a statistic according to which the ROI of using the CMMI model brought considerable added value [RM-006] (see Table 6.3).

An ROI factor of 7 sounds impressive, but the calculation method remains unspecified. The annual error reduction of 45% also seems quite bold, especially since the project phases during which these assessments were made are unclear. The exact assumptions SEI used for this assessment, such as project complexity and types of projects, are not provided. As discussed in relation to Rico's work in the previous section, the added value cannot be reliably calculated.

A more concrete statistical sample was presented by Jürgen Etzkorn at the fifth World Congress for Software Quality in China in 2011, specifically in the context of Automotive SPICE [RM-019].

This study investigated whether a higher maturity level correlates positively with product maturity because bugs are identified as early as possible in the project lifecycle. The author analyzed 26 data points and found a positive correlation between the ASPICE maturity level and product maturity (in terms of the statistical number of product-specific bugs) (see Fig. 6.4, adapted from Mr. Jürgen Etzkorn, ©Mr. Jürgen Etzkorn, BMW). It became apparent that the later the bugs were discovered, the lower the ASPICE-rated maturity was.

The result can be interpreted as follows. The assessment suggests that the most efficient added value of process maturity, as measured by the ASPICE scale, is achieved just before reaching maturity level 2. In other words, an ASPICE maturity level higher than level 2 yields marginal improvements in product quality.

This finding has several implications.

Table 6.3 Added value of a process

Advantages/factors	Value/change
Return on investment (ROI)	7:1
Average productivity gain per year	37%
Annual increase in errors found in the preliminary audit	18%
Reduction of time to market	19%
Annual reduction in reported errors	45%

Fig. 6.4 Correlation between maturity level and product maturity (*Source* [RM-19])

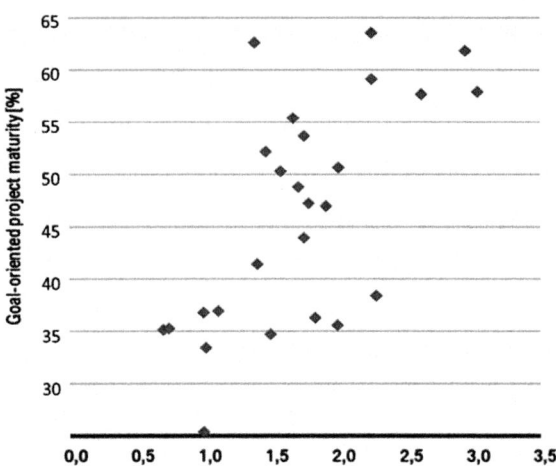

Firstly, as already suspected, ASPICE-based process improvement does not scale linearly. In other words, more is not better.

Secondly, it becomes clear that the significant effort many suppliers invest in achieving higher ASPICE maturity levels must be continually reassessed regarding the effectiveness of process improvement efforts. The merit behind striving for a higher capability level—particularly beyond the second level—appears to be neither scientifically nor empirically proven.

6.3.4   The Limits of Assessing Process Maturity

Various process efficiency measurement models have been developed, many of which have been presented by organizations such as the Software Engineering Institute (SEI) and experts such as David Rico. Their thoughts and methodologies regarding the return on investment of software process improvements offer insight into the value of process improvement in development projects. No theoretical models have been developed to deliver reliable process improvement metrics. The study by Jürgen Etzkorn offers one of the few empirical studies on the direct correlation between the ASPICE maturity level and actual product quality. The analysis suggests that a maturity level 2 offers the optimum balance between process documentation and qualitative results. For higher maturity levels, such as Levels 3, 4, or 5, there is a lack of evidence to show additional added value in the context of product quality.

Against this backdrop, a discrepancy in terms of a correlation between the process quality and the actual product quality becomes apparent. ASPICE is a helpful and popular assessment model. However, our observations indicate that beyond ASPICE maturity level 2, there needs to be more objective evidence of improvements in product quality. This implies that justifying investments in ASPICE maturity levels beyond Level 2 is problematic. It is an important finding that is of enormous significance for our industry. The question arises of whether ASPICE can—or should—be used to achieve higher quality. We will attempt to answer this question in the subsequent subchapters.

6.4   A Constructive ASPICE Critique

The promise of creating higher-quality software with a better process definition has been around for many decades. As objective data is scarce, we have to assume that, as far as we know, there is at least a positive correlation between higher process maturity levels and better product quality. It would be ideal if we could confirm this using industry-wide statistics. However, we are unaware of any such objective and comprehensive evaluation focusing on the automotive industry.

The connection between the causes and effects of ASPICE measures needs to be clarified more than some ASPICE process consultants suggest. The famous proverb "Don't ask a barber if you need a haircut" applies.

At the same time, ASPICE consultants have realized that agility is gaining popularity, and many are trying to convince their clients that ASPICE is agile. David Rico, whom we have already met in the ROI analysis of various process improvement models, created an ROI evaluation in 2008 on a more extensive data basis, whereby he analyzed agile models such as SCRUM in addition to the models already mentioned, such as CMMI, and evaluated the ROI of the respective models (see Table 6.4).

You already know the figures for methods such as CMMI from the previous section. However, the ROI assessment of agile methods such as SCRUM is new. According to this, SCRUM is approx. 3.5 times more efficient than CMMI.

As explained in the previous section, David Rico's method of ROI assessment is fundamentally methodology-agnostic. In this case, the error-based quality assessment is all that matters. This means the ROI assessment is subject to an enormous

Table 6.4 ROI of various processes (data based on [RM-020])

No.	Method	ROI (%)
1	PSP	4.133
2	Inspection	3.272
3	XP	3.103
4	TSP	2.826
5	Agile	1.788
6	TDD	1.607
7	PP	1.499
8	SW-CMM®	871
9	SCRUM	580
10	ISO 9001	229
11	CMMI®	173

margin of error. In his paper, David Rico lists a whole series of limitations that relativize the advantage above in favor of SCRUM.

The question now arises: Why is CMMI so "expensive"? And is this expense justified?

The loss of efficiency when using CMMI compared to SCRUM is no different than with ASPICE; in fact, the opposite is to be feared. Thus, the billion-dollar question is: Why don't we ditch ASPICE and go for SCRUM instead?

Unfortunately, the answer is complex. As we discussed in Chap. 5, agile methods have pros and cons. The standard SCRUM process isn't scalable in environments where safety and security are critical, making its implementation unfeasible. On the other hand, the traditional V-model endures criticism because it is cumbersome, bureaucratic, demotivating, and often viewed as unrealistic.

Automotive SPICE is a tried and tested assessment model, but if ASPICE is naively (literally) implemented as a "process," it causes several inefficiencies and bears unintended risks.

There are many reasons why ASPICE is so expensive. These include the following aspects.

Initial investment: Implementing an ASPICE-compliant process can be a complex endeavor that consumes endless resources. Without a robust roadmap and a generous allocation of well-trained experts and resources, such process improvement efforts can quickly spiral into a resource-intensive nightmare.

Sluggish adaptation: Change is never easy, and resistance to ASPICE can be particularly stubborn. Companies should be prepared to invest more time and energy in change management than expected.

Misunderstanding the nature of ASPICE: ASPICE itself is not designed as a process but rather as a comprehensive checklist. Nevertheless, many process designers interpret ASPICE as a kind of blueprint and develop processes that are a one-to-one copy of ASPICE. Due to the natural redundancies in ASPICE, this can lead to complex, cumbersome process structures that are not only tricky to implement in practice but also problematic to maintain. A revision of ASPICE version 2.4 to version 3.0 made this problem obvious. Whereas in ASPICE 2.4, V-model-relevant processes were numbered from ENG.1 to ENG.10, in version 3.0 a new scheme was defined that included SYS.1 to SYS.5 and SWE.1 to SWE.6. As a result, many suppliers had to adapt all process descriptions, costing considerable effort. A well-designed development process in which the processes were not slavishly reproduced as ASPICE copies would have experienced almost no impact from such a change. Incidentally, a similar upheaval resulted from the switch from CMMI to ASPICE in some organizations in the early 2000s.

Confusing terminology: Some ASPICE concepts may appear too vague, especially in globally operating companies. One example of this is the term "plan." In German-speaking countries, the term is often confused with "schedule". In English, however, a "plan" is more likely to be understood as an "agreement" or "approach". Another example is the terms "change management" and "change request management." "Change management" is more likely to be seen in the integration process. In contrast, "change request management" changes a previously approved baseline. Therefore, a change manager's role is more akin to that of a build master. On the other hand, it is correct for a change request manager to take care of change requests – a role closer to project management than the integration process.

Process maintenance effort: Maintaining ASPICE-compliant documents involves much effort. Companies that shy away from it often face a dilemma: their QA team is overwhelmed, while the mass of documents developed at great effort slowly becomes obsolete.

Excessive process documentation: The flood of different ASPICE-induced documents can become overwhelming. We have often seen process documentation quickly turn from quality to quantity. This is unfortunate because every line of process documentation must be maintained, which directly increases the cost of ASPICE-compliant documentation. The phenomenon of "write-only" documents is often the result—documents nobody reads anymore, but nobody has the courage to remove them.

Team training effort: There needs to be more than a one-off ASPICE training course. Continuous training is essential simply due to natural team fluctuation. However, there is often not enough time or budget for this.

Complex traceability requirements: ASPICE's traceability requirements can be overwhelming and expensive, especially if suitable integrated systems and processes are not in place to enable traceability with viable effort. Using makeshift solutions and Excel sheets to maintain and calculate traceability can quickly become overwhelming and error-prone.

Lack of tool support: Some vendors promise many "all-rounder" tools that are supposed to support SPICE and traceability, but, in practice, only a few keep their promises. Companies such as Atlassian or PTC promise a blue sky, but the reality is often disappointing. In general, software packages that support a complete project lifecycle toolchain are rare, expensive, and complex. Many companies also rightly fear the "lock-in effect," whereby they commit to a toolchain and are then confronted with rising costs, or when new tools come onto the market that promise specific advantages, they have to return to the usual carpet patching without warning.

Skeptical managers: Many managers barely recognize the actual value of ASPICE. The complicated ROI calculation for ASPICE described above already made this clear. Its implementation becomes a tedious Sisyphean task without a rationally justifiable business case.

Position of OEMs: Although OEMs are vocal about the importance of supplier ASPICE compliance, some believe they may not apply these standards as rigorously in-house. This discrepancy raises questions about whether ASPICE is primarily used for supplier control.

Unclear project requirements: Innovation and precise requirements are often at odds. In practice, ASPICE is interpreted by quality assurance as a waterfall. The usual requirement for a "complete specification definition" is unrealistic. Consequently, incomplete requirements are frequently used. The motto seems to be: 'Don't worry too much about that; we will find a way to agree.' Such an attitude can quickly become a source of incalculable risk for suppliers and OEMs alike. If handled overly strictly, full ASPICE compliance is often objectively unattainable. This situation is similar to the change request management process. We have seen customers who generally refuse to use the term "change request." Instead, they prefer alternative euphemisms such as "optimizations" to avoid alarming their QA colleagues or senior management peers on both sides of a development project.

Unrealistic expectations of ASPICE: ASPICE may look good on paper, but many industry insiders see it as a theoretical ideal rather than a practical reality. This is a difficult gap to bridge, especially for smaller suppliers.

Lousy reputation among developers: Critic Volker Bauer sheds light on the supplier situation in his book *The Parasite Principle* (previously mentioned in Chap. 1, [RM-021]). As controversial as his observations may be, industry insiders often confirm them. The author does not explicitly address the topic of ASPICE, but the perception gap between OEMs and suppliers is vast, and this is no different in the case of the ASPICE compliance demands. It is, therefore, not surprising that much of the engineering community appears skeptical of ASPICE.

Unpredictable ASPICE assessment results: Assessment results sometimes vary considerably depending on the choice of assessor. Formal assessment results often widely fluctuate between levels 0 and 2 in the VDA scope.

Expensive ASPICE consulting: ASPICE consulting does not have to be costly, but the reality is often different. The quality of process consulting varies similarly to the results of ASPICE assessments. It cannot be denied that a consultant who works for an assessment and process improvement company may tend to be critical of the quality of his client's ASPICE-relevant processes. The previously mentioned anecdote of a barber and the need for a haircut applies.

Timing conflicts between project and ASPICE: There is a widespread rumor in the industry that the ideal time for an ASPICE assessment can be easily named: never. In our practice, we have often experienced an ASPICE assessment being postponed until the project is on fire. According to the customer contract, this is usually the point at which an ASPICE assessment is overdue. This leads to escalation and frustration on all sides and sometimes increases fluctuation in project teams.

Contractual penalties: ASPICE is often used as a negotiating lever to obtain further concessions from suppliers. These contractual penalties are sometimes quite severe. In addition, there is always the risk of a supplier being blacklisted by the OEMs for lack of formal ASPICE compliance. Such a scenario can be devastating; for smaller suppliers, it can mean the end for good.

This list could go on, but the reader likely already gets the picture. In particular, the topic of traceability deserves further attention. Fine-grained traceability, as defined in ASPICE, not only increases the overall project effort; it becomes increasingly imprecise as the number of traceability levels increases. One of our authors explained this phenomenon at the Berlin Requirements Engineering Symposium in his presentation *SLIMTRACE. How to conquer the traceability monster* [RM-022]. The problem is illustrated in the following Fig. 6.5.

In this hypothetical example, requirement A is "traced" into unit X via the architecture component "Component." At the same time, requirement B is "traced" into unit Y via the same software component.

An alternative path is shown in Fig. 6.6.

In this case, the paths A-Component-X and B-Component-Y apply.

Quantitatively, both alternatives are equivalent. Whether a "component" originally comes from A or B cannot be determined at trace level 3. It is, therefore, unclear which unit covers which requirement because they are figuratively "mixed" via the "component." To achieve clear traceability, a redundant path must be drawn directly from A to Unit X and B to Unit Y. Understandably, such effort is shied away.

In a nutshell, increased formalism does not necessarily lead to objectively better quality. Over the years, ASPICE has become increasingly complicated, costly, and

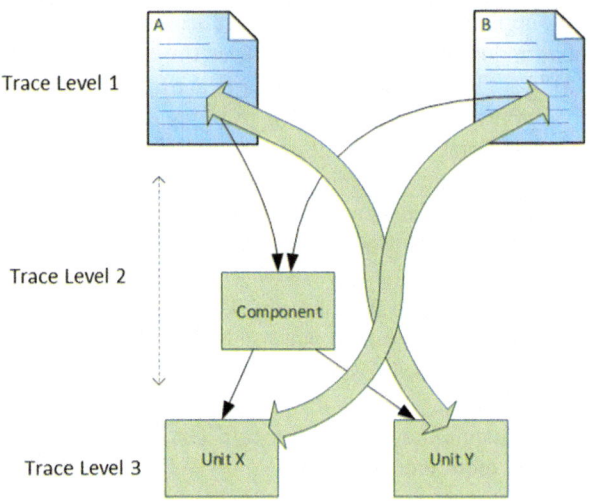

Fig. 6.5 The traceability dilemma (1)

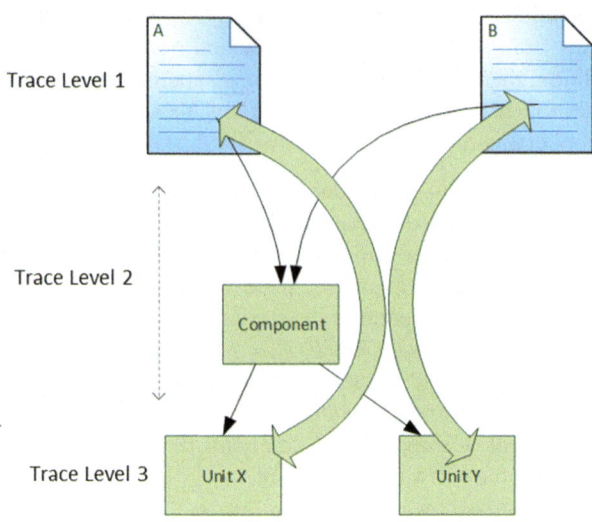

Trace Level 1

Trace Level 2

Trace Level 3

Fig. 6.6 The traceability dilemma (2)

cumbersome. There is a point in a process model's complexity beyond which a further increase in quantity does not improve quality. Some standards, such as ASPICE, have crossed this "break-even" point.

6.5   Advocating for CORE SPICE

The basic idea and motivation behind the development of CORE SPICE is to offer a modern, lean supplement to ASPICE that reflects the current state-of-the-art in the automotive industry. Mind you, we never tire of emphasizing that we are big fans of ASPICE "best practices". It is a beautiful collection of engineering practices that can improve the quality of the resulting development activities when skillfully implemented pragmatically. It is also undisputed that assessment systems such as ASPICE fundamentally promote project quality. ASPICE is arguably the best available process improvement tool in the automotive industry. However, it is also apparent that ASPICE promotes a theoretically perfect ideal, which often remains unattainable.

To illustrate this, we refer to the discredited "waterfall model" (see Fig. 6.7).

According to this oversimplified approach—which is, unfortunately, often assumed to be correct—an idealized process follows the sequence below:

1. Customer requirements are recorded in full. Once complete, they are broken down into system requirements.

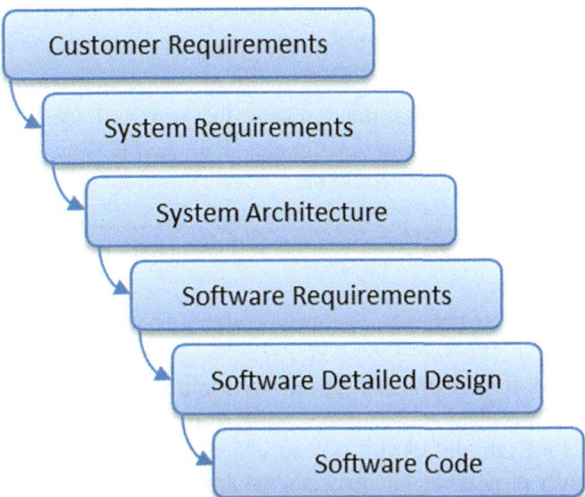

Fig. 6.7 The "pure" waterfall model

2. As soon as the system requirements have been thoroughly checked and agreed upon, the system architect creates a system architectural design based on this result.
3. A software engineer derives the software requirements based on that result.
   … and so on.

This process makes sense in theory. Thus, the customer may agree that such an approach is the way to go. It is, without a doubt, *correct*!

The reasoning is charming. A requirements manager's role is to create a complete and thorough definition of all requirements, not just a rough outline or a partial list. The fundamental expectation *must* be that the analysis results in a flawless, complete set of perfectly defined, well-understood, accurate, and easily maintainable set of requirements until every requirement is fully analyzed and approved. Thus, moving on to the next step—such as creating the system architecture—is strictly off-limits. After all, incomplete requirements cause rework on the architecture level and are therefore undesirable. In this rigid framework, questioning the completeness of this task is nearly impossible, especially during contract negotiations. On paper, allowing any imperfection is simply out of the question. The waterfall model *must* be assumed flawless and logical.

Unfortunately, even quality managers sometimes instinctively or, even worse, deliberately use the assumption of such "waterfall perfection" as a dialectical trick. Their argument typically goes like this:

How do you want to ensure the architecture is correct if the requirements have not yet been fully analyzed and approved?

Since engineers know how development really works—namely, the work is often delivered in an order of tasks that violates the waterfall ideal—the engineers stumble, and that often results in downgrades in an ASPICE assessment.

A logical counter-objection would be:

How can you ensure the project does not fail due to such categorical waterfall imperatives?

To put it more casually, how can we ensure that process quality does not hamper engineering work?

The thoughtless implementation of the waterfall principle has led to countless project disasters. A typical anti-pattern is *analysis paralysis* (for details on analysis paralysis, see, for example, [RM-024]), in which a project that suffers from such impediment never progresses and exceeds the requirements analysis phase.

The "waterfall model" has always been an academic ideal and was never meant to be implemented literally. Leading process experts have been reminding us that an iterative rather than waterfall model should be used for half a century! (See the groundbreaking article by Winston W. Royce on the waterfall model, Managing the Development of Large Software Systems, 1970 [RM-034]).

It is still surprisingly little known that Barry Boehm's V-model is essentially an evolution of the iterative waterfall model. Despite decades of emphasis on iterative processes, this understanding has rarely entered the senior management boardroom. The myth of a sequential waterfall model is probably so persistent because it appears deceptively simple. In contrast, the reality is that every project (especially a software project) is inherently and naturally non-sequential. Specifically, when the pressure in a project's development increases beyond the boiling point—and this is almost always the case in our industry—so-called "schedule compression," as it is called in professional project management (see [RMI-026] for details), is unavoidable. Project managers are forced to drive forward activities at several levels of the development process simultaneously, even at the risk of having to replace some results (known as *rework*).

Another argument for a more pragmatic approach to ASPICE is that ASPICE is not a global standard. While ISO 15504 "SPICE" was an international standard, ASPICE is a German assessment model from VDA (German Automotive Association).

Finally, the agile principle has become increasingly popular since the turn of the millennium. While simple methodologies cannot be fully adopted as they cannot meet the strict expectations of standards such as ISO 26262, there is an urgent need for a pragmatic simplification of development processes. This conflict can be seen across the industry in nearly all the projects we have been involved in.

As a side note, the expectations placed on traditional waterfall-oriented models like ASPICE are increasingly being questioned. Some quality experts might rephrase the previously discussed expectation of a V-model issue as:

How can we ensure that European vehicle manufacturers withstand the growing competition from China?

Ultimately, this is a matter of survival for the European automotive industry (see Chap. 8 for further discussion).

For these reasons, we advocate for CORE SPICE and urge the adoption of realistic process quality targets. This is at the heart of the CORE SPICE agenda.

6.6 Navigating Between Rigidity and Flexibility

CORE SPICE is designed as a lean development concept for software-centric vehicle development. Its primary aim is to support the ISO 26262 standard for functional safety as simply as possible. The aim is for CORE SPICE to come close to the second ASPICE maturity level, establishing itself as a state-of-the-art project coaching model.

We advocate a cooperative approach in project organizations to ensure conformity with CORE SPICE. In contrast with ASPICE, which is based on the principle of control through assessments, conformity in CORE SPICE should mainly be achieved through active project support in safety-relevant projects. The role of the TCC (Team Capability Coach) presented later in this chapter underpins this claim.

CORE SPICE's "process weight" (process complexity) lies between ASPICE and classic agility. Therefore, it is a "hybrid model," as described in [RM-027]. The inventor of the V-model, Barry Boehm, predicted this in 2006 [RM-028].

The "leanness" of CORE SPICE is limited by the requirements of ISO 26262, meaning that CORE SPICE cannot be a variant of SCRUM or similar agile methodologies. Development concepts that support the predictability of automotive projects, such as long-term scheduling up to SOP, must be ensured.

In short, CORE SPICE is as agile as needed but not more agile than necessary.

6.7 CORE SPICE Basic Principles

In this section, we discuss the basic principles of CORE SPICE. First, we introduce the most critical structures of CORE SPICE. We then explain the twelve essential CORE SPICE principles that form the core of this approach. Finally, we discuss the generic practices of CORE SPICE.

6.7.1   CORE SPICE Overview

The CORE SPICE framework rests on these fundamental pillars:

- 12 CORE SPICE Principles.
- 5 CORE SPICE Common Sense Practices.
- CORE SPICE Approaches.
- CORE SPICE Outcomes.
- CORE SPICE Auxiliary Assets.
- ECST (see Sect. 6.14).

The **CORE SPICE Principles** form the foundation of its philosophy, guiding the development strategy for automotive projects.

The ASPICE Process Outcomes inspired the CORE SPICE Outcomes. The idea is based on the observation that goals and results are more important than the analytical decomposition of a project management system into redundant practices, as is the case in ASPICE.

The **Auxiliary Assets** contain miscellaneous elements such as templates, a standard role catalog, checklists, etc.

Figure 6.8 provides an overview of the CORE SPICE structure.

The structure of CORE SPICE is built around a set of project **approaches that encompass** the entire automotive development process. **CORE SPICE Approaches** replace a wealth of traditional ASPICE reference processes.

In CORE SPICE, we intentionally avoid using the term "plan" to prevent potential ambiguity. As previously mentioned, "plan" and "schedule" are often mistakenly used interchangeably in German-speaking regions. Thus, instead of relying on the historically overused term "plan," we adopt the term "approach," which, although still relatively new in the methodology landscape, better aligns with our intent and avoids some pitfalls of traditional terminology.

CORE SPICE should not be understood as a development process in the conventional sense, in which strictly defined sequences of tasks must be executed. Instead, it consists of core concepts, or "approaches," that resemble the mindset of an engineer or project manager. These "core elements" (hence CORE SPICE) must be defined early in a project's lifecycle and accepted by all participants.

Approaches are based on tried and tested standard documents. The project approach, for example, is inspired by the *IEEE Standard for Software Project Management Plans* (ISO/IEC/IEEE 16326).

Approaches are designed as well-commented templates that a project manager can and must complete, review, and use in a project. The template structure may not be neglected or significantly modified; it may be tailored to the project's needs.

It is not essential which tool is used to edit a template. It could be a Microsoft Office document or a Confluence page. The use of tools is defined in the Configuration Management Approach, but CORE SPICE is tool-agnostic, meaning it does not matter which tools are used if they follow the CORE SPICE guideline.

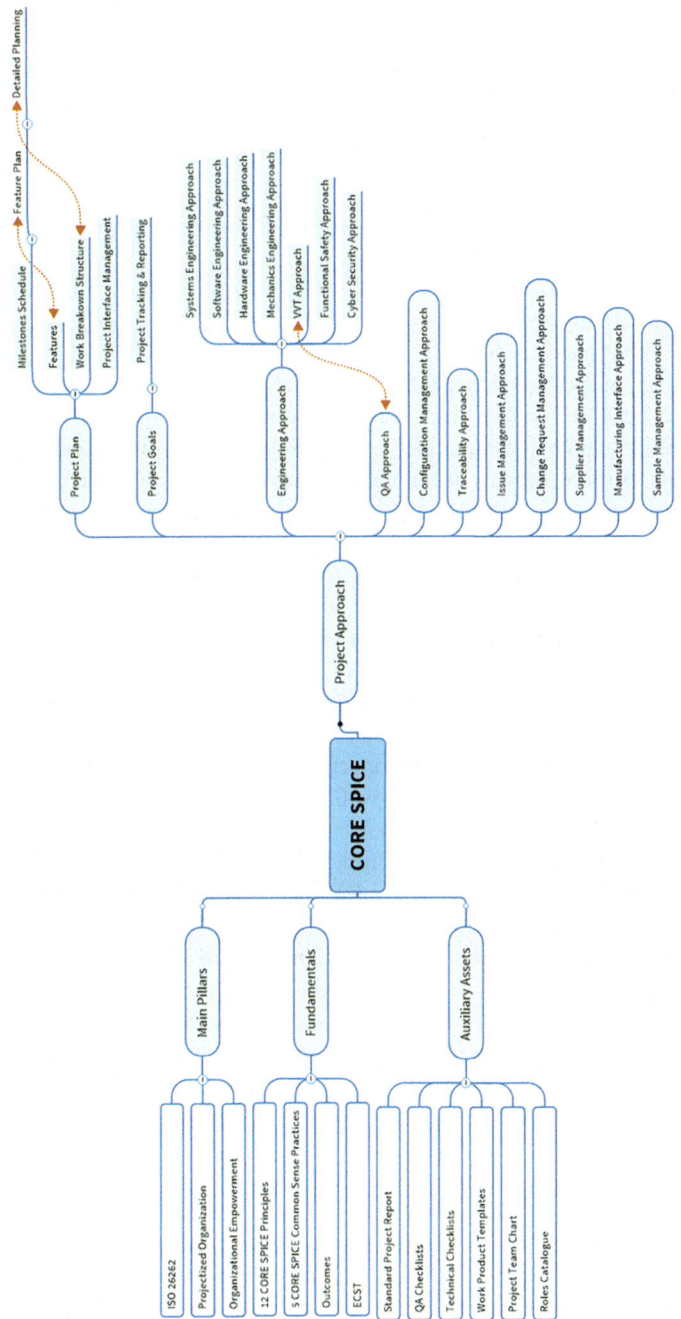

Fig. 6.8 The basic structure of CORE SPICE

6.7.2 The 12 CORE SPICE Principles

CORE SPICE is designed not to replace ASPICE but to enhance the effectiveness of automotive development projects by focusing on practical outcomes within the "Bermuda Triangle" of cost, quality, and time. While ASPICE remains essential for compliance purposes, CORE SPICE aims to reduce the baggage of unnecessary maturity-level discussions.

The CORE SPICE concept is based on the following **basic principles**:

1. **Safety first**: ISO 26262 Functional safety is critical for autonomous driving and increasingly software-based components. Traditional process compliance, on the other hand, is secondary.
2. **Integrity and ethics come before profit**: Disasters like "Dieselgate" did not happen because our QA checklists were too short. Such mishaps occur due to blatant ignorance and notorious "groupthink" in some organizations.
3. **Quality culture instead of formal control**: Quality is not primarily achieved through formal assessments; it emerges from an immersive quality culture consistently prioritizes product quality excellence. A technology-oriented understanding of quality must be seamlessly integrated across all teams, making quality a continuous priority rather than a periodic, formal check.
4. **Approaches instead of sequences of tasks**: CORE SPICE focuses on strategic approaches rather than rigid chains of process steps.
5. **Project over matrix**: CORE SPICE is, first and foremost, project-oriented. A strong project orientation aims to improve efficiency and streamline the project environment so that teams can concentrate on adding value for the customer. For this reason, project roles are well-defined, and the responsibility for each task is always known.
6. **Cooperation instead of confrontation**: Customers must be actively and deeply involved in a project from day one. This helps prevent late surprises and customer escalations.
7. **Risk minimization instead of risk management**: Risks should not be just managed; they must be actively addressed with the goal of risk reduction or elimination. Risk reduction instead of risk management is the motto. For this reason, long-term project planning must always be maintained throughout a project's lifetime.
8. **Ability to deliver instead of final delivery**: Continuous (partial) deliveries to the customer reduce the risk of late delivery.
9. **Engineering skills before consensus**: Teams should be assembled based on skill, and team roles should be defined based on technical abilities. All other factors are secondary. Focusing on fostering a merit culture helps prevent notorious "groupthink."
10. **Responsibility instead of control**: CORE SPICE requires and encourages active participation in problem-solving. The following simple rule applies: if a team member recognizes a need for corrective action, this person remains *personally* responsible for the solution until the issue has been resolved or

the right expert is found who can bring about this solution. Therefore, "the team" cannot be collectively responsible for such measures. Every task needs a concrete assigned person who takes full responsibility for the problem resolution.

11. **Merit instead of bureaucracy**: Less is more. Fundamentally required documents and regulations must be adhered to. Everything else (and everything that can be done) should be automated in the project.

12. **Automated traceability**: To be effective and implemented by the development teams, traceability must be easily calculated. Automating the traceability tasks, at least partially, can achieve this.

The scope of CORE SPICE excludes aspects that can or should be covered by other measures, such as:

– Production (manufacturing), except the PPAP interface.
– Product maintenance (according to SOP).
– Personnel management and HR aspects.
– Contract management.
– Obsolescence management.
– After-market activities.
– IT service management (as described in the ITIL standard).

CORE SPICE is not an "agile ASPICE." However, it does not contradict agile. The following list correlates CORE SPICE with the *Agile Manifesto* [RM-023].

Agile manifesto	12 principles of CORE SPICE
Individuals and interactions are more important than processes and tools	See principle 11
Functioning software is more important than comprehensive documentation	See principle 8
Cooperation with the customer is more important than contract negotiations	See principle 6
Reacting to change is more important than following a plan	See principle 7

The twelve agility principles [RM-024] also do not contradict CORE SPICE. In other words, it is not about inventing a "new agility" or replacing it. However, CORE SPICE principles go beyond the minimalist, agile ideas, as stricter rules must apply in an automotive product development environment. For example, active, long-term project planning and scheduling are essential. OEMs must not be exposed to uncoordinated sub-projects. We are in a business where all three dimensions of the "Bermuda Triangle" are equally critical, and that is what makes our industry so unique and different from large technology groups such as Google, Meta, or Microsoft.

Furthermore, CORE SPICE uses structured "approaches" inspired by conventional plans, such as configuration management plan.

In addition, CORE SPICE defines templates and checklists to help ensure the quality of project work products and facilitate day-to-day project work.

Finally, the role of quality assurance is differently defined. In CORE SPICE, the role of a Team Capability Coach (TCC) is, in a broad sense of this term, a "quality assurance" role essential for successfully using CORE SPICE. This project role is about supplementing the function of a QA manager, in which quality assurance actively participates in projects and, therefore, feels personally responsible for the success of a project. A project-independent role of the QA manager is, of course, still retained. It concentrates on the traditional role, which focuses on a formal review of results about fulfilling the quality criteria of CORE SPICE and related standards.

6.7.3   Common Sense Practices

Every project task is structured consistently in CORE SPICE:

1. **Needs assessment**: What is the need (what needs to be done)?
2. **Capability assessment**: Do we have the resources to meet the demand?
3. **Time and resource plan**: Do we have a time and resource plan? This plan must also be continuously reviewed and corrected.
4. **Determining the status**: What is the project progress status? Are we on track to deliver on time?
5. **Results review**: Was what was initially planned implemented?

The CMMI Generic Practices inspire these Commonsense Practices.
About the individual practices:

1. **Needs assessment:** All project activities must be planned to follow the project objectives defined in the project approach. No resources (including time) may be spent without a formulated need.
2. **Capability assessment:** Good intentions are insufficient; resources must be secured to match the project approach, such as personnel (with skills and hourly quota), licenses, partner commitments, supplier contracts, change requests, agreed specifications, data connections, patents, etc.
3. **Time and resource plan**: Project activities must be planned in the context of the committed resources. This may require a schedule with milestones and resource availability dates.
4. **Status determination**: The progress of each activity must be visible and communicated within the team. Ideally, this is done using an automatic report from a project management system such as JIRA/Confluence. The report, which includes defect curves, resource consumption reports, completed and outstanding work packages, etc., must always be available to the stakeholders defined in the project approach.

5. **Quality check:** When tasks are closed, they must be systematically reviewed. It is, therefore, not enough to close a task ticket or a bug. Usually, the TCC (Team Capability Coach, see Sect. 6.8.5) is responsible for this.

These common practices relate to all work content defined in the configuration management approach. The traceability approach determines the sequence of individual deliveries.

6.8   The Role Model in CORE SPICE

The following discusses the CORE SPICE role model, including role and hierarchy structures. It also presents the project leader and the CORE SPICE-specific TCC roles. This project structure aims to define a robust basis for clear responsibilities and to create efficient reporting and escalation channels.

The Team Capability Coach (TCC) acts as a bridge between engineers and the formal, conventional process quality requirements and actively supports the project work to create a positive working environment for the project team. CORE SPICE represents a hybrid approach, integrating pure agility with the structured rigor of the traditional waterfall methodology.

6.8.1   The Standard Team Chart

In contrast to process assessment models such as ASPICE or CMMI, CORE SPICE predefines roles frequently encountered in the automotive development business.

In CORE SPICE, a role is a project task designated by the Project Lead that is performed consistently throughout the project. In the case of critical roles (such as Project Lead), precisely one role is responsible for the task. Conversely, one person can take on several roles.

Project Team Chart
The Project Lead creates and maintains a Project Team Chart for each project. This defines the project team's roles, which are set out in the role definition. The role holders are integrated into a reporting system. The Project Team Chart is a core document in every CORE SPICE project (see Fig. 6.9).

The vertical relationships between roles might be interpreted as "reporting" lines. Upper role owners support the roles shown below them, and the lower roles report back to those shown above.

The following roles are also required but are not depicted above:

– System Engineer
– Software Engineer
– System Test Engineer

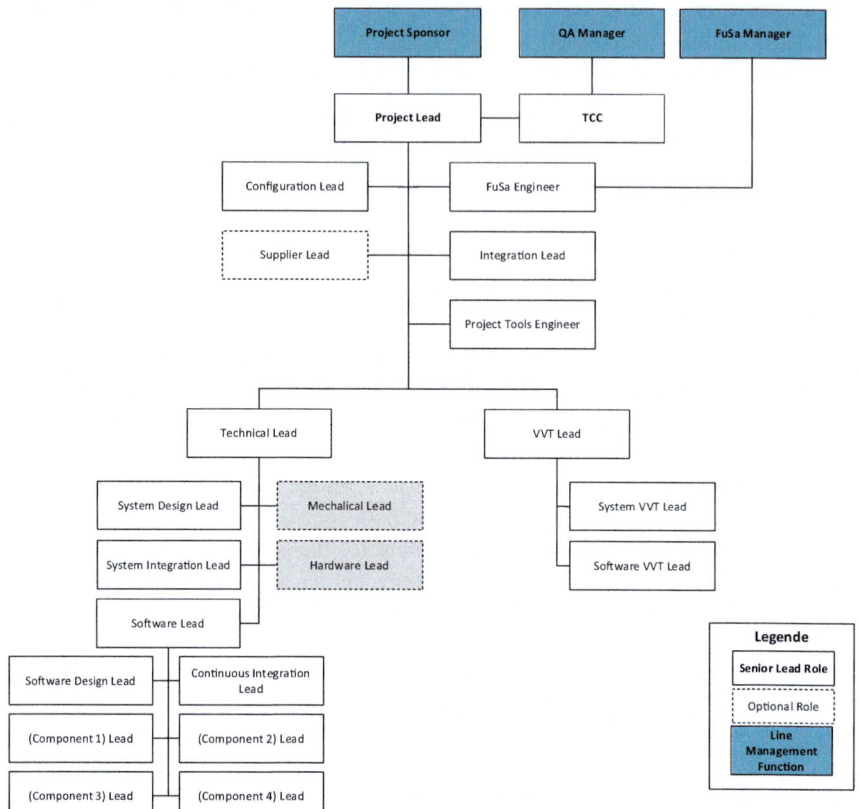

Fig. 6.9 Standard CORE SPICE team chart

– Software Test Engineer

The above roles are assigned to the corresponding lead roles.

Therefore, the project roles in the CORE TEAM are organized in a pyramid. A clearly defined pyramid-shaped team structure creates a clear understanding of roles and is conducive to organizational efficiency. Efficiency and the team structure correlate strongly. In particular, it should be emphasized that—in extreme cases—a self-organized team of a non-trivial size (approx. ten people or more) tends to cause exponential organizational effort. In contrast, a pyramid-shaped team structure has a logarithmic complexity, which is a critical advantage, especially in larger project organizations [RM-029].

Three attributes define each role:

– A specific project role denomination
– Tasks owned by the role
– The required skills

Maintaining a skills matrix aligned with and approved by the HR department is advisable. For example, it could look like this (see Fig. 6.10).

This matrix should always be current and used to plan targeted training and education measures.

Each role's responsibilities in a CORE SPICE project are unambiguously defined. Two roles deserve special attention: Project Lead and Team Capability Coach (TCC). The Project Lead has a singular responsibility for the project. The TCC ensures that quality standards are met and helps the Project Lead minimize risks. These roles are discussed in more detail in the following chapters.

Of course, the definition of each role can be expanded. Role names can also vary, or additional roles can be defined, such as an FMEA Moderator. Roles can also be omitted, such as the System Design Lead role in a software-only project.

It should be emphasized that a role does not necessarily mean 100% capacity utilization. Therefore, the Project Lead must always ask what workload is expected

CORE SPICE SKILL MATRIX

Skill Legend: 1=Competent, 2=Proficient, 3=Expert

#	Role	CORE SPICE	ISO 26262	Specification Methods	UML/SysML Notation	System Design	Software Design	Software Implementation	Hardware Design	Mechanical Design	Test Management	FMEA, FTA Methods	Project Management	Configuration Management	Organizational Talent	Communication Capabilities	Knowledge of the Product Class	Tools: SW Coding Tools and IDEs	Tools: Systems and Networks	Tools: Git Client	Tools: Git Server	Tools: Enterprise Architect	Tools: Microsoft Project	Tools: JIRA	Tools: Software Build Tools	Tools: System Build Tools	Tools: PCB Layout Tools	Tools: CAD	Tools: FMEA/FTA tool
1	Configuration Lead	1					1				1		1	3	2	3				3	3	2	2	2	2	1			
2	Continuous Integration Lead					1	2						3	1	1	1	1			2	2			1	3				
3	FuSa Engineer	2	3	1		2	2	1	2	1	1	3	1		2	3	3		2			1		2					3
4	FuSa Manager	2	3			1	1		1	1		3	1	1	2	3	2							1					1
5	Hardware Lead	2	3	2		2			3	3		1	3	3	3	3		2	1		1	3	1				3	3	3
6	Integration Lead	1			1	1							1	1	1							1		3					1
7	Mechanical Lead	1		2						3		3		3	3	3				1	1	2	1						2
8	Project Lead	3	2	1	1	1	1	1	1	1		1	3	3	3	2		2				3	3						1
9	Project Sponsor	1	1	1								1	3	3	3														
10	Project Tools Engineer	1	1	1	1	1	1		3	1	1		3	1	1		3	2	2	1	2	2	3	1	1	1	1	1	
11	QA Manager	1	3								1				2									1					
12	Software Design Lead	1	3	1	3		3				1	1		3	1		1		3		1								
13	Software Engineer	1	2	2	2	1	2	3	1		1		1		3	3	1	3		3		2	2						
14	Software Lead	1	3	3	3	1	2	2	2		2	3	1	3	3	3		2		3	3	2	1						
15	Software Test Engineer	1	1	1			1				1	3			2	1		1				1	1						
16	Software VVT Lead	2	2	1			1				3	3	1	3	3	3		2		1	2	1							
17	Supplier Lead	2	1	1	1		1				2	1	1	2	1		1		1	2	1								
18	System Design Lead	1	3	2	1		1		2		2	1		3		2			1			1						1	
19	System Engineer	1	1	1	3		1		2				2		2			1			1							1	
20	System Integration Lead	1			1						1			2	1			1											
21	System Test Engineer	1	1	1		1					1	1			1			1				1							
22	Software VVT Lead	2	2	1	2	1	1	2	1		3	1	3	1	3	3	2	1	1	1		2	2	2	1				
23	System VVT Lead	2	2	1							3	1	3	1	3	3	2		3			1	2	1					
24	Team Capability Coach (TCC)	3	2	2	2	2	2	1	1	1		2	1	3	3	2	2	2	2	1	2	2	2	2	1	1	1	1	1
25	Technical Lead	2	3	3	3	3	2	1	2	2	1	3	3	2	3	3	3	1	3	2		3	3	3	1				3
26	VVT Lead	1	2	1		1	1		1	1	3		3	1	3	3		1	1		1	2	1						

Fig. 6.10 Example of a skill matrix

if a role is to be filled effectively. For example, a Project Tools Engineer could be assigned "on call" in a smaller project.

6.8.2   Escalations

One advantage of a hierarchically organized role model is that the escalation strategy is straightforward: every escalation traverses the organization from bottom to top (bottom-up).

An example of an escalation sequence is shown below (see Fig. 6.11).

In this simple example, the escalation starts with the Continuous Integration Lead and, if the reason for the escalation cannot be resolved, ends at the highest level of the escalation ladder: the Project Sponsor.

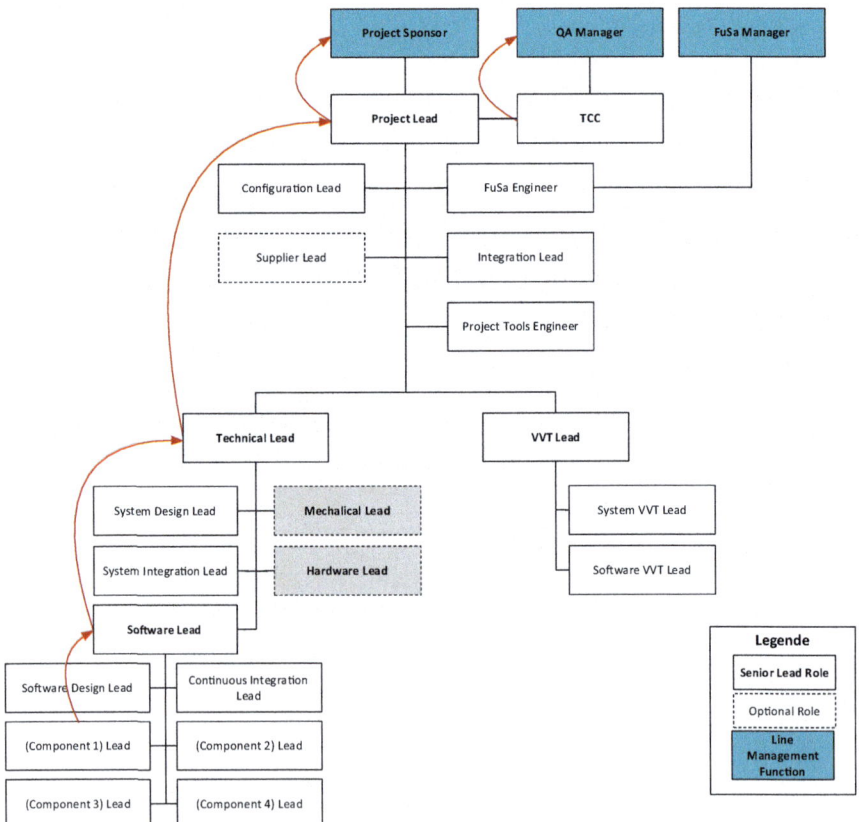

Fig. 6.11 Escalation strategy in CORE SPICE projects

In exceptional cases, the TCC can trigger a fast-track escalation (not shown). This is justified if the situation is critical, for example, in the event of safety-relevant incidents.

6.8.3 The Question of Rights and Team Leadership

A frequent complaint during ASPICE assessments has been that project role-bearers should have formal permission to perform related tasks. Such "findings" stem from confusing responsibilities and role definitions. Such dilemmas are quickly resolved in CORE SPICE. Each role has the right to carry out the activities defined in the role description. If there is any doubt about this, the next highest role in the team chart should be consulted for clarification.

The aim is to guarantee functioning and flexible collaboration in the development team, in which the phenomenon "It's not my job" can no longer occur (see CORE SPICE rule 10).

A similar rule applies to task results and approvals. Approvals are delegated to the next highest level in the team chart.

Each lead has the task of leading their team members, distributing the tasks, collecting the results, and reporting to the higher level of the pyramid. This is not the dreaded "command-and-control" nonsense. The task of each lead is to ensure maximum productivity in the project—no more, no less. In addition, the TCC is active at all levels to ensure that no "silos" form.

While not a "command-and-control" scheme, CORE SPICE is not "agile," either. It is a hybrid approach to project organization that aims to maximize a project team's effectiveness.

6.8.4 CORE SPICE: Project Lead

The Project Lead is always an individual in the project team. Collective lead roles are not permitted in CORE SPICE. A single individual must always be personally responsible for a role task. Of course, one or more "deputies" can be defined for each role, but nominal role holders always bear full responsibility for their direct "deputies."

Tasks of the Project Lead
A Project Lead is responsible for the following tasks:

– Defining, implementing, and maintaining the Project Approach.
– Managing the project scope.
– Assigning and monitoring team roles.
– Motivating and supporting team members to enhance efficiency.
– Organizing, leading, and documenting team meetings.
– Identifying and resolving team conflicts.

- Escalating critical issues to the line organization when necessary.
- Scheduling, organizing, and following up on planning workshops.
- Communicating effectively and promptly with all stakeholders, including line functions and customers.
- Creating, reviewing, and approving project documents such as work breakdown structures and cost estimates.
- Continuously monitoring and adjusting the project schedule as needed.
- Planning, recruiting, and developing project personnel.
- Ensuring adherence to deadlines, milestones, and project goals.
- Managing and controlling the project budget.
- Monitoring project progress and deliveries.
- Ensuring compliance with CORE SPICE requirements and relevant industry standards.
- Maintaining and updating documentation, including meeting minutes and open issues.
- Proactively resolving project-related conflicts and disputes.
- Reporting regularly to the project sponsor and other stakeholders.
- Planning, organizing, and implementing targeted training for the project team.
- Assessing and mitigating project risks and problems.
- Maintaining and updating project knowledge management.
- Building and maintaining relationships with external stakeholders, including suppliers and partners.
- Leading by example to inspire and motivate the team.

This long list of responsibilities suggests that being a Project Lead takes a unique skill set. The skills of a Project Lead include the following:

- Certified in-depth knowledge of project management (e.g., PMP).
- Strong understanding of the automotive industry.
- Ability to develop detailed strategies and plans with timelines, milestones, and resource allocation.
- Competence in identifying potential risks and creating mitigation strategies.
- Familiarity with test and validation methods.
- Expertise in managing expectations and communication across all project participants.
- Proficiency in budget management and cost control.
- Ensuring project quality meets industry standards and specifications.
- Leadership and motivational skills for managing technical teams.
- Understanding of vehicle systems, components, and their functions.
- Knowledge of both V-model and agile processes.
- Familiarity with automotive standards like ISO 26262, Automotive SPICE, and CORE SPICE.
- Awareness of required automotive certifications (e.g., ISO 16949) and AIAG standards (e.g., PPAP).
- Understanding of vehicle electronics, mechanics, and software.

- Relationship management with suppliers and knowledge of the automotive supply chain.
- Familiarity with automotive design tools.
- Ability to communicate complex information to diverse stakeholders.
- Problem-solving skills for unforeseen challenges.
- Knowledge of emissions and environmental regulations.
- Commitment to continuous learning and staying updated on industry trends.
- Ability to inspire and lead with charisma.

Leadership skills are critical for a Project Lead. They must communicate the project mission as effectively and credibly as Steve Jobs did when he unveiled the first iPhone in 2007. Of course, not everyone can have Steve Jobs's charisma, but they should at least become a role model worth striving for.

A passion for technology and relentless conscientiousness are essential traits of an ideal Project Lead. This might seem like an exaggeration, but it's far from just an ideal. Our industry demands more than mere rule-followers and rigid bureaucrats; it requires leaders who thrive on innovation and action. A Project Lead who clings to rigid principles will likely struggle in this dynamic environment.

A Project Lead must also fully internalize a project's meaningfulness and be able to communicate this meaningfulness to the team and other stakeholders fully. This is also a vital leadership quality.

Then, resorting to hyperbole that conveys the importance even better, a Project Lead must not only grasp the true meaning of a project in depth but must also be able to ignite the fire of this vision in the team and all those involved.

After all, almost every car system development project involves new technologies, innovative solutions, prototypes that have never been used in practice, borderline laws of physics, and "mission impossible" novel products in which new technological territory must be explored and conquered. It is an open secret that many such developments are incredibly daring. In our praxis, we have experienced product developments that bordered on madness but were successfully delivered and are now used in hundreds of thousands of vehicles daily. For an excellent Project Lead, such challenges are motivating and an incentive.

The motivation myth
In addition to academic and technical skills and the often-emphasized experience, we are firmly convinced that an almost fanatical determination and motivation are indispensable in our profession. An innovative development project always involves a high level of personal risk. The entire development team, but above all, the core role of the Project Lead, must be willing to take the risk *consciously*.

However, a high level of intrinsic motivation is a fragile asset. One wrong statement or bad day can sometimes be enough to permanently demotivate the entire project team. Nobody can be "motivated"; demotivating, on the other hand, is easy.

Therefore, a Project Lead who is competent in work psychology is indispensable. However, since our technical universities rarely offer courses in leadership, it isn't easy to find good Project Leads. Self-study and reading inspiring books are often all that is available.

An innovative way to address this shortcoming, evident across industries but particularly pronounced in ours, which is so hungry for effective leadership, would be to implement systematic training opportunities. However, extensive leadership training requires time—something Project Leads often lack. This is where coaching and 'training on the job' come into play—see Project Coach [RM-030]. A Project Coach could support Project Leads directly, guiding them and enhancing their project management skills in real time.

The role of the Project Lead must be well-established
At the same time, it is critical that the entire organization (project and the line management) properly understands the role of the Project Lead. In particular, the Project Sponsor must accept and actively support the role of the Project Lead as described above. Effective project management is unrealistic if a Project Lead acts only as a project coordinator. Complex projects under the "Bermuda Triangle" pressure cannot succeed within a conventional matrix organization.

6.8.5   CORE SPICE: TCC (Team Capability Coach)

Quality managers usually accompany traditional car development projects. Their role is often limited to participating in peer reviews, carrying out spot checks (short process quality checks), creating and maintaining a quality management plan, and carrying out or supporting audits and assessments.

In our practice, we realized that more intense and, above all, **continuous** project support from positively **motivating** quality experts is required so that project teams not only survive sporadic assessments but can be actively supported by these experts in their daily project work. We have often observed that teams request and appreciate such support. This is mainly the case when the help of QA experts is not only perceived as an additional burden or annoying means of formal control in stressful times but is systematically offered as qualified participation in QA problem-solving. Such support helps reduce the frustration caused by strict process compliance requirements and maintains team motivation.

The effectiveness of maintaining team motivation while ensuring the quality of the development process is always a challenge. Some OEMs have established dedicated quality support roles for this purpose. For example, a so-called SQIL (Software Quality Improvement Leader) was introduced in the VW Group as a support measure for troubled supplier projects (as already discussed in Chap. 4 on VDA Automotive SPICE). Unfortunately, experience has shown that SQILs often only focus on following specific checklists and strictly controlling extensive quality compliance demands such as KGAS (Group Basic Software Requirements) and a literal interpretation of ASPICE as a "process." Furthermore, SQIL is a costly foreign body within a supplier project because the goal of such measures is not product quality but conformity with rules and formal process compliance as close to the letter as possible—and not much more. SQIL is often just an additional burden in a typically troubled supplier project. In theory, an SQIL is supposed to "help"—but, as the saying goes, with "friends" like these, who needs enemies?

A TCC, on the other hand, is an internal project role responsible for achieving project goals, including maintaining process quality. Rather than merely overseeing formal process quality tasks, a TCC should serve as a driving force within the project, supporting effective processes and creating a close feedback loop between team members and the TCC. As a direct interface to the QA Manager, the TCC also plays a crucial role in "translating" formal quality expectations into practical engineering terms. The TCC acts as a bridge between engineers, the Project Lead, and formal quality standards, fully understanding both "languages"—"engineering" and "quality"—to ensure seamless collaboration and alignment.

The role of a Team Capability Coach was first formulated by one of our authors in 2018 [RM-031]. The name TCC—Team Capability Coach—already suggests that this role embodies more than the usual tasks of a process quality engineer. Also, the term "coach" might sound similar to the role of an agile coach, but this doesn't fully convey the meaning of the TCC's responsibilities. Instead, TCC is a synthesis of team coach, process quality expert, FuSa expert, development engineer, and CORE SPICE expert.

Tasks of the TCC
The tasks of a TCC are defined as follows:

- Supporting the Project Lead with decision-making, escalations, and relevant QA reporting.
- Acting as the interface between customers and suppliers regarding process quality.
- Establishing a CORE SPICE-compliant Project Methodology.
- Implementing CORE SPICE training courses and providing practical coaching.
- Conducting QA audits and assessments, including CORE SPICE peer reviews.
- Developing and maintaining the QA Approach.
- Participating in technical peer reviews and project status meetings.
- Identifying improvement opportunities and leading "lessons learned" sessions.
- Monitoring the Issue Management System (IMS).
- Moderating escalations and assisting with internal and external assessments.
- Ensuring effective communication between the project team and stakeholders.
- Promoting a positive and efficient working environment.
- Collaborating in risk identification and management.
- Introducing continuous improvement initiatives.
- Providing consultation in areas of functional safety (FuSa) and cybersecurity.
- Taking active measures to minimize risk.
- Reporting to the QA Manager.

A TCC must have a wide range of skills. These include:

- Broad engineering knowledge, particularly in software development.
- Expertise in process quality, functional safety (FuSa), cybersecurity, and testing methodologies.
- Strong knowledge of quality standards such as Automotive SPICE, ISO 9001, ISO/TS 16949 and ISTQB.
- Specialized knowledge in the automotive domain.
- Deep understanding of relevant project methodologies.
- Negotiation and persuasive communication skills.
- Experience in working with external assessors.
- High emotional intelligence (EQ).
- Ability to continuously learn and adapt in relevant domains.
- Effective communication skills across all project levels, from testers to senior management.

Ideally, a TCC will have spent ten or more years in automotive product development projects, worked in various project roles themselves—most notably as a project manager, quality manager, developer, and test manager or engineer—and have sufficient seniority and competence to communicate effectively and credibly at all levels of a project. The software competence of a TCC is crucial, as this core competence is expected and required. Software is the future of the automobile. Therefore, a TCC must have experienced the challenges of software development firsthand to support product quality effectively.

The traditional function of a quality manager (or SQIL or similar), frequently perceived as a tormentor, is thus replaced by the TCC, which effectively supports the QA manager role. A TCC is a trusted person in the team and, therefore, an internal team role—so it is not independent. The independent entity that oversees the project in terms of traditional project quality remains the role of the independent QA Manager. The TCC maintains a good relationship with the QA Manager, to whom the TCC reports the quality status. This ensures constructive, goal-oriented, efficient, and quality-oriented teamwork. A TCC is the missing link that solves the "Bermuda Triangle" dilemma.

6.9   CORE SPICE: Approaches

In CORE SPICE, the relevant activities are integrated into the associated approaches. Only fundamental processes are defined therein. The principle is that getting the right results is more important than monitoring the order in which these results (work products) are generated.

Process complexity in car development projects poses a significant safety risk that needs to be addressed. Engineers are not merely cogs in a machine; they have dedicated years of their lives to mastering their craft. They must never be overburdened by having to tick boxes on endless checklists or adhere to rigid process quality specifications that may contribute little to product quality. By enforcing such unnecessary process compliance demands, we risk undermining their ability to focus on what truly matters: designing safe, high-quality vehicles. We must balance necessary process discipline and the freedom for engineers to apply their expertise where it counts most. A demotivated engineer is a high safety risk that is notoriously overlooked. As a lean process approach, CORE SPICE helps to increase product safety and reduce product risk while preserving team motivation.

The misconception of the "waterfall" process was discussed in Sect. 6.5. CORE SPICE's outcome-centric approach effectively addresses this. Instead of rigid, deterministic task chains, CORE SPICE emphasizes *outcomes*, offering developers greater freedom in interpreting and meeting process demands. This approach, detailed in Sect. 6.9.12. CORE SPICE Traceability recognizes that critical processes in requirements and design do not always follow a strict, linear path. Each Project Approach in CORE SPICE focuses on these strategic outcomes and includes only those core outcomes that truly add qualitative value to the project based on industry best practices.

This approach-based structure aims to strike an optimal balance between structured processes and work efficiency, allowing for less formal rigidity and more opportunities for self-motivated engineering work to thrive.

In the following chapters, we briefly describe the individual CORE SPICE approaches. For each approach, we provide templates on our corespice.org page. These annotated templates also represent the quality specifications that must be used for a project to be CORE SPICE-compliant.

6.9.1 Project Approach

A project approach aims to establish a project management system for system development in which pragmatic management of project deadlines, product quality, and budget aspects are agreed upon. Effective communication and risk minimization, team leadership, and stakeholder management are also among the core objectives of approaches, as is the strategy and monitoring of the project scope (scope management).

A project approach is based on a relatively comprehensive and well-annotated template. The key chapters include aspects such as:

– Project purpose.
– Project scope.
– Team organization.
– Project methodology.
– Schedule and feature plans.
– Resource management.
– Risk reduction.
– Outcomes.
– Human resource.

- Stakeholder management.
- Project metrics.
- Project closeout.
- Acronyms and glossary.

The Project Lead defines the project approach at the start of the project on behalf of the Project Sponsor, coordinates it with key stakeholders, and always maintains it. This includes identifying stakeholders, defining internal customer milestones, agreeing on the budget, planning resources, defining a risk minimization strategy, establishing communication channels, defining fundamental project requirements (e.g., ASIL classification), developing procurement planning, etc.

The Project Schedule and the Feature Plan are integral components of the Project Approach. The Featured Plan is defined as the packaging of system functions by the state-of-the-art feature-driven systems engineering procedure (based on the principle of Dr. Granrath et al. [RM-034], derived from the ISO 15288 systems approach, which is a fundamental basis of CORE SPICE). When understood in this way, the FDD principle (feature-driven development) is anchored in the systems thinking approach (systems thinking is discussed in Sect. 6.14) and is suitable for consistent, customer-oriented planning and end-to-end variant management. Scheduling (project schedule) is seamlessly linked to the FDD concept.

CORE SPICE follows the V-model approach around the traceability concept analogous to ASPICE. Alternative lifecycle models, such as agile methodologies, could be more robust from the point of view of CORE SPICE, as the approach model is flexible. Therefore, an FDD-based approach can be used based on object-oriented system development principles, such as INCOSE's OOSEM.

These aspects must be defined in the Project Approach. It should be designed sensibly and pragmatically so it can be handed out to all new project participants and provide an overview of the project content and tasks.

Furthermore, the Project Approach references all other approaches. For example, the Configuration Management Approach should exist as a separate document and not be described redundantly in the Project Approach. Freedom from redundancy in all documents is a high priority in CORE SPICE.

The Project Lead reviews the project approach with the Project Sponsor, the TCC, the QA Manager, and—if the project is FuSa-relevant—the FuSa Manager and all "leads", such as the Technical Lead.

6.9.2   Systems Engineering Approach

The systems engineering approach aims to formulate a holistic view of the product to be developed, in which stakeholder requirements are implemented and system risks are minimized. This requires an interdisciplinary, professional approach that documents and quality-assures the product, considers scalability and sustainability, and creates state-of-the-art architecture.

The Systems Engineering Approach is based on the conventional SEMP (Systems Engineering Management Plan), generally based on the IEEE 1220 standard. However, the Systems Engineering Approach is not a copy of this; it is a modified version that is also primarily suitable for the Software Engineering Approach and is compatible with the IEEE standard for software management plans.

The table of contents of the Systems Engineering Approach comprises the following chapters:

– Introduction
– Systems Team Organization
– Training
– Systems Engineering Process Planning
– Systems Engineering Outcomes
– Major Systems Deliverables and Results
– System Breakdown Structure
– Standards and Procedures
– Resource Allocation
– Constraints
– Systems Specification
– Interface Management
– System Design Verification
– Technical System Reviews
– Systems Schedule
– Systems Status Report
– Systems Risk Reduction Strategy
– Systems Issue Management and Root Cause Analysis
– Systems Engineering Tools
– Related Approaches
– Acronyms and Glossary

No approach is maintained with redundant content. This is why, for example, traceability is not listed, as there is a dedicated Traceability Approach for this in CORE SPICE.

The Technical Lead's project role involves defining the Systems Engineering Approach. As usual in CORE SPICE, the approach is checked by the next highest role in the project, in this case, the Project Lead, and approved by the TCC.

6.9.3   Software Engineering Approach

A software engineering approach aims to ensure the quality, consistency, reusability, scalability, and maintainability of the software developed for a given product.

The Software Engineering Approach is structured similarly to the Systems Engineering Approach, but the content of the individual sections is software-specific. In some respects, however, the structure of the Software Engineering

Approach differs from that of the systems engineering approach, particularly regarding the outcomes.

The table of contents of the Software Engineering Approach comprises the following chapters:

- Introduction
- Software Team Organization
- Training
- Software Engineering Process Planning
- Software Engineering Outcomes
- Major Software Deliverables and Results
- Software Breakdown Structure
- Standards and Procedures
- Resource Allocation
- Constraints
- Software Specification
- Interface Management
- Software Design Verification
- Technical Software Reviews
- Software Schedule
- Software Status Report
- Software Risk Reduction Strategy
- Software Issue Management and Root Cause Analysis
- Software Continuous Integration
- Software Engineering Tools
- Related Approaches
- Acronyms and Glossary

The Software Lead is responsible for defining and maintaining the Software Engineering Approach. The individually defined outcomes include work results such as the approach, scheduling, software-specific milestones, root cause analysis strategy and team organization, software-specific stakeholder analysis, etc.

The Software Engineering Approach is reviewed and approved by at least the Technical Lead and the TCC.

6.9.4 Configuration Management Approach

The Configuration Management Approach is inspired by the IEEE-828 standard for configuration management plans but adapted and extended by sections defined separately for systems and software. This is due to the different requirements regarding methods, tools, and the granularity of the processes in these areas.

- Purpose
- Scope

- Introduction
- Configuration Management Organization
- Training and Resources
- Configuration Management Outcomes
- Configuration Identification
- Configuration Control
- Configuration Status Accounting
- Configuration Audits
- Configuration Management for Interfaces
- Configuration Management Tools and Infrastructure
- Supplier Configuration Management
- Merging and Branching Strategy
- Configuration Management with Variants
- Plan Maintenance
- Acronyms and Glossary

The list of configuration items (CI) is a critical configuration management component. In the simplest case, it is a spreadsheet with all the data objects that must be managed systematically (see Fig. 6.12).

It should be noted that in complex development environments where several projects are managed in a program and different variants must be worked with, managing the configuration items consistently is always challenging.

The Configuration Management Approach and the CI List must constantly be updated, as they are essential for consistent baseline creation. The CI List must

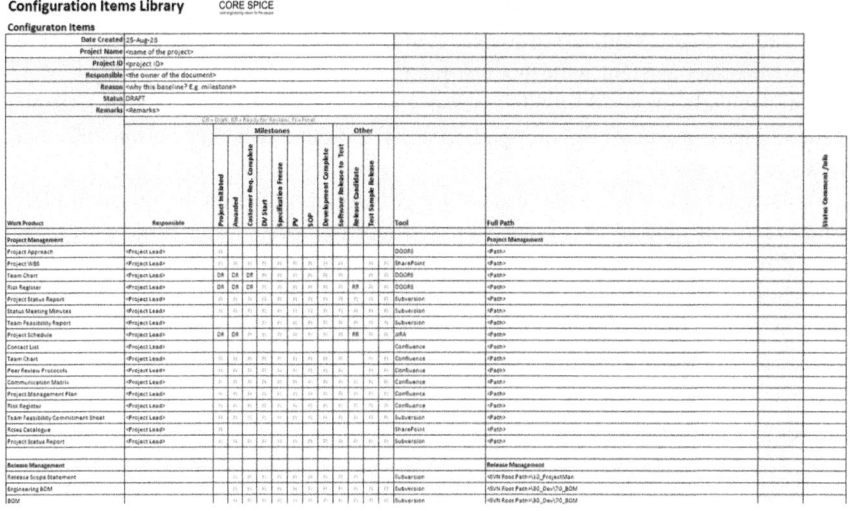

Fig. 6.12 Excerpt from a CI library realized with excel

include the suppliers' configuration items if the project also involves collaboration with external teams (suppliers).

The Configuration Lead is also responsible for release management.

The role of a Configuration Lead is labor-intensive and involves a high level of responsibility. Although it is often overlooked, it is vital as achieving FuSa compliance would be impossible without it.

The Configuration Management Approach is created by the Configuration Lead and reviewed with at least the Technical Lead and the TCC. For projects classified as A to D on the ASIL scale, the FuSa Manager also participates in peer reviews or inspections.

6.9.5   VVT Approach

The VVT Approach (Verification, Validation, Test) is a practical method that ensures a product can be validated, verified, and tested, providing a robust framework for product development.

The VVT Approach is derived from the ISO 29119 standard. A lot of confusion has been caused by the way test plans have been handled in the past. For example, "test plan" and "test strategy" were always seen separately (depending on the consultants' opinions). We have seen cases where the test engineers were forced to write a separate, highly redundant test strategy in addition to the test plan. Such disputes are a demotivating waste of time and mostly simply dialectical, frustrating hair-splitting. However, even hard-core ASPICE dogmatics have realized that such wasteful discussions should finally be resolved. Moving "plans" to Level 2 in ASPICE 4.0 from BPs can be viewed as a somewhat reluctant way to fix this problem. We have also seen overly extensive test plans containing hundreds of pages successfully evaluated in formal ASPICE assessments, while the actual test results were inconsistent with this test plan.

Unfortunately, many ASPICE assessors lack hands-on experience in test management, leaving them ill-equipped to truly evaluate whether testing has been conducted "correctly."

While several test levels are crucial for ensuring product quality, oversimplified phrases like "testing = checking if the product is *correct*; validation = checking if the product was implemented *correctly*" offer little clarity. Such coffee-kitchen catchphrases don't do justice to the complexities of the testing world. These discussions can be frustrating for those with firsthand experience in test management. Instead, we aim to provide a practical overview that clearly outlines the different types of testing.

Validation: "Validation" aims to ensure that all problems known in our experience have been detected.

Verification: The aim is to ensure that all problems derived from the valid product specification can be prevented.

Testing: The aim is to ensure that all test cases derived from the specification have been successfully executed.

In principle, the terms could be visualized as follows (see Fig. 6.13).
These three aspects could be summarized as follows:

– If we cannot test it, we can only verify it.
– If we can neither test nor verify, we can only validate it.

A car component that recognizes traffic signs (TSR—Traffic Sign Recognition) can exemplify this challenge. The specification states, for example, that a TSR shall automatically recognize 95 percent of stop signs. Further requirements may include the rate of false positive detections, weather conditions (rain, hail, snow, fog, etc.), lighting conditions, national regulations, age and condition of a traffic sign, recognition of temporary (provisional) traffic signs, etc. The main challenge is to define realistic verification criteria for such requirements. In some cases, only a few such specific test cases can be included in the test catalog. A testing method could implement a traffic sign displayed on the test stand and then automatically check whether a pattern recognition algorithm works. Verification can also be statistical. For example, the pass/fail criteria could include statistically evaluating the hit rate after several traffic signs.

In contrast, the TSR system can be validated by "trying it out" under real-life conditions. You go for a drive with the customer and systematically or even spontaneously try out how the system behaves in different situations. The validation is rated as "successful" if the customers like what they see.

The VVT Approach covers all these approaches, plus test team organization, resources, test equipment, test demarcation, definition of test levels, training of the testing team, risk considerations, etc.

The following topics are included in the template for the VVT Approach:

– Purpose
– Scope
– Introduction
– Test Policy
– Unit Testing
– Integration Testing

Fig. 6.13 Validation, verification and testing

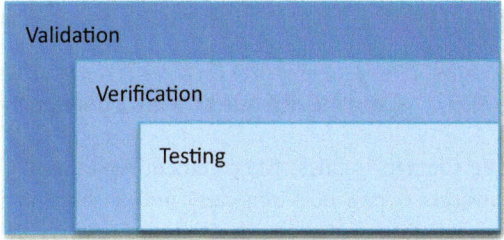

Validation

Verification

Testing

- Qualification Testing
- System Integration Testing
- Acceptance Testing
- DV and PV
- End-of-line Test
- Test Coverage
- Test Cases and Procedures
- Test Data
- Test Environment
- Risk Management
- Test Execution
- Test Reporting
- Outcomes
- Resource Allocation
- Review and Approval
- Acronyms and Glossary.

The detailed specifications for the respective chapters are set out in the VVT Approach template.

The VV Lead also leads the subordinate leads, such as the System VVT Lead and Software VVT Lead, so the testing team acts as an overall testing organization that covers the right side of the V-model and collaborates with the left side (Technical Lead, Software Lead, Mechanical Lead, and Hardware Lead).

It is also vital to prevent a wasteful discussion about the correct definition of test levels. Particularly in the traditional ASPICE world, the distinction between software integration testing and software qualification testing is clear on paper. However, there are often heated discussions about whether a test case is a SWE.6 or SWE.5 test case. A precise definition and differentiation of the test levels in the plan is required to avoid ambiguity. This becomes even more complicated when "hardware-software-interface" (HSI) is involved in the discussion as an additional layer. In these instances, it becomes evident that such disputes do not contribute to product quality or test completeness. Instead, they lead to hair-splitting and may degenerate into dogmatic ASPICE debates. Only an experienced test manager—in our nomenclature, a VVT Lead—with the right instincts can bring the discussion under control and end it without emotional collateral damage.

The VVT Lead is responsible for creating and maintaining the VVT Approach. All leads (Project Lead, Configuration Lead, etc.), the TCC, the FuSa Manager (if the project is FuSa-relevant), and the QA Manager must review it.

6.9.6   Change Request Management Approach

The Change Request Management Approach (CRMA) aims to control the scope (project scope). It is generally not possible to manage a development project in the automotive domain in an agile, reactive way, as the impact of a recklessly decided change can be devastating across the entire spectrum of the project management "Bermuda Triangle" during project implementation. While we may want to provide the customer with "everything," it's equally important to protect them from potential pitfalls by systematically scrutinizing any scope deviations within the already agreed-upon project scope. This is not necessarily about financial constraints. Instead, it concerns risks relating to system safety, system and software architecture stability, and team capacity. Prevention of the notorious "scope creep," where a project's scope expands uncontrollably, should be considered in the CRMA.

The following overview shows the table of contents of a CRMA:

- Purpose
- Scope
- CRM Overview
- Roles and Responsibilities
- Key CRM Outcomes
- Change Request Identification
- Change Request Submission
- Change Request Impact Analysis
- Impact on Schedule
- Cost Impact
- Risk Assessment
- Priority Considerations
- Change Request Approval
- Change Request Implementation Tracking
- Change Request Implementation Confirmation
- Change Request Closure
- Monitoring and Reporting
- Acronyms and Glossary.

An essential premise of CORE SPICE is always to minimize project risk. The CRMA is a critical component of the concept. Therefore, it should be carefully designed and reviewed. In smaller projects (up to approximately 30 people), the Project Lead can create and maintain the CRMA.

Furthermore, a Change Control Board (CCB) is defined for non-trivial projects. The Project Lead appoints this board, comprised of a panel of experts who review the change requests, decide on their implementation, and check the resulting realization.

In larger projects, a separate auxiliary role of a Change Request Lead may be defined (not included in our standard team chart), which can handle the work of handling change requests. Such a role always reports directly to the Project Lead.

The Project Sponsor, QA Manager, and TCC check the CRMA.

6.9.7 QA Approach

The QA Approach aims to guarantee product quality, minimize risk, communicate opportunities and risks in quality assurance, and maintain consistency throughout the project while continuously upholding the CORE SPICE approach. Team motivation in the context of product quality and the proactive efficiency of quality assurance is also essential. A key goal of this approach is to ensure that quality assurance supports the team without becoming an obstacle to project progress.

Quality assurance in CORE SPICE is fundamentally focused on the efficiency and effectiveness of the project organization so the product quality that OEMs and end customers expect can be achieved. Product quality assurance has already been discussed in the VVT Approach, so the QA Approach is limited to checking the standards, monitoring the project methodology, minimizing risks in the project, and QA reporting.

The QA Approach is inspired by the IEEE 730 "Standard for Software Quality Assurance Plans". Other aspects of the QA Approach relate to security-related quality requirements.

- Purpose
- Scope
- Quality Policy and Objectives
- QA Roles and Responsibilities
- Audits, Assessments, and Continuous Improvement
- QA Outcomes
- Related Approaches
- Quality Risk Reduction
- QA Issue Tracking Verification
- QA Competence
- QA Metrics
- Supplier QA
- FuSa and Cybersecurity Interface
- Acronyms and Glossary.

The preference is for the QA Approach to be created and maintained by the TCC. It is reviewed at least by the Project Lead, the QA Manager, and—if the project is safety-relevant—the FuSa Manager.

6.9.8 Functional Safety Approach

The Functional Safety Approach is only created if a project is ISO 26262-relevant, i.e., classified at ASIL levels A to D.

The Functional Safety Approach aims to minimize the functional safety risks in a product that contains electronic and software components. To this end, a consistent safety lifecycle must be defined to ensure product safety to an acceptable degree from the first day of development to obsolescence. A *safety case* (safety justification) must be created at the latest SOP time, which can and must also withstand an audit.

As explained in Chap. 4 on functional safety, a safety-relevant project must meet stringent requirements for formal documentation. For this reason, this role is an integral part of the CORE SPICE team structure.

The table of contents of a Functional Safety Approach comprises the following chapters:

- Purpose
- Scope
- Roles and Responsibilities
- Functional Safety Management
- Safety Goals and Safety Requirements
- Safety Lifecycle
- Overall Safety Lifecycle
- Hardware Development Lifecycle
- Mechanic Development Lifecycle
- Software Development Lifecycle
- Activities and Measures in the Safety Lifecycle
- FuSa Outcomes
- CORE SPICE Compliance
- Relationship with Cybersecurity Management
- Related Approaches and Plans
- Safety Training and Competence
- Measurement and Analysis
- Safety Documentation Management
- Supplier Safety
- DIA
- Safety Audits and Reviews
- Acronyms and Glossary

A Functional Safety Approach mustn't be seen as a one-off "assessment fix" but as a safety-centric mindset. For this reason, the FuSa Engineer, the entire project team, and the Project Lead must have internalized FuSa principles and lived them through the whole project lifecycle. This is a "safety culture," and it must be practiced consistently. "Safety first" is and remains the top priority in the automotive sector.

The FuSa Engineer creates the Functional Safety Approach, which is reviewed at least with the FuSa Manager, the Technical Lead, and the VVT Lead.

6.9.9   Cybersecurity Approach

The Cybersecurity Approach aims to minimize data security and privacy risks in a car system containing electronic and software components.

The table of contents of a Cybersecurity Approach follows below:

– Purpose
– Scope
– Cybersecurity Organization
– Cybersecurity Training
– Cybersecurity Policy and Rules
– Cybersecurity Goals and TARA
– Cybersecurity and Privacy Requirements
– Outcomes
– Cybersecurity Risk Reduction
– Tools
– Dependencies on Other Approaches
– Special Considerations for Threat Scenarios
– Suppliers' Involvement
– Cybersecurity Escalation Approach
– Cybersecurity Reporting
– Acronyms and Glossary.

The Functional Safety Approach and Cybersecurity Approach are closely linked, as hacker attacks can also jeopardize the operational safety of a vehicle. It is enough to imagine a hacker using a cell phone connection in the car (WAN) to inject code through a stack overflow that interferes with the CAN bus via the MOST/IP bridge. A vehicle with a "steer-by-wire" feature moving 130 km per hour on the highway can initiate a sudden 90-degree turn. It's best not to imagine the result. In this case, the system reacted correctly regarding functional safety but incorrectly regarding cybersecurity.

For this reason, cybersecurity is becoming increasingly important, just like functional safety. One day, when cars have Level 5 autonomy, cybersecurity will become paramount. As with safety culture, a cultural shift is indispensable for cybersecurity so that engineers find and possibly resolve cybersecurity threats during development, not only when cars hit public roads.

CORE SPICE currently does not provide a separate role for cybersecurity. Instead, the System Lead and Software Lead are responsible for this topic. The System Lead or the Software Lead designs the Cybersecurity Approach and has the result reviewed at least by the TCC, Technical Lead, FuSa Engineer, Technical Lead, and VVT Lead.

6.9.10 Issue Management Approach

Alongside the Configuration Management Approach, the Issue Management Approach is one without which a non-trivial project cannot function. The Issue Management System (IMS) described therein enables the traceability and activities required to plan and resolve tasks and problems. The IMS defined in the Issue Management Approach is related to the project schedule specified in the Project Approach, the Change Request Management Approach, the Configuration Management Approach, and, in fact, all other approaches.

The Issue Management Approach aims to manage project tasks systematically and comprehensibly. This CORE SPICE approach goes beyond the SUP.9 reference process defined in ASPICE as integrating tasks and problems.

The Issue Management Approach includes:

- Purpose
- Scope
- Responsibilities
- Issue Management Tools
- Outcomes
- Issue Identification
- Issue Classification
- Issue Analysis
- Issue Resolution
- Issue Tracking and Monitoring
- Feedback Mechanism
- Documentation and Record Keeping
- IMS Training and Competence
- IMS Audits
- Issue Communication and Escalations
- IMS Reporting
- Dependencies with Other Approaches
- Acronyms and Glossary.

As defined in the Issue Management Approach, the IMS is a technology-heavy solution used in the Change Request Management Approach. However, other aspects of the Issue Management Approach are also defined, such as the organization of root cause analyses (including topics such as 5D/8D analysis and documentation) and a comprehensive, powerful reporting system in which, for example, the much-noticed *defect curve* and *burndown charts* are implemented.

The Project Lead formulates the Issue Management Approach. The Project Tools Engineer helps implement the technical side of the IMS and realizes the technical aspects of all reports, such as burndown charts and defect curves.

The Project Lead has at least reviewed the Issue Management Approach with the TCC.

6.9.11   Supplier Management Approach

A Supplier Management Approach must be described if the project has hired suppliers to implement parts of the project. In this case, these tasks are regarded as sub-projects. For example, "time-and-material" tasks—small packages that are completed by external experts in small batches on an ongoing basis and remunerated on a time-and-material basis—are not dealt with in the Supplier Management Approach. Such activities are instead included in the regular Project Approach.

The purpose of a supplier approach is to manage existing supply contracts, define the work packages to be delivered, and determine them together with the suppliers to ensure correct acceptance of deliveries.

The following is a table of contents of a Supplier Management Approach:

- Purpose
- Scope
- Supplier Management Responsibilities
- Supplier Compliance Criteria
- Supplier Agreements
- Supplier Performance Monitoring
- Supplier Risk Reduction Strategy
- Communication and Escalation Procedures
- Supplier Development and Training
- Supplier Audits and Assessments
- Documentation
- Review and Continuous Improvement
- Outcomes
- Dependencies with Other Approaches
- Acronyms and Glossary.

The associated role is that of the Supplier Lead, who is a Project Lead "light," as the tasks include many topics that also arise in the domain of project management, such as stakeholder analysis, communication strategy, supplier meetings, supplier escalations, quality requirements for suppliers, acceptance of deliveries, definition of the DIA (Development Interface Agreement), etc. The Supplier Lead must also prevent the dreaded "sandwich syndrome", whereby a supplier makes agreements with the customer that are not in line with the Project Lead's intentions.

The Supplier Lead works closely with the Project Lead and the TCC to minimize the risks associated with external suppliers.

The Supplier Lead reviews the Supplier Management Approach with the Project Lead and the TCC.

6.9.12 Traceability Approach

The Traceability Approach defines how traceability should be implemented, as
ASPICE describes. ASPICE traceability looks as follows (see Fig. 6.14).

This means that at least 19 bidirectional traceability must be managed. This
figure does not include the traceability aspects that change requests must also
account for. They further increase the traceability complexity of ASPICE.

The Traceability Approach defines how the individual traceability paths are
technically implemented, how a coverage report is structured, which tools are
used to implement the connections between the related work products, and who
is responsible for compliance with the traceability rules. In CORE SPICE, traceability evaluation is expected to be mostly automated, so compliance with the
traceability rules makes development more straightforward.

The Traceability Approach includes:

- Purpose
- Scope
- Responsibilities
- Training
- Traceability Objectives
- Outcomes
- Traceability Concept
- Traceability Metrics and Reporting
- Traceability Peer Reviews

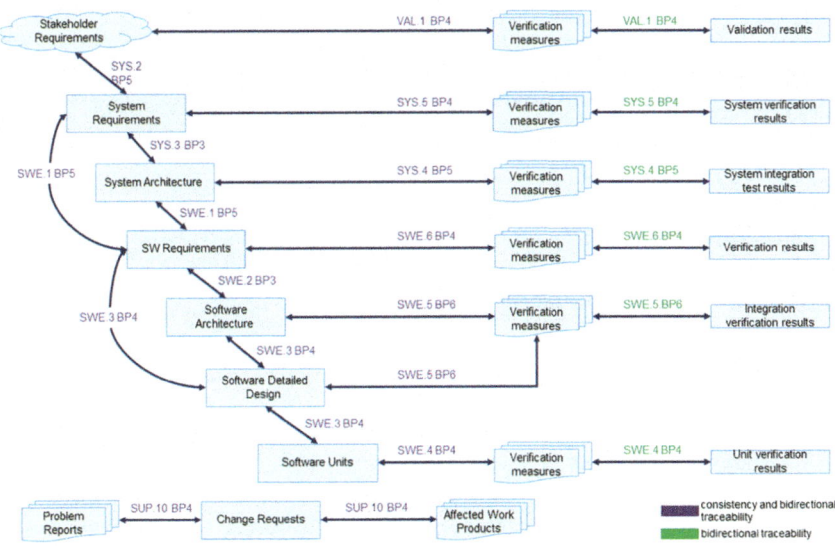

Fig. 6.14 ASPICE traceability (according to PAM/PRM v4.0)

– Acronyms and Glossary

The TCC creates the Traceability Approach, which is reviewed at least with the Project Lead, the Configuration Lead, and the Technical Lead.

6.9.13 Manufacturing Interface Approach

The Manufacturing Interface Approach is particular to CORE SPICE because it is not directly linked to system development. However, our experience shows that teams are often suddenly confronted with the topic of PPAP, for example, and nobody seems to know what PPAP is and why the interface to producing car parts is so important.

PPAP stands for Production Part Approval Process. It is an American standard developed by the AIAG (Automotive Industry Action Group) to ensure the proper delivery quality of vehicle parts. The German VDA has developed a similar standard called PPF (Production and Product Release).

AIAG [RM-032] describes the PPAP process in great detail, so there is no need to replicate that content here. Instead, the purpose of the Manufacturing Interface Approach is to serve as a framework for the PPAP process defined by AIAG (or the PPF process required by the VDA). This framework ensures that the responsibilities for the interface between car development and product delivery in production are clearly understood and function smoothly. The table of contents of the Manufacturing Interface Approach includes the following chapters:

– Purpose
– Scope
– Responsibilities
– Introduction
– Objectives
– PPAP/PPF Summary and Motivation
– PPAP/PPF Documents
– Approval Levels
– PPAP/PPF Procedures
– Communication Rules
– Documentation and Document Retention
– Outcomes
– Training
– Certification and Approval
– Related Approaches
– Acronyms and Glossary.

Many of the PPAP or PPF documents are created during development anyway (such as DFMEA or design records (system design). Others have to be planned additionally, such as a master sample.

The Project Lead is responsible for creating the Manufacturing Interface Approach. It depends on the help of colleagues who work outside the project scope, for example, in the planning and implementing of a production facility. Therefore, these experts, the TCC, the Technical Lead, and the Project Sponsor review the Manufacturing Interface Approach.

6.10   CORE SPICE: Standard Assets

CORE SPICE assists with and requires the existence of certain standard documents, which are listed in Table 6.5.

CORE SPICE offers templates and raw checklists that are outside the scope of this book. These assets are available in the CORE SPICE area at www.coresp ice.org.

6.11   CORE SPICE: Reviews

Peer reviews are at the heart of CORE SPICE. They increase product quality, ensure more effective communication, and, above all, ensure continuous improvement of project processes.

Reviews take place on several levels, see Fig. 6.15.

The following summarizes the most important types of reviews that should be planned and carried out in a CORE SPICE-compliant project.

Table 6.5 CORE SPICE standard assets

Asset	Remark
Standard project status report	CORE SPICE provides an Excel example. The actual status report is to be drafted by the Project Lead
QA checklist	Checklists must be available for each result and created before the start of the project
Technical checklist	Organization-specific, technical checklists that are created as required
Work product templates	Templates for work products that need to be instantiated during development
Outcomes	CORE SPICE specifies the list of outcomes anchored directly in the approaches as specifications. The list is then checked in a CORE SPICE peer review
Project team chart	CORE SPICE provides a raw version of the standard team chart. The Project Lead completes the chart with the TCC and other relevant stakeholders
Roles catalog	CORE SPICE provides a standard role catalog for expansion by the Project Lead

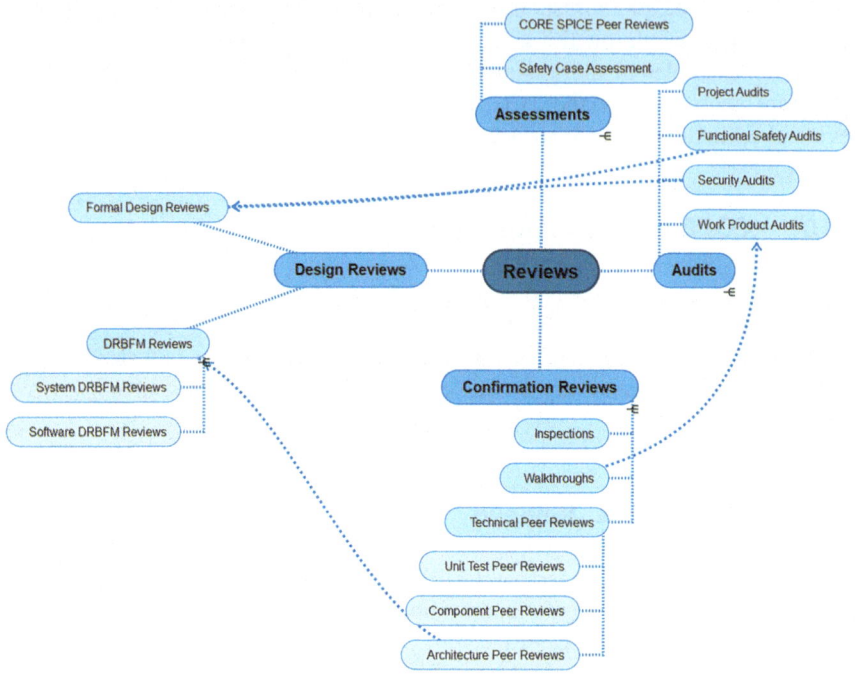

Fig. 6.15 Review types

Confirmation reviews aim to check the correct implementation of the system or software specification.

- **Inspections** are explicitly recommended in FuSa-relevant projects on ASIL level B onwards. They are carried out to identify defects, errors, or non-conformities in work products, regardless of whether they are documents or other work products. Inspections are formal and follow a strictly defined process.
- **Walkthroughs** are less strictly defined than inspections. They aim to find errors and bugs in work products and can replace work product audits.

Technical **peer reviews** are conducted to validate a work product (or group of work products). These reviews are carried out at several levels, for example:

- **Uni test peer reviews are performed before new code is checked in. Component peer reviews** are analogous to this but at the architecture level.
- **Architecture peer reviews** are holistic design reviews typically carried out after creating the decisive elements of architecture (at the system or software level). If a change to the existing architecture is involved, the DRBFM review is a suitable option.

Assessments are systematic evaluations of processes or particularly critical work products.

– **Safety Case Assessment is a formal review of the Safety requirements of the safety cases** in the context of Safety Goals for the formally defined Safety Item Definitions. This review is typically the basis for road approval for a vehicle.

CORE SPICE Reviews

– CORE SPICE reviews are evaluated to assess the suitability of a new supplier (also as part of the "potential analysis") or to confirm that the project is still following the CORE SPICE principles.

Audits check compliance with standards and (possibly legal) regulations.

– **Project audits**, such as the DP (Design Validation), are typically carried out before a significant milestone.
– **Functional safety audits** are carried out in the context of the safety lifecycle, often at the end of a significant project phase (such as DP or PV). Such audits can also be requested by external stakeholders or at the senior management level, such as by the QA Manager.
– **Security audits** are carried out in a similar way to functional safety audits.

Design reviews can take the form of formal design reviews or DRBFM reviews.

– **Formal design reviews** are usually carried out to check the fulfillment of customer requirements. They can be combined with functional safety audits and security audits.
– **DRBFM reviews** are carried out to validate a significant change in the system or software architecture.

We will now explain the DRBFM review type in more detail. DRBFM—a Design Review Based on Failure Mode—is a review method developed in the Japanese automotive industry as a structured design review process. It is used everywhere throughout the Japanese industry. DRBFM is a form of "critical thinking" review (for more on critical thinking, see Sect. 6.14). The focus of DRBFM reviews is on proactively preventing potential problems by analyzing and dealing with possible failure modes during the design phase. The philosophy behind this is that many production problems can be avoided by identifying weaknesses early in the design stage by scrutinizing possible defects or failures, even if they are not explicitly included in the requirements specification. DRBFM affects all technical discipline levels: electronics, mechanics, and software. The difference with other review types lies in the following aspects:

- DRBFM focuses on design changes.
- DRBFM is discussion-based. It is intended to be an expert discussion, not one conducted "top-down" by management. It is pure, 100 percent peer review.
- DRBFM is an integrative approach: the review process should be firmly integrated into the development process.

The DRBFM review procedure is based on the following GD^3:

- **Good design:** Design should be robust, consistent, transparent, and user-friendly (in terms of the end customer, but also in terms of internal design).
- **Good discussion:** a well-prepared expert review that includes crucial experts and is realized through a systematic discussion of design changes, but also open discussions and brainstorming.
- **Good design review:** systematically examining all findings using checklists and objective metrics.

The principles of this process are outlined as follows:

Visualization: Identification and visualization of the design parts discussed in the DRBFM review.

Training: Anyone without DRBFM training should do it immediately.

Preparation: Establishment of an evaluation system:

   Rank A: Error is very likely
   Rank B: Error is possible
   Rank C: Error is unlikely.

A standard DRBFM worksheet lists the components and facilitates functional discussions.

Technical review: Based on the identified components, a team of technical experts systematically reviews all component changes and functions. The ranking above system documents concerns, risks, and potential errors in writing.

The tried-and-tested "5 Whys" principle is systematically applied to (potential) sources of error. This allows all *root causes* to be recorded and understood in detail.

Discussion of the findings: The technical review's results are now discussed with a broader group of experts and stakeholders. The aim is to gain as diverse an insight as possible into the future system.

Recursion: This discussion is often organized over several sessions so all components and expert groups can participate.

Final review: A meeting in which the results and measures are summarized

Open discussion, critical thinking, and creativity are essential in all these sessions.

The integration of DRBFM into the CORE SPICE approach has several advantages:

- The method provides a thorough check (**validation**) that ensures products and systems meet their requirements.
- DRBFM promotes teamwork and communication. In this sense, DRBFM is an excellent **team-building activity**.
- DRBFM is an outstanding method for **minimizing risk**.

Reviews are checks that help us ensure that product quality can be proactively increased. We use different types of reviews to examine the system and software design from all perspectives. It is important to emphasize that reviews should not be treated as "box-ticking exercises." Therefore, reviews cannot and must not be issued as external work packages – such reviews are often pointless. Reviews are the best way to connect diverse brains and find the best solutions to existing and potential challenges that may arise during development. It's about getting better together.

6.12 CORE SPICE Peer Reviews

No complex reviews or assessment rules are defined for CORE SPICE, as with CMMI or ASPICE. The existence of the relevant approaches is a determining factor in assessing a supplier's suitability for the project award. If all relevant approaches are systematically defined under the CORE SPICE template, a supplier is evaluated by presenting the representative outcomes described in the CORE SPICE template. CORE SPICE maturity resembles ASPICE Level 2 in the VDA scope (ASPICE 4.0) without needing a further formal assessment framework.

6.13 CORE SPICE: Further Considerations

Beyond the CORE SPICE aspects outlined earlier, we will now discuss additional key aspects: scalability, application within the OEM environment, and essential tips for successfully implementing CORE SPICE.

Scalability
The question of the scalability of a project management system such as CORE SPICE arises for larger projects. CORE SPICE's scalability can be implemented using the conventional, organically growing team structure. Let us focus on scalability.

Systematic team organization is essential for non-trivial projects. Groups with more than about eight developers often find it difficult to organize themselves

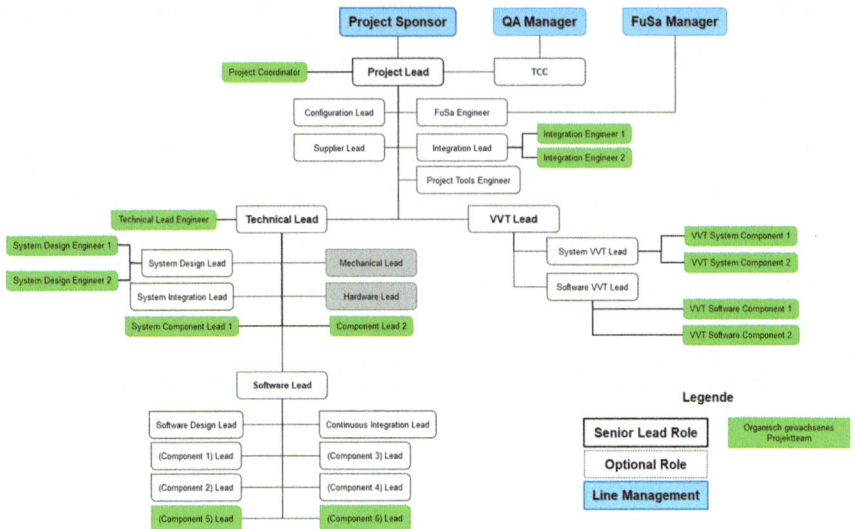

Fig. 6.16 The organically grown project team

[RM-029]. The concept of roles is of central importance for structured team growth. The concept described in Sect. 6.8 is expanded in this case according to the principle of a *fractal*. The structure then grows vertically and horizontally. The horizontal team structure can be effectively managed by segmenting tasks, such as breaking down the overall system into smaller, coherent system elements (components or technical teams). A good practice is to align project teams with the system or software architecture, ensuring communication occurs through the most streamlined functional interfaces rather than traditional departmental boundaries. Meanwhile, vertical growth can be facilitated by introducing specialized, role-specific positions within the team structure. (Fig. 6.16).

The horizontally grown number of components in the hypothetical example above (system and/or software components) is subdivided by Component Leads. These report to the corresponding lead (e.g., Technical Lead). Specialists and additional team members are engineers who report to the corresponding lead roles, such as the System Design Engineer or the System Design Lead. The senior lead roles cannot usually be easily split, but auxiliary roles can be created, such as coordinators, Project Lead Coordinators, or Technical Lead Coordinators.

A project team that has grown in this way can work effectively with up to 200 members. When the team's natural limit is reached, more advanced organizational concepts are required. A more complex program management system usually handles this. However, CORE SPICE focuses on project management systems; program management is a more advanced discipline. Topics such as program and platform management are not covered in CORE SPICE.

OEM Level

Every new vehicle development is inherently a program management task. The complexity of modular systems and platform development goes beyond what can be addressed by project management practices alone.

One of the first tasks in platform development is developing an overall vehicle architecture. CORE SPICE becomes more applicable if the overall program is divided into individual projects. Thanks to the traceability automation required in CORE SPICE and a focus on outcomes (instead of mere activities lined up in activity chains), CORE SPICE provides a sound basis for multi-project management.

CORE SPICE Success Factors

The twelve CORE SPICE principles (see Sect. 6.7.2) are crucial for successfully implementing CORE SPICE. The following success factors should be highlighted (CSP [number] refers to the corresponding CORE SPICE principle):

- **Develop a functioning team structure** (CSP5, CSP9, CSP10) early. A team chart with skill-based roles is fundamental.
- **De-program the team.** This means team members who bring ideologies and process biases from outside must resolve these issues early. The project mission, a sensible pace of work, and the twelve CORE SPICE principles must be communicated and "sold" to the team (buy-in) so that no wasteful discussions need to take place within the team (in the sense of "ASPICE wants ..." or "Agile is different" etc., etc.) (CSP9).
- **Develop the approaches early** and ensure they are subject to meaningful reviews (CSP4). Peer reviews should be applied at all development levels, including systems and software.
- **Train the team in DRBFM review techniques.** Before a DRBFM session, every invited team member should be familiar with and have internalized the review principles.
- **Establish a *constructive* peer review culture** (CSP3, CSP7). Small changes in peer reviews should occur at the software level before checking in to the main branch. The rule is: No peer review—*no merge*. Such rules should be established at all project levels.
- **Ensure that the team has in-depth FuSa knowledge** (CSP1). The FuSa Engineer must work with the team from day one. This is especially crucial when developing the system and software specification. Ensure that the TCC and FuSa Engineer work closely together.
- **Automate everything you can** (CSP11, CSP12): Status and progress reports, software builds, traceability reports, burn-down/burn-up charts, test scripts, baseline, automated peer review reports, etc. All of this must be implemented as early as possible. This is why the Project Tools Engineer's role is so essential.
- **Install an effective Team Capability Coach early in the project** (CSP3, CSP10). This helps ensure quality is integrated at the right time, preventing it from becoming an inconvenient afterthought or an excessive early burden.

- **Involve suppliers early**. Suppliers should work *on* the project and not just be involved in the next interim delivery (CSP6).
- **Ensure that CORE SPICE traceability is in place from day one**. Traceability developed later is often pointless.
- **Deliver incrementally as often as possible**. For software, this could be a daily build (CSP9) frequently shared with the customer. Here, too, automate what you can.
- **Build an efficient test team at an early stage**. This ensures early and consistent delivery capability between releases (CS7, CSP8).
- **Always work to minimize project risks**. This concerns technical and ethical risks, especially product safety-related (CSP2). Systematically use and encourage Team Chart-based escalation.

The twelve CORE SPICE principles are the foundation of successful CORE SPICE implementation. Critical success factors include the early establishment of effective team structures, clear communication among teams and stakeholders, transparent principles, process automation, and consistent minimization of technical and ethical risks. A clear project mission is also essential, led by an inspiring team and capable project leads. Success becomes elusive without a convincingly presented and effectively communicated goal; unclear goals cannot be achieved.

6.14 Effective Critical Systems Thinking—A Broader Perspective

CORE SPICE is a project management approach in the automotive industry that emphasizes the practical aspect of effective car systems development. CORE SPICE presents essential steps towards the next phase of the transition to new technologies, such as electrification and autonomous driving so that those transformative changes can be implemented successfully, effectively, and—perhaps more importantly—as efficiently as possible. However, like the traditional Automotive SPICE, CORE SPICE only represents a part of what a product development organization such as a car manufacturer needs to master. In addition to a well-considered project management system, a holistic view of the organization, understood in terms of the entire company, is required to make short-term and long-term business decisions. At the same time, it is essential not to be distracted by often chaotic and externally induced ideas and ideological trends manifesting as "hypes," such as the rigid waterfall model or business process re-engineering.

It is sometimes difficult to think clearly in the hustle and bustle of our business, but these questions are crucial in a project: What needs to be done *next*? What do we need to do *in the long term*? How do systems like CORE SPICE fit in with the electrification challenge? Are we even doing the right thing? How are we supposed to know? How should we do it? And are we even allowed to do it? These and other questions can and must be asked and answered so that projects and the broader corporate strategy can be defined and successfully implemented.

These questions should be considered in the context of a broadly defined "corporate philosophy". Two popular schools of thought are relevant in our context.

- Critical Thinking
- System Thinking

Critical Thinking is an approach that combines the analysis of facts, evidence, and arguments to form a judgment through rational, skeptical, and unbiased evaluation. It requires a willingness to think independently, with discipline and self-criticism. Effective communication and problem-solving while overcoming egocentrism and bias are critical success factors. Curiosity, open-mindedness, a systematic approach, analytics, and engineering maturity are essential for Critical Thinking. It is an inner dialog that can be used to find better, new solutions to complex and traditional problems.

Systems Thinking is a method of seeing the world as a whole and, simultaneously, as a set of internally interconnected aspects to better understand it. It helps find solutions to complex scenarios and interrelationships so that a more effective solution can be found to complex issues. It is used to analyze problems in complex scenarios.

In our business world, the two concepts are closely interwoven. For example, long-term perspectives postulated in Systems Thinking can only become successful if the fact-based foundation stemming from Critical Thinking is in place. The principle of continuous learning derived from Systems Thinking only makes sense if the principle of Critical Thinking is applied so that the *right* things can be learned, not things that are nonsensical, effectively useless, or even harmful.

Furthermore, our industry must expand both approaches to incorporate a project-oriented mindset. This includes time and planning components, a *sense of urgency*, and a measure of system and software development effectiveness. In addition, software must be integrated into the strategy so it is no longer perceived as a foreign body in the corporate environment. Combining these aspects with Critical Thinking and System Thinking results in **Effective Critical System Thinking (ECST)** [RM-129].

The following basic concepts help explain this approach:

Holistic view: The automotive industry is dynamic and networked. Changes in one area can impact the entire company and its environment beyond the automotive industry.

Analytical problem identification: All activities must consider specific aspects of our industry, from the challenges of supply chain disruption to sustainability. This must be driven by rigorous, data-driven analysis.

Interdisciplinary cooperation: Close cooperation between different teams and disciplines, such as hardware, mechanics, software, marketing, design, HR aspects, environmental aspects, etc., must be considered holistically. The principle of rationality should be upheld to safeguard the entire organization from being swayed by hype and fads.

Continuous learning: This is essential to staying one step ahead of evolving technologies, regulations, and market requirements.

Feedback loops: Continuous feedback in the direct and immediate environment of all teams and operational activities must be cultivated to gain insights, refine processes, and make data-driven decisions.

Innovation and creativity: Creative thinking must be encouraged to find solutions that increase efficiency at all company levels and offer direct and indirect benefits for the company, customers, suppliers, and society.

Strategic balance: There must be a balance between short-term results and long-term planning. Day-to-day decisions must align with the organization's long-term vision while maintaining an openness to new opportunities. Active risk minimization must become integral to all operational and strategic activities.

The ECST approach, a fusion of Systems Thinking, Critical Thinking, and project orientation principles, is a philosophy that the fast-paced automotive world needs to successfully manage the transition to the post-ICE era. It promotes a holistic, adaptive, data-driven approach that considers immediate and long-term impacts while encouraging collaboration and innovation.

CORE SPICE is an ECST approach that embodies the success models of companies like Apple and Tesla. It is holistic, organic, and systemic, guiding organizations through the challenges of the automotive transition while capitalizing on opportunities.

6.15   CORE SPICE: Project Restructuring with CORE SPICE

CORE SPICE is designed as a minimum requirement for safety-related projects focusing on system and software development. It covers ASPICE in the VDA scope up to the second level, including safety-related aspects. This feature can be used to remediate non-performing projects in this category by using the CORE SPICE approaches and the twelve CORE SPICE principles with the ECST mindset.

A "non-performing project" (or "project in distress") is in trouble because it is not delivering the desired results or does not meet customer expectations or requirements due to problems such as budget overruns, schedule delays, unclear responsibilities, poor planning, scope creep (unexpected expansion of the project scope), or lack of promised functionality. Such projects are often too critical to fail because they are usually part of a broader car platform that promises a competitive advantage for the customer (OEM).

Project restructuring with CORE SPICE is usually carried out when a project has encountered severe difficulties, and the customer may even have threatened to take contractual action. Sometimes, such a restructuring can be carried out using conventional measures. These measures address the challenges of managing the "Bermuda Triangle":

- Adjustment of the project scope
- Change in quality expectations
- Stretching the schedule
- Adding or replacing project resources

In our industry, these strategies are only partially applicable, as adjusting all factors simultaneously is usually not feasible. For instance, quality requirements cannot be compromised in a safety–critical project classified ASIL D. However, selectively increasing resources is often unavoidable. Just as an insolvency administrator is essential during insolvency, a restructuring expert is crucial for project realignment.

For project restructuring, CORE SPICE can be used as a blueprint. This includes aspects such as:

- Redesigning the project organization: structuring the project through a narrow interface between the project (Project Lead) and the line (Project Sponsor)
- Adaptation of responsibilities in the project according to a Team Chart structure
- Traceability concept with automation
- Establishing the approaches based on the twelve CORE SPICE principles and five generic CORE SPICE practices
- Setting up regular reviews and status updates to monitor progress
- Establishment of an active risk minimization approach
- Improving communication through effective and secure interaction with all stakeholders along the communication and escalation channels defined in the team chart
- Active involvement of customer management in the restructuring process
- Establishment of short and long-term planning based on the FDD (Feature Driven Development) principle, defined in the Project Approach
- Active project coaching by an experienced Project Coach who can also take on the TCC task
- Automation of processes (e.g., reporting). It is never too late to increase efficiency, especially regarding daily reporting.

Strong leadership is crucial in a distressed project. If the Project Coach takes on the task of restructuring, the project can be managed with the proper sense of time criticality and customer needs.

Inspired by a conventional project restructuring process, CORE SPICE is used to stabilize a distressed project in this way:

Diagnosis: The Project Coach conducts an in-depth analysis of the current project status, identifies weaknesses, and assesses compliance with the CORE SPICE principles.

Planning: Creation of a remediation plan focusing on the CORE SPICE principles and the elimination of identified problems, with the subsequent approval of the project remediation concept by the client

Implementation: The Project Coach leads the implementation of restructuring measures, promotes cooperation, and ensures that the team members are actively involved in solving the problem.

Monitoring and control: Monitoring project progress, adjusting the remediation plan and project priorities as necessary, and ensuring compliance with the CORE SPICE principles.

Completion and handover: The restructuring task is completed either when the project objectives have been achieved, or the project has been successfully restructured to the extent that the Project Coach can hand it over to the Project Lead.

If the handover takes place before completion, the Project Coach will typically remain in contact with the Project Lead to step in and provide advice and support as required.

6.16   Comparing Effort: CORE SPICE and ASPICE

If developers, customers, project managers, and QA engineers find ASPICE to be challenging, it is because the complexity is built into the very foundations of ASPICE. ASPICE is based on more than 200 individual BPs and GPs, plus generic work products, traceability requirements, and the process hierarchy, not to mention further interpretation props and books required to understand the assessor's mindset and to pass an ASPICE assessment as required by OEMs. These elements, such as SWE.1, SWE.6, GP2.1.x and GP 2.2.x (at level 2 of ASPICE), are closely interrelated. If you were to draw a graph for SWE.1, SWE.6, and the associated BPs, the relationship graph would look something like the one shown in Fig. 6.17.

Of course, this is only a symbolic image, as not *all* these relationships are meaningful. However, there is a grain of truth in this exaggerated representation. Experts familiar with complexity theory will instantly recognize that ASPICE is a graph, and graphs are known for their complexity. Indeed, the way ASPICE is implemented and handled – namely as a network of relationships between BPs, GPs, WPs (generic work products), and traceability elements – means ASPICE is a concept that is inherently exponentially complex. In complexity theory, we speak of "NP-completeness". This is the highest possible level of complexity. The human brain is not designed to comprehend exponential complexity, so we routinely underestimate ASPICE complexity, and the associated effort required to implement an ASPICE-compliant process – primarily when implemented literally. By its very nature, it is almost impossible to master this complexity, even with a reasonable amount of effort.

Another weakness of ASPICE lies in its strict completeness requirements for process elements. If even one process falls below 85 percent completion for Level 1 or 50 percent for Level 2, it is considered non-compliant. Failing ASPICE compliance can have far-reaching consequences (see also the discussion in Chap. 4

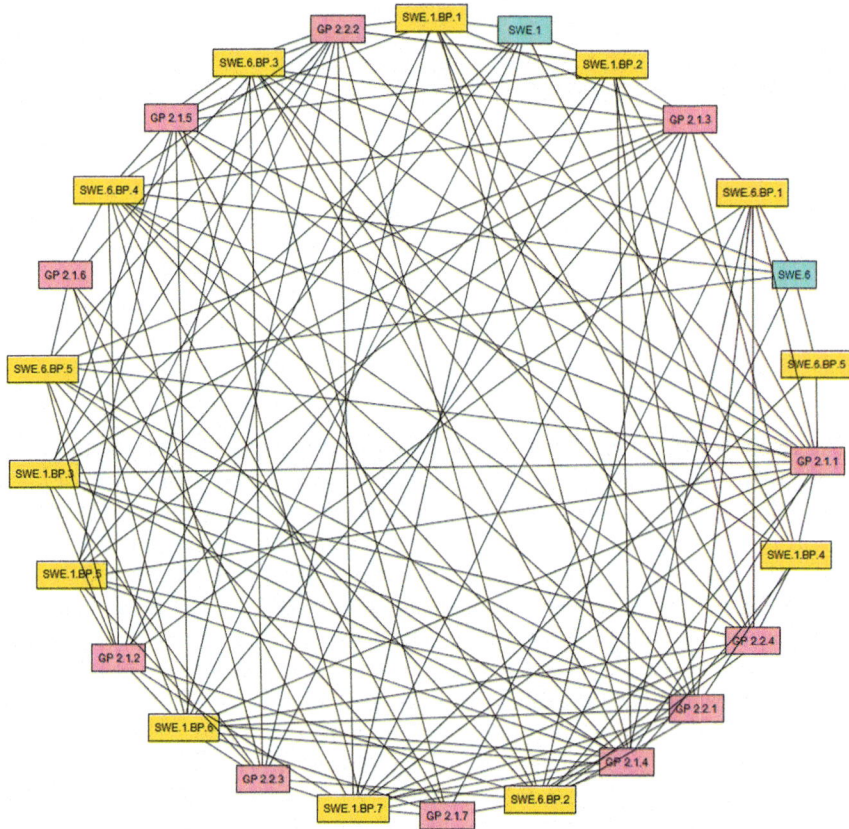

Fig. 6.17 Visualization of ASPICE complexity

on VDA ASPICE). We have already discussed other challenges associated with ASPICE in Sect. 6.4.

Added to this is the effort required to achieve ASPICE maturity level 2. It is not uncommon for the initial assessment to show that the team (or department, depending on the scope), i.e., the organization, is at ASPICE level 0 in most (or even all) reference processes. A process improvement–i.e., a full adaptation of all required reference processes by the development teams–places a heavy financial and time burden on an organization. A sample calculation illustrates this fact (Fig. 6.18).

This estimate is based on a hypothetical approach using the number of generic work products (126 in ASPICE 3.1). The resulting numbers are consistent with our own experience in ASPICE projects. In this example, a development team of approximately 80 people is involved, with a timeframe of 36 months from start to successful completion of a formal ASPICE assessment. An ASPICE consultant (usually required) was not included in the effort. However, this is often not the

ASPICE Assessments					
ASPICE Assessment	Hourly rate	Effort (hours)	Factor	Subtotal	Remarks
Team - Assessment preparation	80	6	16	7.680 €	Inclusive Dry Runs and Assessment Training
Lead Assessor - Assessment preparation	350	16	1	5.600 €	
Lead Assessor - Performing the assessment	350	6	8	16.800 €	6 days x 8 hours
Lead Assessor - Follow-up, report, queries	350	16	1	5.600 €	
Expenses (Hotel, trip)	3.000	1	1	3.000 €	
Team - Assessment participation (3 persons in	80	6	16	7.680 €	3 persons per process in each session
Opening & results presentation (80 develope	80	80	2	12.800 €	Opening: 1hr, final presentation: 1hr
Total for 1 assessment		1	1	59.160 €	
Usually, at least two assessments are performed for a project			2	118.320 €	2 assessment in all
Process improvement from CL0 to CL2					
Process development for ASPICE + trainings	80	120	16	153.600 €	3 weeks x 16 processes (VDA scope in PAM v3.1)
Work Products	80	60	126	604.800 €	Average value, incl. reviews, templates, checklists, variants, etc.
Setting up of a traceability concept	80	80	19	121.600 €	19 traces; coverage, effort per trace: appr. 80 hours
ASPICE - consultant	250	1200	1	300.000 €	Spread over 3 years
Process implementation in the project	80	337,5	16	432.000 €	16 processes (VDA scope in PAM v3.1), reviews, CM, SUP8/9/10, etc., 3 years x 1 hour/week
Spot Checks: QA Lead	80	40	16	51.200 €	1 Spot-check every 3 months; 1 hour preparation + 2 hours for performing + 1 hour post-treatment
Dry runs: team	80	40	16	51.200 €	5 developers/team, 4 hours/Dry run, 1 Dry run before each assessment
Total costs:				1.560.800 €	
ASPICE TOTAL COSTS (Process + Assessment)				1.679.120 €	

Fig. 6.18 Example calculation achieving ASPICE L2 in VDA scope

case; some ASPICE improvement projects have been supported for several years by external consultants.

Development projects are not linear; they are *networks*. Regardless of how simple we try to design and document them, the interconnectivity of teamwork always carries the risk of complexity getting out of hand. Therefore, the "moving parts" time in the system must be reduced. That's why CORE SPICE focuses on *outcomes* instead of basic practices (BPs). Of course, several cross-connections are also required. However, since far fewer outcomes are defined than ASPICE BPs and the approaches have the function of encapsulating the complexity of the process documentation, the focus shifts more from BPs to a few dozen approaches and auxiliary assets such as templates, which also tend to be required for a specific approach. CORE SPICE, therefore, resembles a tree, which implies a logarithmic complexity. The complexity is encapsulated in the approaches, but this does not mean that the project tasks are any less demanding. Formulated somewhat more abstractly, one could say that ASPICE is a syntactic, analytical variant of a project management system, while CORE SPICE corresponds to a semantic equivalence of the task. So, if different processes explicitly require the equivalent of several BPs, in CORE SPICE, the tasks are bundled via the approaches, which gives the development team more freedom of movement.

Similar to ASPICE, a general cost estimate for the regular development of a project management system with CORE SPICE is shown in Fig. 6.19.

Let's go back to our previous example of a team of 80 people. The team is required to work together for one and a half years. This is possible thanks to the

CORE SPICE Efficient Process Optimization					
CORE SPICE Process Coaching	Hourly rat	Effort (hou	Factor	Subtotal	Remarks
Customizing CORE SPICE approaching and work	80,00 €	80	56	358.400,00 €	14 approaches x 56 assets
TCC	180,00 €	2500	1	450.000,00 €	1.5 years
CORE SPICE peer reviews - TCC + Co-TCC	180,00 €	40	2	14.400,00 €	1 week peer review + 1 week report, certificate, summary report
CORE SPICE peer reviews - TCC + Co-TCC	80,00 €	24	5	9.600,00 €	3 persons on average over the team span of the peer reviews
Total				832.400,00 €	

Fig. 6.19 CORE SPICE effort estimation

Table 6.6 CORE SPICE versus ASPICE–decision table

Goal	Recommendation	Context
Regular development	CORE SPICE	Pay attention to customer contract design!
Project renovation	CORE SPICE	The aim is delivery capability
ASPICE preliminary stage	CORE SPICE, then ASPICE	Start-up, smaller organizations, ASPICE optional
Customer demands ASPICE L3	ASPICE	Conventional ASPICE process improvement
Project efficiency	CORE SPICE	The objectives are team efficiency and team motivation, with customer projects potentially relevant in the future

simplified structure of CORE SPICE. Finally, a CORE SPICE peer review occurs with another CORE SPICE expert not involved in the project. As the TCC is setting up the CORE SPICE system with the team right from the start, the usual ASPICE "re-assessments" are no longer necessary.

Fortunately, CORE SPICE is compatible with ASPICE at level 2 of the VDA scope. Safety-relevant aspects are covered so that a CORE SPICE-compliant project can be ISO 26262—to ASIL D-compliant. The Safety Manager is supported in checking the safety case from the start of the CORE SPICE transition until the CORE SPICE TCC checks the safety case.

Ultimately, the decision on the team's path rests with the customer. CORE SPICE allows for more direct and intensive support from the TCC, streamlining processes that are typically burdened with the formalities inherent in ASPICE. Either CORE SPICE or ASPICE should be used depending on the strategic and operational objectives (see Table 6.6).

Finally, the following comments are relevant to avoid unnecessary misunderstandings:

- Projects already on the brink of failure or considered "practically failed" can still be saved if the "Bermuda Triangle" project management is relaxed.
- Project turnaround is a personal responsibility. While "all-hands" meetings, committees, and group decisions may have their place, they are no substitute for individual accountability.
- Typically, teams are familiar with the customer's ASPICE requirement, which has often already been announced and signed in the development contract. At the same time, most team members understand that such a requirement is unrealistic in a project that is already struggling. Attempting to introduce a new process model under these circumstances would likely fail. However, CORE SPICE is not a new "process"; it is a set of rules that should cover the customer requirements through the approaches and ensure ISO 26262 compliance.

Should the customer decide to implement the full ASPICE model later, the CORE SPICE approaches already created are a perfect basis for this purpose.

Project restructuring places a heavy burden on all team members, which can only be endured if there is a realistic chance of a successful project restructuring. The first step is, therefore, to take stock of the current situation. It is rare, but possible, that the Project Coach recognizes after the initial assessment or at the latest in the planning phase that the project cannot be successfully restructured. In such cases, the decision to discontinue CORE SPICE support is made in consultation with the customer.

6.17   Conclusion

CORE SPICE is not a replacement for Automotive SPICE; it is a project management framework that supports ASPICE compliance. However, it can be a first—and, sometimes, sufficient—step towards improving the ASPICE-compliant development process, especially in safety-related projects.

CORE SPICE emphasizes team effectiveness and supports efficiency in these projects. While ASPICE is an *assessment* scheme, CORE SPICE is a project *coaching* concept that pursues the same goal: quality in project management. However, CORE SPICE also promotes the balanced agility and efficient development work required for OEMs and suppliers to compete in the age of rapid system development.

CORE SPICE is a project coaching approach that helps to implement ISO 26262 and cybersecurity requirements. These are the most critical requirements in the automotive industry, where software has become the core business. Also, CORE SPICE offers an alternative for smaller car part suppliers struggling with the growing complexity of assessment models like ASPICE's increasing demands.

CORE SPICE is a critical component in the corporate framework of a vehicle manufacturer. (Fig. 6.20).

Embedded in this industry context, CORE SPICE embodies a project-centered way of thinking and, with its approaches, provides a sound basis for effective project management. The approaches place a strong focus on the project objectives and results, thereby increasing the project's efficiency. Compared to traditional methods such as ASPICE, CORE SPICE enables a faster response to changes and a more targeted allocation of resources.

A defining strength of CORE SPICE is the ongoing project support provided by the Team Capability Coach (TCC). Unlike an external consultant, the TCC is a core team member deeply embedded in the project's daily operations. The TCC's role is crucial in effectively communicating and implementing the CORE SPICE principles, ensuring that ISO 26262 (functional safety) and ISO/SAE 21434 (cybersecurity) are seamlessly integrated into the project. With the TCC's dedicated guidance, teams are equipped to meet quality standards and continually enhance their processes, reducing project risks and encouraging a culture of continuous improvement.

The core responsibility of the TCC is reducing project risks. By integrating a TCC into the project cycle early, risks are continuously and actively minimized. This contributes to greater project confidence and promotes trust among project stakeholders.

CORE SPICE offers project coaching for projects that need more intensive support. This service goes beyond the tasks defined in the TCC role. A Project Coach provides in-depth, project-specific support at the project level and strengthens the interface with stakeholders such as the customer and suppliers. Project Coaches are tasked with guiding projects out of customer escalations, mentoring Project Leads to excel in their roles, and ensuring the successful delivery of the final product to the customer.

While CORE SPICE originated in the automotive industry, its methodology is adaptable across various sectors. Industries such as aerospace, medical devices, rail, and defense—where stringent quality and safety standards are indispensable—can also use CORE SPICE to enhance development effectiveness and ensure safety compliance.

It gets better from now on

CORE SPICE is more than a project management framework; it encourages more risk-aware and consistently results-driven project management. Integrating CORE SPICE into a development project, with ongoing support from a TCC and an optional Project Coach, addresses the challenges of the automotive industry, which can no longer bear the burden of bureaucratic and overly formal traditional development processes in the face of growing pressure to transition towards electrification and software-defined cars.

The path to the future of automotive systems development with CORE SPICE promises increased efficiency and effectiveness and—maybe even more essential in the age of a notorious lack of experts—better talent retention. Project organizations

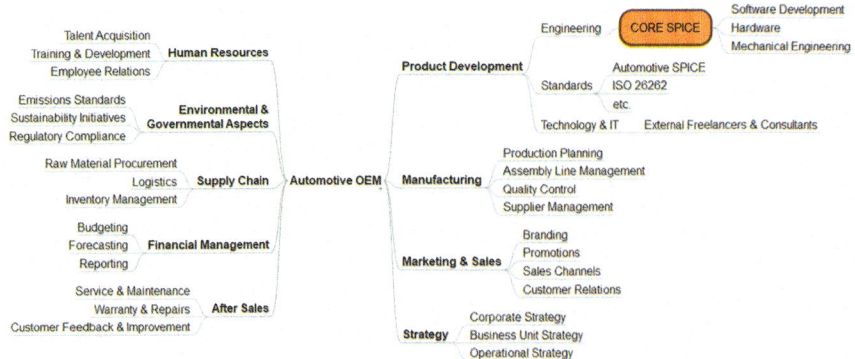

Fig. 6.20 CORE SPICE context

that embrace this philosophy can shorten their development times, reduce costs, and improve the quality and safety of their products by consistently complying with essential standards such as ISO 26262 (functional safety) and ISO 21434 (cybersecurity).

CORE SPICE is not a methodology; it's a game-changer in approaching automotive development. In a world where electrification, autonomous driving, and software-defined cars reinvent the industry, CORE SPICE is the key to unlocking leaner, more agile, fiercely goal-oriented project organizations. The challenges of today's rapidly evolving market demand nothing less. Let us join this transformative approach and take the lead toward a future where innovation is not just a goal but a way of life. The road to success with CORE SPICE is ahead of us. Let's drive it forward.

References

[RM-006] Benefits of CMM-Based, SEI/James Herbs Leb, Anita Carleton, et al, https://resources. sei.cmu.edu/asset_files/technicalreport/1994_005_001_16310.pdf, accessed 09.07.2023

[RM-013] ISO 9000:2005 Quality Management Systems, international ISO Standards

[RM-016] Automotive SPICE Guidelines, https://vda-qmc.de/wp-content/uploads/2023/06/BGB_ A-SPICE-Guidlines-2.0_V8.pdf, accessed on 11.08.2023.

[RM-017] Automotive SPICE™ in Practice: Interpretation Guide for Users and Assessors, Markus Müller/Klaus Hörmann/Lars Dittmann (Author), Jörg Zimmer (Author), ISBN-10?: 9783864903267

[RM-018] ROI of Software Process Improvement, David F. Rico, ISBN 1–932159–24-X

[RM-019] 5th World Congress for Software Quality - Shanghai, China, November 2011, details available from Jürgen Etzkorn, BMW

[RM-020] What is the ROI of Agile vs. Traditional Methods?, David Rico 2008, https://davidfrico. com/rico08g.pdf, accessed 14.08.2023

[RM-021] Das Schmarotzer-Prinzip - Wie deutsche Automobilhersteller ihre Zulieferer ausbeuten, Volker Bauer, ISBN-10 3940445843

[RM-022] SLIMTRACE. How to conquer the traceability monster, Roman Mildner, Berlin Requirements Engineering Symposium 2012

[RM-023] The Agile Manifesto, https://www.agilealliance.org/agile101/the-agile-manifesto, accessed 18-Aug-2023, 18-Aug-2023

[RM-024] The 12 Principles behind the Agile Manifesto, https://www.agilealliance.org/agile101/ 12-principles-behind-the-agile-manifesto, accessed on 18.08.2023

[RM-026] A Guide to the Project Management Body of Knowledge, Sixth Edition, ISBN: 978-1-62825-382-5

[RM-027] Balancing Agility and Discipline, Barry Boehm/Richard Turner, ISBN: 0–321–18612–5

[RM-028] A View of 20th and 21st Century Software Engineering (Key Note ICSE 2006), Barry Boehm, https://isr.uci.edu/icse-06/program/keynotes/boehm.html, accessed 22.08.2023

[RM-029] Effective Adaptive-Agile Team Structure, Roman Mildner, https://projectcrunch.com/ efficient-adaptive-agile-team-structure, accessed 22.08.2023

[RM-030] Project Coach, https://projectcrunch.com/project-coach-adaptive-agile, accessed on 23.08.2023.

[RM-031] The TCC, https://projectcrunch.com/tcc, accessed 23.08.2023.

[RM-032] (PPAP) Production Part Approval Process, AIAG, https://www.aiag.org/quality/automo tive-core-tools/ppap, accessed on 27.08.2023

[RM-034] Managing the Development of Large Software Systems, https://dl.acm.org/doi/pdf/
https://doi.org/10.5555/41765.41801, accessed 31.08.2023.
[RM-129] ECST, ProjectCrunch.com, https://projectcrunch.com/esct, accessed 10/16/2023

Strategic Roadmap: Shaping the Future of the Automotive Business

Abstract

This chapter examines the strategic landscape of the automotive industry, focusing particularly on German manufacturers' responses to electrification challenges. It analyzes the strategies of major players, including Volkswagen, BMW, Mercedes-Benz, and Tesla, using SWOT analysis to evaluate their positions. The chapter discusses critical challenges, including software strategy, battery technology, raw material sourcing, and competition from China. It explores the industry's struggle with software development, talent acquisition, and the need for new approaches to PR and marketing. The chapter emphasizes how traditional manufacturers must adapt their strategies to remain competitive in the rapidly evolving automotive market.

Our industry is at a critical crossroads where the path of tradition intersects with the necessity for bold innovation and a transformative future. Traditional car manufacturers face new global competition, changing geopolitics, and new technologies. In Germany, the diesel strategy once seemed like a safe bet. However, electrification is now forcing an unprecedented change in strategy. This chapter discusses the critical aspects of new automotive business strategies.,[1] focusing mainly on the German OEMs. In addition to providing an overview of each OEM's activities, we will conduct a simplified SWOT analysis (Strengths, Weaknesses, Opportunities, and Threats) [RM-077], a methodology previously employed in Chap. 5 for project management.

This discussion aims to develop and summarize an understanding of the critical strategic factors shaping our industry. We will then consolidate these ideas and proposals in Chap. 8.

---

[1] For the term "business strategy", see [RM-077].

© The Author(s), under exclusive license to Springer Fachmedien Wiesbaden GmbH, part of Springer Nature 2025
R. Mildner et al., *Car IT Reloaded*, https://doi.org/10.1007/978-3-658-47691-5_7

This chapter will focus on the three major German car manufacturers: VW, BMW, and Mercedes.

7.1　Current Strategy Review

Since the invention of the combustion engine, the car industry has experienced a boom that has lasted for decades. American and European, especially German vehicle manufacturers, grew steadily. When Volkswagen AG became the largest vehicle manufacturer in the world in 2015, "Deutschland, Inc." became the "automotive world champion."

This circumstance becomes evident within the context of the European economy. Today, the automotive industry is one of the most important economic sectors in the entire EU. It directly and indirectly supports nearly 14 million jobs, corresponding to 6.1 percent of the workforce in the EU [RM-074]. Seven percent of the gross national product is generated in this industry. In addition, around 30 percent of European R&D budgets flow into the automotive industry.

7.1.1　Strategies of Selected Vehicle Manufacturers

Demand for EVs and plug-in hybrids is surging. In terms of new registrations, car technology has developed as follows (Fig. 7.1).

In 2022, many customers were still buying combustion engines.

Percentage:

- Gasoline: 33%
- Diesel: 18%
- BEV: 18%
- Hybrid: 18%
- Plug-in: 14%

An evaluation from June 2023 shows that German car manufacturers are successfully chasing Tesla despite pessimistic expectations in the German auto press (Fig. 7.2).

Discussing plug-in cars as a major key factor in the long term is not very helpful, as gas vehicles will disappear in the coming years. The long-term future belongs to pure BEVs.

German car manufacturers are meeting this challenge with different EV strategies, although they also have some similarities. In the following, we examine the strategies of the three major German car manufacturers, plus the Tesla strategy for comparison, and then summarize the similarities in the chap. 7.1.1.5.

7.1.1.1　Volkswagen

Volkswagen AG stands out for its unique brand diversity:

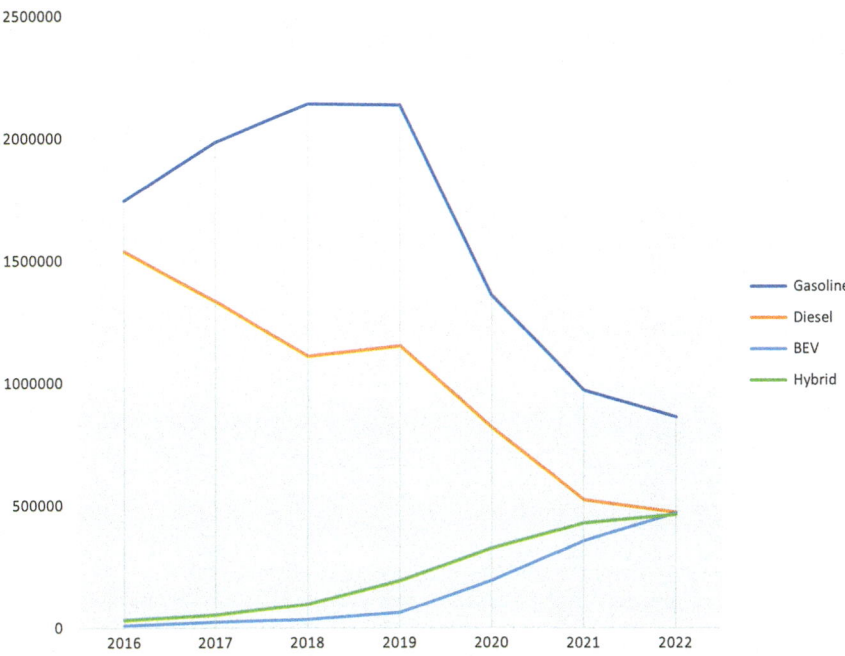

Fig. 7.1 Registrations 2016–2022 (data source: KBA)

- Audi (acquired in 1965 – at that time, the company was called "Auto Union")
- SEAT (acquired 1986)
- Skoda (acquired 1991)
- Bentley (acquired 1998)
- Bugatti (acquired 1998)
- Lamborghini (acquired 1998)
- Porsche (finally acquired in 2012)
- Ducati (taken over 2012)

Further majority shareholdings: MAN (2012) and Scania (2014).

Many models accompany this diversity. Counting all models and their variants, including engine options, results in several hundred, excluding discontinued models like the Phaeton. Multiplying this by the number of different equipment variants, paint finishes, packages, and options returns a five-digit figure.

The Volkswagen Group is also experimenting with innovative brand initiatives such as Cupra (founded in 2018), a SEAT spin-off in Spain, which offers niche opportunities as an unusual alternative to high-priced market segments such as BMW or Mercedes.

In recent years, Volkswagen has developed and implemented several strategies that build on one another:

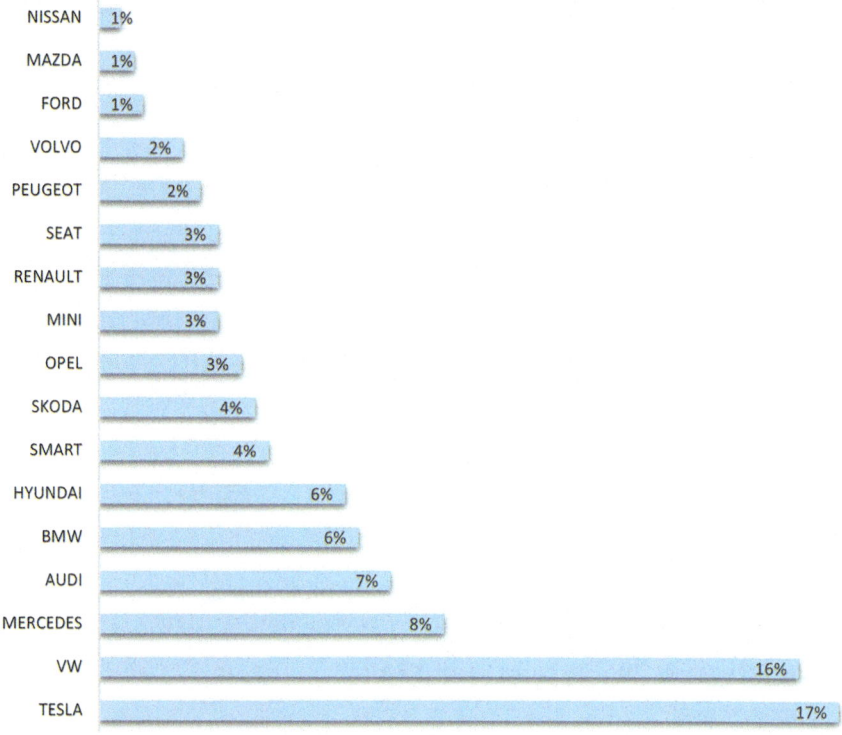

Fig. 7.2 EV registrations January-June 2023 (data source: KBA)

- 2016: **TOGETHER 2025**—A comprehensive strategy with an emphasis on sustainable mobility
- March 2021: **ACCELERATE**—Accelerating the transition to e-mobility. Volkswagen's vehicles will be 70 percent electric in Europe and 50 percent electric in China and the USA by 2030.
- July 2021: **NEW AUTO**—This is a comprehensive strategy for the entire Group. The time horizon is the year 2030.
- May 2023: **ACCELERATE FORWARD**—Limited to the core VW brand. The goal is to achieve a 6.5 percent return on sales for the core brand (currently approx. 3 percent). The aim is to achieve savings of 10 billion euros, which will be invested in electrification.

The **NEW AUTO strategy** remains the Group's leading strategy. Its twelve pillars were published in September 2021 [RM-107] and are briefly described below:

1. **Mechatronics–Backbone and Scalable Systems Platform (SSP)**: Enables maximum synergy effects, reduced investment costs and regular technology updates
2. **Software—CARIAD: One E3 2.0 Platform and Autonomous Driving (AD) Stack**: Development of software platforms for connectivity and automated driving
3. **Battery—Cell and Battery Strategy**: Complete value chain, including battery management, cell production and recycling
4. **Charging—Charging and Energy Services**: Comprehensive charging infrastructure
5. **Mobility Solutions:** Future-oriented mobility solutions taking into account the opportunities associated with autonomous driving
6. **ESG, Decarbonization and Integrity**: Sustainability, decarbonization, integrity and responsibility towards employees and society
7. **Business Model 2.0**: Integration of services and integration of brands, customers, retailers and markets
8. **North America (NAR) region**: Focus on the North American region, including e-mobility and the market launch of attractive EVs
9. **China region**: Expanding market share in the EV segment and securing market share in the ICE segment
10. **Group Steering Model**: Accelerated decision-making to achieve the objectives of the Group strategy
11. **People & Transformation**: Recruiting top talent and supporting employees with extensive training to manage the transition to electrification
12. **Financing the transformation**: Extensive investments in electrification and e-mobility with a focus on costs and efficiency to secure financing.

The initiatives can be interpreted as a four-pillar strategy Fig. 7.3.

The Trinity Project, which combines the new platform, an optimized supply chain, and more efficient production approaches, expresses this strategy. As an extraordinary measure, CARIAD was founded as a separate entity to bundle software-related activities. The German car industry plans a massive investment of 180 billion euros in electrification over the next five years.

Volkswagen is focused on electrifying its current lineup and modernizing traditional, older models. For example, the VW Scout, which was once popular in the USA, is set to roll off the production line as a new EV in 2026. At the IAA 2023, Volkswagen consistently announced that it would build on the traditional design of the Polo and Golf [RM-110]. The strategy appears to make perfect sense, as the average age of a German car buyer has recently risen to 53 [RM-111]. This initiative is expected to appeal to the tastes of more conservative customers.

Global battery production facilities will be established, and Volkswagen plans to build battery factories in North America. Volkswagen's approach to battery technology is unified cell technology, which enables efficient battery production. In addition, efforts are being made to secure the raw materials required for EVs, including in Germany, Australia, and China.

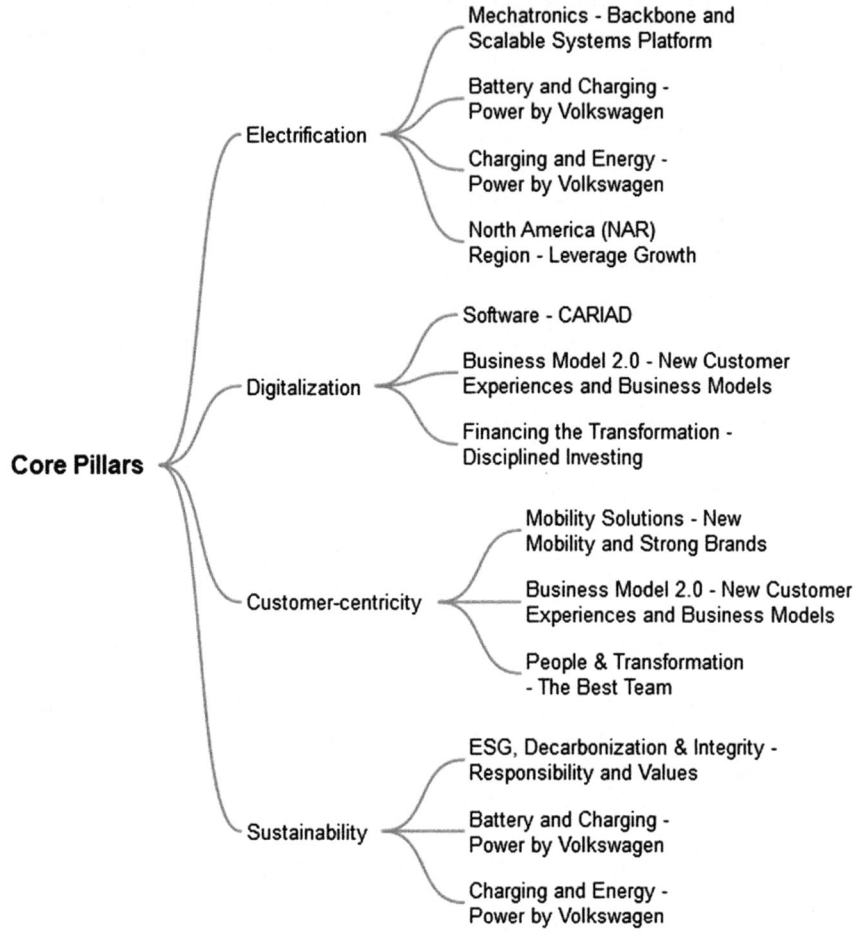

Fig. 7.3 The new auto pillars (own interpretation)

At the same time, Volkswagen has invested 300 million dollars to work with QuantumScape on solid-state batteries. Bill Gates has also invested in QuantumScape.

In the field of autonomous driving, Volkswagen has expanded its partnership with Mobileye, which began in 2016 after the cooperation with Argo failed [RM-121].

The CARIAD software division plans investments in the E^3 platform 1.2 and the future 2.0 vehicle platform. The Audi Q6 e-tron initially previews the new software generation E^3 1.2. The *Software Defined Vehicle Hub* will be launched shortly and will drive the development of the future software architecture E^3 2.0. One of CARIAD's aims is to simplify and accelerate software development processes [RM-109].

Table 7.1 Volkswagen—SWOT

Strengths	Weaknesses
High brand awareness	Damaged PR ("Dieselgate")
Stable finances	Low market share in the USA
Broad portfolio	Relatively high price level
Diversified brands	Complex corporate structure
Diversified production sites	Software strategy is still being developed
Enormous R&D budget	Defensive in the area of autonomous driving
	Low profitability per employee
	Inflexibility due to VW law
	Dependence on global suppliers
	High barriers for suppliers
Opportunities	**Threats (risks)**
Increase in market share in the EV segment	Intense competition (China, Tesla)
Rising price of fossil fuels	Further fines, if necessary
Acquisitions in the technology sector	Increasing burden due to new laws
Strong partnerships in the service sector	Recession in Germany and worldwide
Autonomous driving	Ageing customer base in Germany vs. globally
Improvements in the image of the eco segment	Rising debt burden
Government subsidies	Rising cost of raw materials
Expansion in China and North America	Euro-dollar currency fluctuations
	War in Europe

VW is also working on an improved electric drivetrain. This will be followed in 2025 by the improved modular electric drive matrix MEB +, which will have an extended range, shorter charging times, and new models with entry-level prices below 25,000 euros. The plan is to present eleven new electric vehicles by 2027.

A strategic SWOT matrix rounds off the overview of the VW strategy: Table 7.1.

In summary, Volkswagen is pursuing an intensive, multi-phase strategy to accelerate e-mobility and digital transformation. It has made significant investments in global electrification, innovative battery technology, software development, and the introduction of new electric vehicles. Volkswagen's priorities are increasing efficiency and strengthening its customer image as a traditional brand.

7.1.1.2 BMW

"Branding" is vital to BMW as a brand with a powerful emotional emphasis and is therefore carefully cultivated. The BMW, Rolls-Royce, and MINI brands enjoy high recognition in Germany and globally.

As with other German OEMs, electrification has been an abstract idea for decades. Initially, BMW favored the hydrogen cell as the environmentally friendly technology of the then-distant future, even though it was well-known that lithium-ion batteries are more efficient than hydrogen cells (see the efficiency discussion in Chap. 1). Currently, fuel cell technology is seen as a niche market with an unclear future in the passenger vehicle sector. However, buses and trucks might still adopt this technology down the line. In the meantime, BMW has now jumped on the lithium-ion bandwagon.

In 2012, even before Volkswagen and Mercedes, BMW presented the i3—the first BEV. BMW can, therefore, boast of being a pioneer in the German EV market. However, the i3 did not fit the BMW image, which is known worldwide as a sleek, fast luxury brand. The MINI brand, bought in 1994, may have been better for BMW's first EV model, yet the company only introduced the first MINI BEV in 2019. Despite the i3's brand alignment challenges, BMW has continued refining its EV market approach. This evolution is reflected in the BMW marketing strategy, which is based on the following cornerstones ([RM-112]):

1. **Focus on innovation**: BMW's marketing strategy emphasizes innovation and cutting-edge technology.
2. **Individualization**: BMW offers its customers individualization options.
3. **Strong emotional bond**: BMW nurtures performance and sleek car design as part of a modern lifestyle.
4. **Partnerships**: For example, in the field of battery technology (Envision AESC) or in the style/fashion sector with Louis Vuitton.
5. **Performance**: BMW vehicles are performance-oriented. Speed, strength, and agility are essential aspects of the brand.
6. **Focus on social media**: Appearances on social media platforms such as LinkedIn, Instagram, Facebook, YouTube, and TikTok cultivate customer loyalty and maintain customer contact.
7. **Exceptional customer service**: Exceptional customer service is essential to BMW's marketing strategy.
8. **Artificial intelligence (AI)**: BMW uses AI to identify customer preferences.

The prefix "i" (for "innovation") is used consistently at BMW in a design-oriented manner. Vehicle types such as i3, i4, i7, and iX are examples of this consistent design language. This consistency also extends to areas of development and production, where the "i" prefixes are also integrated:

- iFactory (production strategy)
- iDrive (infotainment system)
- iPerformance (plug-in cars)
- iNEXT (autonomous driving)
- iVision (concept cars)
- iVision Circular (an example of a concept car)

BMW emphasizes the elusive strategy element called "The BMW "New Class,"" which summarizes the electrification efforts intended to make ecological sense and still meet customers' high expectations. The "Circularity" principle emphasizes the reusability of all components.

BMW recently announced a revised BMW Operating System (BMW OS), an integrated platform that maps the entire vehicle. Until now, BMW has used the free Linux operating system as the basis for vehicle software. In the future,

Android (OOAS—Android Automotive OS) will also be integrated into the operating system, which, since 2018, can be updated via OTA (over-the-air), similar to Tesla.

The new interior design, iDrive, will include panoramic displays, 3D head-up displays, central display design, and more.

The iFactory follows BMW's plan to make its entire production environmentally friendly and efficient, following the "circularity" principle. The slogans "Lean, Green, Digital" indicate this strategy. Artificial intelligence is being researched and applied in manufacturing, logistics, development, and quality assurance. At the same time, BMW is struggling with high production costs per vehicle, which is why the Munich-based company decided to achieve a cost reduction of 25 percent per vehicle in 2021 to secure BMW's long-term competitiveness [RM-117].

Efforts in the field of autonomous driving and connected cars are bundled in the iNEXT initiative (also known as "iX"), under the acronym D-ACES (Design, Autonomous Driving, Connectivity, Electromobility and Services). BMW previously worked with Mobileye on developing autonomy but has not pursued this partnership since around 2021. Instead, BMW will rely on Qualcomm [RM-122] in the future. The most advanced models in this context are the i5 and i7, which offer a "Level 2 + " hands-free feature up to 130 km per hour.

Production of "New Class" vehicles based on the iFactory principle will begin in 2025 with a car the size of the current 3 Series [RM-115].

BMW is now on course to push the electrification strategy seriously, but the company does not want to commit to a date when it will no longer produce ICEs. Environmental activists are, therefore, often skeptical as to whether BMW is taking the Net Zero strategy to heart. However, this is a logical conclusion drawn from their consistent orientation towards the premium customer segment. Nevertheless, BMW is launching EVs in both the MINI and the high-end segment. After all, according to BMW, all factories are powered by green electricity [RM-114].

BMW is producing lithium-ion batteries in the USA in partnership with AESC, a Nissan and NEC joint venture. BMW currently has no plans to develop battery cells itself; instead, it is working with partners. In addition to the USA, BMW produces batteries at several locations in Germany and China (for the "New Class" vehicles).

BMW is working on further developing the new solid-state batteries as the next, more efficient battery technology (Table 7.2).

BMW has established itself in the premium segment. With initiatives like the "New Class" strategy and the iFactory concept, BMW is committed to electrification. Even though the Group initially favored the hydrogen cell as the future, it has changed its strategic direction towards lithium-ion battery technology. Further development of the BMW Operating System and partnerships in the battery sector are the cornerstones of BMW's electrification strategy.

Table 7.2 BMW—SWOT

Strengths	Weaknesses
Strong global brand presence	High customer prices
Successful global partnerships	High production costs per vehicle
Highly innovative quality products	Dependence on global suppliers
Geographically diverse production	
Well-defined long-term strategy	
Strong finances	
Roadmap for autonomous driving	
Attractive employer	
Opportunities	**Threats (risks)**
Fast-moving EV market	Increasing competition (Tesla, China)
Expansion in North America and China	Recession
New types of digital services	Increasing number of legal requirements
Increasing production efficiency	Euro-dollar currency fluctuations
New technologies (e.g., solid state)	War in Europe

7.1.1.3 Mercedes-Benz Group

Mercedes-Benz Group AG, known as Daimler-Benz AG, DaimlerChrysler AG, and Daimler AG until 2022, is one of the world's most traditional car manufacturers. Mercedes is a vehicle brand with a leading global position in the luxury segment. The popularity of this brand extends far beyond the automobile business. Mercedes, for example, is by far the most sung-about car brand in pop music (far ahead of Cadillac), which testifies to the popularity of the Mercedes brand [RM-118]. Mercedes is a cornerstone in the consumer market, embodying automotive luxury and pop culture icon status.

After merging all divisions into one Mercedes-Benz Group, the commercial vehicles and passenger car divisions have been integrated into the overall strategy. Commercial vehicles such as the Vito, Springer, and Citan are particularly profitable, so this area will likely remain in the overall portfolio as a valuable contribution margin in the long term. Both divisions increasingly focus on the modular principle, an approach that has been common practice at Volkswagen, among others, since the early 1990s.

The fundamental strategic approach is fully integrated into the marketing concept. The slogan "Deliver the most desirable vehicle" sums up the strategy in a nutshell. Mercedes advertises the "Economics of Desire", which is emphasized intensively at all communication levels. This applies to all passenger car brands and sub-brands: Mercedes, AMG, Maybach, G, and EQ.

The following key points outline this strategy:

– **Profitability**: Focus on profitable segments (at least 75 percent of all products).
– **Sales growth**: 60 percent growth in the premium segment by 2026.
– **Full electrification by 2030**: From 2030, Mercedes will no longer offer ICE-powered products.

- **EV range**: Dedicated EV platforms are MB.EA for medium to large models, AMG.EA as a dedicated performance EV platform, and VAN.EA for light commercial vehicles. In 2025, these three new architectures, developed exclusively for electric vehicles, will be introduced.

Mercedes names four cornerstones of the new strategy: focus on efficiency, performance, off-road technology, and *next level*:

- **Efficiency**: The best electric range, powerful performance, a silent ride, and easy charging.
- **Performance**: This is the sporty dimension of the AMG.EA platform is used as an e-platform.
- **Off-road technology**: The off-road drive is consistently electric.
- *Next level*: Prioritization of new technologies, for example, in the battery sector.

The Mercedes software strategy, under the claim "Lead in Car Software," is based on three aspects:

- **Autonomous driving**: The aim is to offer full autonomy (level 5).
- **New customer functions**: Innovative technologies and services will be used to implement new features that offer customers added value.
- **Customer convenience**: Mercedes prioritizes convenience to provide a pleasant and stress-free customer experience.

Mercedes boasts that it launched the Vito E-CELL–a commercially available, modern German electric vehicle – in 2010. In contrast to these early attempts at modern electromobility, Mercedes has already progressed several generations. Mercedes' architecture, called VAN.EA, is currently being redesigned as part of the overall MB.OS landscape. The aim is to create an overarching architecture – hardware and software – that is standardized for all. Mercedes prides itself on not doing things by halves; the entire platform architecture has been completely redesigned specifically for the needs of EVs. The architectural principles consist of four pillars:

1. **Purpose-built architecture**: An architecture that remains open for Mercedes partners.
2. **Personalized services**: Driver data is managed via a central data key (similar to single sign-on).
3. **Privacy by design**: The MB.OS architecture is designed from the ground up to protect privacy and ensure consistent data protection.
4. **Full over-the-air updatability**: Mercedes has been using the OTA feature since 2013 (for the EQS in the USA). Now, it is being applied platform-wide.

A holistic view of the architecture is emphasized, which is to be implemented realistically and pragmatically. This makes it clear that Mercedes wants to avoid making any ill-considered mistakes. The balance between luxury and function will remain a Mercedes design principle.

Mercedes is pursuing a partnership with Nvidia in autonomous driving, which was announced in 2020 [RM-119]. Nvidia is developing special hardware for autonomous driving called "Thor" (the predecessor chip was called "Orin"), which is part of the NVIDIA DRIVE Thor SoC (system-on-a-chip) architecture. The partnership has already borne its first fruits: Mercedes announced in 2022 that the first Level 3 feature was already approved in Germany. In 2023, Mercedes received road approval for the DRIVE PILOT in the US state of Nevada. The system will be installed as an option in the EQS model and is designed to take over from the driver in slow-moving traffic at speeds up to around 37 km per hour. Nvidia has announced that it is already working on the Level 4 SoC for the "DRIVE Odin" system to be released in 2025.

In the field of batteries, Mercedes is working with the Californian company Sila (Sila Nano), in which the Stuttgart-based company has acquired a stake. Sila Nano uses titanium-silicone technology, using silicone instead of graphite. It promises a 20 to 40 percent higher energy density. Mercedes has set the time horizon for the new cell technology at around 2025.

The Group is working on establishing a global network of charging stations in Europe, North America, China, and other markets. It plans to invest 1 billion dollars in the USA to build 2,500 charging stations by 2027 [RM-124].

The Mercedes Group is proud of the company's profitability—Mercedes currently achieves a net return of around 10 percent.

Additionally, Mercedes is actively sourcing new talent by recruiting over 3,000 top software engineers [RM-123]. The "Re-Skill" initiative will prepare the company for the electric future. To this end, Mercedes plans to invest 1.3 billion euros in employee training by 2030. Employees will be involved in motivation and monetary terms via profit-based bonuses.

A strategic overview in the conventional SWOT matrix is shown Table 7.3.

In summary, Mercedes-Benz's strategy consistently focuses on electrification, luxury, and improved profitability. For its successful implementation, it will be crucial to skillfully expand and intensify electrification with Mercedes' successful marketing strategy. Mercedes appears well-positioned to continue leading and expanding globally in the luxury segment.

7.1.1.4 Tesla

In over 130 years of automotive history, Tesla is still considered a newcomer. Since the company was founded in 2003, Tesla has experienced ups and downs (see Chap. 1). Since Elon Musk took over in 2008, the start-up has led the way in modern electric vehicles and become an industry leader, drawing both admiration and skepticism. Elon Musk is a striking figure who contrasts starkly with the typical profile of established automobile industry leaders.

The first Tesla—Tesla Roadster—succeeded in 2008, followed by several other breakthroughs and EV advancements:

Table 7.3 Mercedes – SWOT

Strengths	Weaknesses
Excellent global brand in the premium segment	Intense competition
Innovation leader (function and safety)	High brand reputation risk
Future-oriented R&D budget	
Strong marketing	
Strong commitment to racing	
Global production presence	
Leader in autonomous driving	
Stable finances	
Opportunities	**Threats (risks)**
Hybrid market opportunities	High fuel prices (ICE)
New markets (emerging markets)	Burden due to many legal regulations
Strong PR impact	War in Europe
High growth in the luxury segment	
Rising fuel prices (EV)	

- A Roadster with a range of 1,000 km
- Several models, such as Model S, Model M, Model Y and Cybertruck
- Tesla Semi—a fully electric truck
- Constantly improved lithium-ion cells (4680 cells, in cooperation with Panasonic)
- Solar panels, solar roofs, battery cabinets (Powerwall)
- Gigafactory and Gigacasting: A revolutionary approach in which car bodies are produced in a single step, saving time and resources.

Products that have been announced or are rumored in the industry include:

- Model 2: An EV that should be cheaper than the Model 3
- Model V: A van
- Tesla Bot: An android robot
- A Tesla smartphone

Tesla's EV production is rushing from one record to the next. Tesla produces the most BEVs worldwide and ranks first in EV production in Germany (see Chap. 1 for more information). Tesla has no hybrid PHEVs in its range, but this would not be expedient in the long term, as the production of combustion engines will be banned almost entirely by law in the foreseeable future.

The list of innovative Tesla features is getting longer and longer. As Tesla primarily manufactures software products, new features, and updates are constantly installed OTA (over-the-air) directly into Tesla cars, as is common in PCs or smartphones. These features can be objectively useful, such as improving range through enhanced energy management. Other features are more sophisticated or simply cute, such as "Dog Mode," which allows Tesla owners to monitor their beloved pets via cell phone. Many critical system fixes can be provided remotely, which would otherwise result in expensive recalls.

Tesla's system platform is unique. Having chosen not to partner with companies such as Mobileye or Nvidia, Tesla is developing its own chip and software. Tesla does not outsource its software because it is integral to its core operations. The entire platform is fully integrated from a single source, which enables agile development and the development of further innovations and improvements at short notice without changing the hardware.

Tesla's marketing and PR strategy is also unique. The company is known as one of the most innovative car manufacturers in the world and uses this for inexhaustible PR capital. In addition to Elon Musk's famous dance interludes, he is omnipresent in the media. Tesla and Musk are so closely linked that it is hard to imagine what will happen if Musk is no longer at the helm one day. Tesla is a vertically integrated vehicle manufacturer, also reflected in its marketing strategy. Vehicle, person, and media presence have merged into one overall media product. This strategy can be outlined as follows:

Emotional attachment at all levels: The Tesla community is almost fanatical. Tesla drivers occasionally risk their lives to demonstrate that Tesla's Autopilot works. This is also evident in the sales aspect: pre-orders for the Cybertruck amounted to more than 1.6 million. This unique emotional connection could perhaps only be compared to the iPhone and Apple in their prime.

No or only sporadic advertising: Tesla does not seem to consider the usual advertising in conventional media to be necessary. Instead, Tesla relies on Elon Musk's X.com (formerly Twitter).

Influencers: Tesla works with well-known influencers who act as Tesla's mouthpiece. One example is the YouTube channel of Marques Brownlee, also known as MKBHD. He demonstrates the latest and "coolest" Tesla features with expert commentary. His YouTube videos receive millions of views.

The Tesla community: Tesla fans love Tesla and publicize it at every opportunity. Countless posts on social media platforms such as X.com and YouTube are full of benevolent examples that show how attractive Tesla is.

Elon Musk as a person: Musk likely acquired Twitter to use the platform without any limitations for his PR purposes. This motive alone could justify the purchase despite widespread criticism of the high acquisition cost. Musk also enjoys giving interviews, is involved in various social activities, and never misses an opportunity to promote Tesla.

Environmental protection: Tesla sees itself as a pioneer and proverbial "white knight", saving the world from climate change. This gives the company a powerful moral superiority that can hardly be surpassed by traditional car manufacturers – at least as long as they still sell ICE cars. An army of influencers also work for Tesla for free because they see themselves as the world's saviors.

As long as Tesla remains the leading supplier of electric vehicles, focusing 100 percent on electric products and not "dirty" combustion cars, traditional car manufacturers will find it hard to fight this PR strategy effectively.

Tesla also goes its own way when it comes to sales. There are hardly any car dealerships selling Teslas; Tesla drivers buy their cars online. Buying a car from Tesla is not much more complicated than buying a more expensive smartphone. Unless Tesla makes the same mistake as many other manufacturers by offering hundreds of variants, they will probably not need a distribution network any time soon.

The situation is similar with repairs. Since Tesla cars do not require regular emissions testing and remain low maintenance compared to ICE vehicles, regular maintenance or inspections are unnecessary. EVs do not require oil changes and rarely need new brake pads, as the EVs cause less wear and tear due to the regenerative braking system. If a minor repair is required, a mobile Tesla team can perform repairs while the owner works in the office.

A key aspect of Tesla's marketing strategy is the Supercharger charging network. The concept of widely distributed fast charging stations in the USA and other countries is a unique innovation and particularly attractive in countries with high fuel costs. Tesla offers charging at these stations at comparatively low prices. In the initial phase, Tesla even offered free charging for life at the Superchargers for specific vehicle models, which was a strong incentive to buy. However, Tesla no longer offers this service on new purchases.

Tesla is also striving for more environmentally friendly products and production. For example, the company wants to develop cobalt-free batteries. Like Mercedes and other manufacturers, Tesla is working on silicon anode batteries, but there are no concrete announcements regarding the practical implementation of this new technology. Officially, Tesla is still working on the 4860 lithium-ion cells. It is unknown whether, like Toyota or Mercedes, Tesla is working on solid-state batteries. For that reason, solid-state strategies are not included in the "strengths" section of Tesla-SWOT (Table 7.4).

A particular risk for Tesla is that Musk's growing political presence and unusual public statements increase the risk of being politically "undercut", which could affect the company brand and its ability to innovate. Growing competition from China could also jeopardize Tesla's market position, especially if Chinese EV manufacturers offer "cooler" alternatives and Musk underestimates the threat.

7.1.1.5 Strategic Summary: The German Perspective

German car manufacturers are pursuing e-car strategies with a similar focus. The main difference lies in the target customer segments and marketing strategies.

Car manufacturers face similar challenges and share notable similarities:

- **Cannibalization**: ICEs are currently the primary source of revenue, so they have to subsidize EVs heavily. Economies of scale with EVs are still a dream.
- **Software talent is in demand**: All manufacturers claim to have embraced software as a core business, yet arguably, not a single CEO has a background or education in software. This discrepancy between software's apparent importance and management teams' makeup is cited as one reason software experts do not find the automotive industry attractive.

Table 7.4 Tesla—SWOT

Strengths	Weaknesses
High profile	Dependence on one person: Elon Musk
Elon Musk as a "brand"	Quality problems
Excellent innovation efforts	High price of EVs
Profitability	
High growth	
Vertical integration	
Attractive employer	
Supercharger network	
Sources of income through carbon credits	
Sales, production, and development agility	
Advanced battery technology	
Opportunities	**Threats (risks)**
Utilization of profits for price reductions	Growing competition from China
Expansion into further markets	Supply bottlenecks (especially for batteries)
Partnerships	Falling EV subsidies
Use of climate protection PR	Hydrogen cars
E-bikes	Uncertainty regarding autonomous driving
Additional services (car and third-party)	New laws and restrictions
Integration of other products (solar roofs etc.)	War in Europe
Licensing of Tesla technology	
Rising price of fossil fuels	

- While software solutions are increasingly becoming critical factors in this sector, there is often a lack of understanding of these areas at the executive level. This leads to growing frustration among software developers, who frequently feel misunderstood. When talented software experts are hired from outside to fill management positions, they often fail due to embedded structures and processes that leave little room for innovative and agile software development practices. This ongoing frustration usually results in these talents leaving the company, weakening car manufacturers' innovative strength. In Sect. 7.2.2 we look at this problem in detail.
- Extensive **standards** and legal regulations complicate the e-transition.
- German OEMs have been caught cold by the **Chinese EV offensive**, and it is unclear how this new challenge will be met.
- The **electrical supply chain is fragile** and heavily dependent on foreign suppliers, particularly for battery technology, mainly sourced from China.
- All German brands enjoy a **strong global brand presence**. The task is to defend these brands against EV newcomers (Tesla, Chinese manufacturers).
- **Low vertical integration** leads to slower adaptation of new technologies. Vertical integration is not seriously attempted.
- Strong dependence on suppliers leads to complex interdependencies through numerous **fragmented system architectures**.
- The **poor charging station coverage** is slowing down EV adoption.
- **Autonomous driving**: Mercedes has launched its first L3 feature, but autonomous driving remains largely a vision of the future.

- **Gigacasting**: German manufacturers do not regard Tesla's production strategy highly.
- **Lack of profitability in EVs**: Tesla has recently become the most profitable vehicle manufacturer. While Volkswagen now has to cut back and save, Tesla can use the net surplus for price reductions and innovative products.
- **ICEs are still the primary source of income**: All three major German vehicle manufacturers "live" through ICE vehicles, which are cross-subsidizing the development of new EVs. They are running out of time to boost EV sales quickly enough that profitability does not suffer. It will probably be a tightrope act until EV mass production becomes profitable.

Business figures can be used to observe future trends. An important business metric that indicates a company's profitability and financial strength is the company's net margin (Fig. 7.4, reproduced with permission from Seeking Alpha Ltd).

After a typical start-up dry spell, Tesla became quite profitable, which explains its ability to lower prices for end customers.

However, despite the enthusiasm of Tesla fans, Volkswagen remains clearly in the lead in terms of total sales (Fig. 7.5, reproduced with permission from Seeking Alpha Ltd).

Another key figure that measures employee efficiency is the net contribution margin per company employee (Table 7.5).

Volkswagen plans to fund essential EV investments through cost-saving measures.

A further challenge for German vehicle manufacturers will be that in the future, around 30 percent fewer components will potentially be needed to build EVs compared to vehicles with combustion engines – an estimation that will probably not be without consequences for employees. This problem applies to many suppliers,

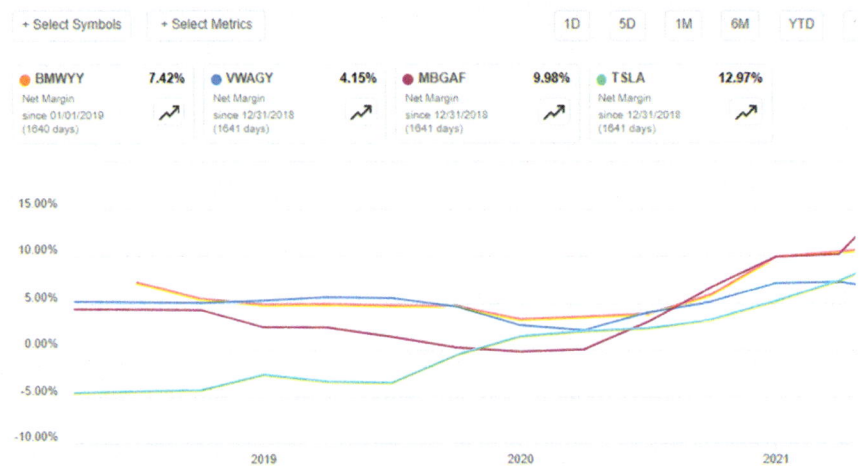

Fig. 7.4 Net margins of BMW, Volkswagen, Mercedes, and Tesla. *Source* Seeking Alpha)

Fig. 7.5 Gross sales, German manufacturers and Tesla, 12-month moving figures, in US$ (source: Seeking Alpha)

Table 7.5 Profitability per employee (in US$, reproduced with permission from Seeking Alpha Ltd)

Volkswagen	20,323
BMW	81,654
Mercedes-Benz	100,269
Tesla	95,381

meaning it can quickly take on economically relevant dimensions. The Volkswagen Group (including the core brands such as Audi, etc.) employs over 600,000 people worldwide. A socio-political crisis would be inevitable if a third of these jobs suddenly became redundant.

7.2   The New Software Strategy

In our industry, software has often been accepted as a necessary evil, which has consequences for its status with car manufacturers. As software was frequently perceived as a foreign body, software or software-heavy ECUs were often outsourced. This strategy was quite successful for a long time. Software plays a crucial role in autonomous electric vehicles. Once a problem child, software must now transform into the industry's poster child.

7.2.1   Software as the Problem Child

Since the 1980s, it has become clear that software is not simply a variant of computer hardware, which was the domain of electrical engineers. Popular terms such as "bugs" originally came from the electrical engineering domain. Back then, bugs were a problem because if an insect got between contacts in an electromechanical computer of the early years, the short circuit created by the charred corpse of the insect caused an error that had to be laboriously searched for and rectified (hence, "debugged"). Software did not exist as a separate discipline.

With the ever-growing number of software functions in practically every indus-try, the realization slowly seeped through that software is not just special hardware with compilers on top of it. Compared to software, hardware is a very stable, slowly changing discipline. Software, however, is a volatile, fast-moving, con-stantly evolving field that thrives on the fact that it is inherently *agile*. While developing a new hardware function can take weeks or months – or even years, if new semiconductor modules have to be developed–a new software function that an end customer demands or requests can sometimes be implemented within hours. This is probably why some managers perceive software as a straightforward task. Software coding may appear deceptively simple but issues can arise when dozens or hundreds of these "quick-and-dirty" features are added. The exploding number of software features comes at a price:

– New software functions can impact the functionality of other features. These consequences often only become apparent under specific, rare circumstances.
– More features mean more system load and resource consumption, such as computing time or main or NVM memory (non-volatile memory). As these features have variable resource requirements, the maximum requirement must be calculated. This is often not done automatically; software developers must consciously search for these "overhangs".
– Often, software resource requirements cannot be determined rationally, so assumptions and worst-case scenarios must be used. Only a thorough valida-tion of the resource consumption can reduce the risk of unexpected failure at runtime.
– The test phase becomes longer and more complex as the interactions between features have to be tested. It is sometimes surprisingly long, and it has to be repeated every time the software is changed ("regression test") at all test levels (qualification test, integration test, unit test) along the nearly 20 traceability paths required by ASPICE.
– Maintaining and updating software is becoming more challenging.
– Software documentation must be constantly updated to maintain an overview of all features and their dependencies. Otherwise, software development devolves into a trial-and-error process rather than a systematic approach.
– There may be an increased defect rate as the complexity of new software grows. As some algorithms are non-deterministic (e.g., ADAS features based on neural networks), more tests must be carried out, some of which can only be done manually. Under certain circumstances, errors can paralyze the entire software platform.
– The training period for new developers can increase considerably as they have to familiarize themselves with an increasingly complex code base. Therefore, global outsourcing can be a strategic mistake. Outsourcing teams suffer from higher turnover, which exacerbates this issue.
– As the number of features increases, so does the risk that not all of them will ever be used at runtime, leading to the phenomenon of "dead code." Under-standably, dormant "dead code" is explicitly prohibited in the FuSa standard

ISO 26262. However, code cleanup can be costly; this effort must be planned and paid for.

- The user experience can suffer if the interface is overloaded with too many features or confusing ("featuritis").
- Constantly adding features can mean the software architecture becomes unsustainable. Redesigning it can be extremely costly.
- Users may require more training when confronted with additional functions and options.

In many traditional companies, there is a surprising lack of knowledge about these software phenomena. Thus, OEM managers often succumb to the temptation to outsource the troublesome "problem child" software. Previously, it was easy to control the situation by outsourcing software-heavy parts of the vehicle platform. If the software didn't meet expectations, it was easy to blame the supplier. Another common approach was outsourcing the entire platform's integration to external service providers. Of course, the integrator cannot take responsibility for the content of the various sub-suppliers of the software-heavy components, as the individual ECUs represent a "black box" for the integrator. However, there is sometimes the illusion that integration outsourcing can eliminate software problems simply by performing thorough integration testing. Such a strategy is occasionally ridiculed as "magical thinking."

These examples demonstrate that the issue is not merely a "software crisis" but a crisis within the software supplier industry. Jim Farley, Ford's CEO, summarized this software problem in an interview in 2023 [RM-135]. In his opinion, traditional car manufacturers like Ford face integration challenges when software-heavy components are outsourced. Historically, manufacturers have outsourced various vehicle modules (e.g., seat and engine control) to hundreds of suppliers to reduce costs. The problem is that different companies create the software for these modules, which in turn must seamlessly communicate with each other. Module changes could require approvals from several external companies. The increasing number of ECUs requires seamless cooperation between software strategies, architectures, and IP rights, which often compete, making software updates and integration more challenging.

It is, therefore, challenging to implement a seamlessly integrated car platform with the traditional multi-supplier approach in an acceptable amount of time. Software "from a single source" can only be developed efficiently if a vehicle manufacturer actively assumes responsibility for the platform. This requires the entire management team to (want to) understand the fundamental principles of software development, and this seems to be precisely where the difficulty lies. So-called vertical integration, which enables companies like Tesla (and others such as Rivian or Chinese vehicle manufacturers) to develop efficiently, appears difficult to embrace for the traditional automotive industry. This challenge potentially applies to both manufacturers and their suppliers.

A truly holistic systems thinking would involve less rhetoric about software being the core business, as often touted in company brochures and at annual general meetings, and more about fully embracing the realignment towards a software-centric approach. However, this would also mean that software manufacturers would have to go much deeper into software development and thus increase the vertical range of manufacture again.

7.2.2 What Makes Software Engineers Tick

Some vehicle industry veterans seem puzzled by software. Let's take a closer look at the issue because this is the real sticking point in the new world of autonomous EVs.

One particular aspect of the problem is the nature of the "phenomenon of software" itself. Academics sometimes dismiss computer scientists and software engineers as "unworldly nerds." The common stereotype of a software developer is that of a pale, overweight, unshaven hacker who stuffs himself with fast food and rarely crawls out of the corner of his basement, let alone in daylight. Of course, there are those, but they are exceptional cases. At the same time, some of these "nerds," such as Bill Gates or Steve Jobs, to name just a few examples, are almost fanatically worshipped. Perhaps underlying envy also plays a role, as the profitability of many software giants is often light years above the notoriously low profitability of the automotive industry. Some managers may find it frustrating to deal with these "weird nerds" just because software has suddenly become a key competence in our industry. Software is the proverbial "new kid on the block". However, a thorough understanding of the nature of software engineering must be internalized if OEMs want to survive in the tough new mobility software-driven business.

Therefore, it makes perfect sense – especially for those of our readers who cannot or do not want to deal with the subject of software – to take a sympathetic look at the "software engineer species". We can actually learn a lot from these supposed eccentrics.

On the well-known social network Reddit, there is an interesting discussion about the article *There Will Be Blood*, which offers deeper insights into the soul of a software developer [RM-147]. It's no secret that Reddit readers are predominantly software-heavy, so it makes sense to listen when they're having a rational discussion. The Reddit thread is extensive, but we'll summarize some key insights for our readers:

Integration and Software Approach

- Tesla works with vertical integration, seamlessly connecting hardware and software via a common API.
- Older OEMs do not understand the importance of APIs.
- The concept of a "car operating system" is emphasized, suggesting that cars should have a unified operating system, similar to smartphones.
- Outsourcing software, especially to other countries such as India, is viewed extremely critically due to potential quality losses.

- Tesla's approach has been compared to the "androidization" of the automotive industry, which points to a change in values: that software must have a special status in automotive engineering.

Industry Comparisons and Talents

- The demise of Nokia is a sad example that suggests legacy car manufacturers could suffer a similar fate if they do not adapt.
- The old OEMs are assumed to lack the necessary talent to survive in the evolving industry.
- The enthusiasm of developers for the integration of smartphone software in cars is an indicator of the direction in which the industry is developing.
- AI is recognized as a cutting-edge technology with applications that extend far beyond self-driving cars.

Project Management and Development

- Increasing the number of developers working on a troubled project does not necessarily finish it faster. Reference is made to the book *The Mythical Man-Month*, which emphasizes that hiring new developers in troubled projects actually tends to decelerate the project's progress.
- Older OEMs are criticized for their user interface, which many consider to be below average.
- There is a belief that hardware managers do not understand software requirements, leading to inefficiencies.
- The importance of small, talented teams is emphasized, with the iPhone development team cited as an example (it is known to have consisted of only around 100 engineers).
- Legacy OEMs are criticized for being reluctant to pay appropriately for top software talent. Instead, the principle of "mass instead of class" is favored.

Hardware and Software Development

- Tesla's approach is to build the car around the software, which contradicts the traditional OEM approach in which software is viewed as a pesky add-on.
- Many German cars are criticized for having several modules from different suppliers, which makes them difficult to maintain.
- Over-the-air (OTA) updates are praised as a disruptive feature of Tesla that enables quick correction or improvement of car functionality.
- The hardware should be designed to adapt to the software, not the other way around.

OEMs' Understanding of Software

- While the traditional OEMs always emphasize how important software is, there is skepticism regarding whether the old OEMs actually understand the *criticality* of software.
- The user interface (UI) is a key selling point, and practical examples are used to criticize the fact that traditional OEMs are weak in this area.
- Several times, it has been mentioned that not everything in software development has to be perfect at market launch, as over-the-air (OTA) updates can fix problems later. This is a culture shock for traditional vehicle manufacturers.

Creativity and Flexibility

- It is believed that the development of electric vehicles requires more than just structured planning. It demands creativity, a spirit of innovation, and flexibility.
- One humorous comment is that the Germans could excel at developing an electric vehicle if a comprehensive checklist existed. However, the real challenge lies in being innovative and adaptable.

This Reddit discussion provides many insights that can offer strategic added value for traditional car manufacturers. The community of software engineers appears to have grasped some crucial insights that have yet to fully resonate with the automotive industry – or, perhaps, particularly with some seasoned managers. But computer scientists are not that different from the "old" engineers. Just think of Ferdinand Piëch, with his passion for technology, who honed his skills at ETH Zurich and was relentlessly innovative. ETH has been a software engineering, AI, and machine learning leader for decades. If Ferdinand Piëch were 18 years old today, he would most likely choose computer science rather than mechanical engineering as his main subject.

The bottom line is that we urgently need more Software-Ferdinand-Piëchs with a passion for software in automotive engineering. The future of our industry, and that of traditional vehicle manufacturers, depends on it. The former German President Roman Herzog once said: 'Germany needs a jolt. We have to say goodbye to cherished vested interests. Each individual must perform better than before. I know that this is uncomfortable. But there is no other way. We have to act now so that Germany doesn't fall behind.' At the time, he was talking about the German economy in the 1990s, but today, this quote seems particularly applicable to the traditional automotive industry. The "jolt" that Roman Herzog spoke of must now come from the automotive community to ensure our legacy OEMs don't lose the race and get left in the dust.

7.2.3   The Industry's Call for New Software Talent

The idea that software is just a fifth wheel on the car—an attitude popular in the past—belongs in the dustbin of automotive history. Adopting a systems approach that places software at the core of vehicle development is urgently necessary.

The first key finding is that software is not cheap. It is expensive because strategically important software components, such as software vehicle platforms (vehicle operating systems, car OS), cannot usually be effectively outsourced globally to cheaper regions with lower wages ("offshoring").

In other words, software development must become a central component of the company's core business. Software must not only become a top priority; it's imperative to have a software engineer at the C-level of a vehicle manufacturer.

Another observation is the radical change in the perception of vehicle design. The "software-defined vehicle"–a conceptual approach in which a software-heavy system design represents the first step in the conception of a new car platform–has recently become a popular phrase [RM-137]. So, while thermodynamics (combustion engine) and mechanics (transmission, body, and chassis) were once the sticking points, software and battery have now become the leading differentiators for car design.

A final insight is the observation that the software guild itself must also change to work more efficiently in car construction. While pure software engineers are sometimes sufficient, what's needed are software engineers who also understand the semiconductor industry and have a basic knowledge of mechanical engineering. Just as electrical engineers once learned software on the side, software engineers must now learn electrical and mechanical engineering on the side. Only then can traditional car manufacturers withstand the upcoming electrification storm.

7.3   The "Fleet Consumption" Challenge

Environmental regulations that the automotive industry must comply with are constantly increasing. These regulations aim to reduce the environmental impact of vehicles. EVs play a vital role as they enable the ultimate goal of "zero emission vehicles." This target is not always followed voluntarily; manufacturers face significant penalties in case of non-compliance. Since "Dieselgate," it has become clear that this is a real threat, so car manufacturers are investing billions in EV technology. At the same time, many countries are offering EV subsidies, as EVs generally cannot be developed and sold profitably at the moment—with the notable exception of Tesla.

However, no subsidies can compensate for the massive investment costs in EV development. The pressure is enormous: this "carrot and stick strategy" means that traditional car makers cannot offer EVs profitably for years, possibly even decades to come. At the same time, the legal thumbscrew is being tightened ever further. For example, stricter CO_2 limits have been in force in the EU since 2021. The exact limits are complex and are calculated individually for each brand based on the average weight of their vehicle fleet. This means heavier vehicles may have slightly higher CO_2 emissions than lighter ones. The 95 g/kilometer is an average target value, but the specific targets vary depending on the manufacturer and vehicle fleet. If the specified values, which are constantly decreasing, are exceeded, this results in extreme penalties: 95 euros x the sum of exceeding the fleet target in grams per kilometer x the number of vehicles sold. This can quickly add up to millions or even billions of euros of penalties for a vehicle manufacturer.

In the long term, EVs will eventually become profitable for the manufacturers involved. With ongoing development in the areas of battery technology and charging infrastructure, as well as strict cost reductions, this positive balance is expected to one day be achieved. The chances of this happening are promising; as observed in the first chapter, the number of EVs is rising rapidly while the number of ICE-powered vehicles is declining. It appears to be only a matter of time before ICE vehicles are phased out completely, leaving only EVs on the market.

7.4   Strategic Partnerships

During the supply chain crisis caused by the coronavirus pandemic, most car manufacturers suffered from a chip shortage. The shortages were so severe that new cars were occasionally delivered without the necessary chip, with these components being retrofitted later [RM-140]. The reason for these difficulties was the length and nature of the international supply chain. For example, fewer vehicle components were supplied during China's "Zero Covid" strategy. As many semiconductors are manufactured in China, these components could not be shipped.

The number of semiconductors in ECUs is constantly increasing. Furthermore, existing discrete chips sometimes cannot be efficiently integrated into more complex architecture. For instance, Tesla realized that systems with many discrete components were not feasible due to high-performance requirements and problematic energy and heat management. As a result, Tesla terminated several collaborations with specialized chip manufacturers, including Mobileye [RM-120], in visual pattern recognition. Instead, Tesla developed its own SoC – System-on-a-Chip. Tesla's SoC implements several functions in a single ECU with two redundant processors (redundancy postulated by FuSa ISO 26262 standard [RM-0136]), including:

– Front Collision Avoidance
– Lane Departure Warning
– Lane Departure Avoidance
– Emergency Lane Departure Avoidance
– Side Collision Avoidance
– TACC (Traffic-Aware Cruise Control)
– Autosteer, Accelerate, and Brake
– Car Park

- Auto Lane Change
- Read Speed Signs
- Summon
- Smart Summon
- Navigate on Autopilot
- Respond to Traffic Lights and Stop Signs
- Full Self Driving (planned)

Before integrated SoCs existed, such solutions were developed using separate ECUs provided by different suppliers. Coordinating between suppliers and integrating these ECUs into the car architecture via various communication protocols and buses involved a great deal of effort. The SoC approach, therefore, offers considerable savings.

Tesla continues to use discrete chips (e.g., from Intel or AMD) for less safety–critical components, such as infotainment. Other semiconductor components are increasingly part of this "ecosystem": CPUs, GPUs, DSP, ASIC, FPGA, etc.

Meanwhile, cooperation between suppliers and car manufacturers has entered a new phase (see Fig. 7.6).

These collaborations are intended for long-term partnerships, but no official announcements have been made about closer cooperation between traditional OEMs and other partners in system-on-chip (SoC) development. Several car manufacturers still lack a convincing SoC strategy.

Specialized hardware is essential for autonomous driving at levels 3 to 5. The current strategy, which relies on multiple separate ECUs connected to various sensors (lidar, radar, camera, ultrasound, etc.), is too energy-intensive and strains conventional architecture. While Tesla has been pursuing an innovative strategy

	Front-end enablers	Operating System (OS)	Compute Platform
GM	**Android** *(platform usage of Android Auto from 2024)*	**Ultifi** *(GM-owned software platform),* **Redhat** *(strategic collaboration on SDVs)*	**Qualcomm** *(strategic collaboration on digital cockpits, telematics systems and ADAS)*
Toyota	**Lexus** *(Toyota-owned multimedia system since 2021)*	**Arene OS** *(Toyota-owned real-time software platform)*	**Oracle** *(offboard - strategic collaboration on high performance workloads)* **No partner for onboard** announced
VW	**Android** *(future usage of Android Auto & Google Apps)* **Thundersoft** *(collab. for Chinese market)* **Harman** *(supply of open ecosystem for Apps)*	**vw.OS** *(VW-owned OS platform)* **Blackberry QNX** *(supply of safety-certified embedded software integrated in vw.OS)* **Vector** *(supply of automotive ethernet)*	**Qualcomm** *(strategic collaboration with focus on automated driving functions up to L4)*
BMW	**Android** *(platform usage of Android Auto in BMW OS8)*	**Android** *(platform usage for infotainment, App store designed for BMW)* **QNX/ Blackberry** *(supply of safety-certified embedded software)*	**Qualcomm** *(strategic collaboration with focus on automated driving functions)* **Intel/ Mobileye** *(strategic collaboration for autonomous driving platform)*
Mercedes	**Unity** *(strategic partnership to enable infotainment applications on new MB.OS)*	**MB.OS** *(announced Mercedes-owned OS platform)* **QNX/ Blackberry** *(supply of safety-certified embedded software)*	**Nvidia** *(strategic collaboration on SDVs, future focus on AI applications)*
Renault	**Android** *(strategic collaboration on SDV offboard & onboard applications)* **Valeo** *(supply of on-board application software)*	**Android** *(announced strategic partnership digital architecture for SDVs)*	**Valeo** *(Supply of SDV components incl. HPCs)* **Qualcomm** *(strategic collaboration with focus on next-gen. vehicle architecture)*
Stellantis	**Amazon** *(strategic collaboration on software solutions starting in 2024)*	**Mobile Drive** *(strategic collaboration together with Foxconn on in-vehicle user experiences)*	**Qualcomm** *(strategic collaboration on digital experience)*

Fig. 7.6 Strategic partnerships in the area of software and operating systems (PwC Ltd., [RM-0138])

in this area for years, other manufacturers must catch up and aim to surpass Tesla in these efforts.

7.5 The Chinese Challenge

It won't be surprising to learn that China's agile automotive industry has been massively underestimated. As recently as 2020, Volkswagen wanted to help China with mobile electrification. Now, it seems the strategic tables have been turned.

The idea that German manufacturers can sell more cars in China and thus cushion the losses in the USA and Europe is based on a misunderstanding. China has long since ceased to be the geopolitical Cinderella needing assistance. The current international balance of power is diametrically opposed to the situation in which China was an import market for Western luxury goods such as BMW or Mercedes. The new Chinese assertiveness is swiftly changing the geopolitical and economic balance. At the same time, Western industrial leaders don't seem to be able to get to grips with the Middle Kingdom. Chinese power acts like a monolithic entity that can react more agilely to opportunities and risks than, for example, the EU.

The Chinese government is grappling with numerous challenges, many of which receive little attention in Western media. Among these challenges is a significant rise in youth unemployment, now exceeding 21 percent. Given the automotive industry's traditional role as a key driver of employment, China naturally seeks to capitalize on the global demand for EVs. In Europe, there is a noticeable shortage of affordable EVs. China aims to capitalize on this gap to create new growth opportunities for its economy and skilled workforce – a legitimate move in the context of national interest.

Previously viewed as a developing country in the EV sector, China has emerged as a global player, with some even referring to a "Chinese EV invasion" in Europe (see EV figures in Chap. 1). In response, EU authorities have investigated potential illegal subsidies for Chinese EV manufacturers [RM-106]. The outcome of this inquiry remains uncertain.

There is increasing discussion about the need for a "new China strategy". Suggestions range from import tariffs on Chinese EVs to features developed specifically for the Chinese market, such as karaoke integration in new EVs. However, the gap between Western and Chinese manufacturing and development costs is insurmountable. Chinese manufacturers will enjoy a significant price advantage in the foreseeable future. One approach could be to offset the price disadvantage of European manufacturers by focusing on quality and effective PR. However, achieving this would require strong software expertise to deliver the necessary features, which still seems rare in Europe. While around 45 percent of employees at car manufacturers in the USA and China have software expertise, the figure in Europe is a meager 20 percent [RM-140]. The development speed (time-to-market) would also have to be doubled. **Creating a new model in the EU takes around four years; for some Chinese vehicle manufacturers, it takes two years.**

Cost savings alone cannot overcome these disadvantages. For years, there have been murmurs that the car industry needs new approaches and skills. The recent challenge posed by China has exposed a concerning structural weakness in European vehicle manufacture. Yet, despite this, the need for structural change has not been fully recognized. Without acknowledging and addressing this issue, it will be challenging to respond effectively to China's growing competition.

7.6   Public Relations: The Silent World Power

The term "public relations" was coined in the 1920s by Edward Bernays, the nephew of Sigmund Freud, because the synonym "propaganda" had a negative connotation. Bernays is celebrated as the "father of public relations" and is an omnipresent figure in the marketing industry. In one of his most famous books, *Propaganda* [RM-141], he describes a strategy to control public opinion based on emotions. One of his favorite examples of successful PR work was an assignment from a tobacco manufacturer to make smoking socially acceptable among women. At the time, it was unusual for women to smoke in public in the USA, and the tobacco industry suffered as a result. Bernays wanted to change this. He organized a demonstration on a sunny public holiday, during which he invited various female celebrities. The press was also invited. At his signal, the women present lit cigarettes, after which Bernays pointed to them and shouted into the crowd that this was the "torch of freedom". Within a few weeks, women were smoking at will in every public space.

This story might seem trivial, but doubling the revenue of an entire industry is nothing short of extraordinary. Achieving such a breathtaking result with minimal investment is a testament to the overwhelming influence and power of the PR and advertising industry – a global titan that shapes economies and perceptions with unparalleled might.

This fact becomes apparent when looking at sales figures. The global PR and advertising industry generated over 100 billion dollars in 2023, and advertising generates around 760 billion dollars annually. The advertising industry is expected to exceed *1 trillion dollars* in two years, and the PR industry is growing similarly [RM-142]. One of the world's most prominent advertising and PR companies is Google (Alphabet). A deep-rooted intertwining with business and politics means nothing happens without and certainly not against these PR juggernauts.

The significance of effective PR strategies becomes clear when scandals and crises occur, such as "Dieselgate." Entire industries can rise or fall based on their PR approach. For example, Tesla's ascent to becoming an automotive giant was due mainly to its skillful PR use.

The events of recent years have shown that public opinion, when backed by a clever PR strategy, can often outweigh objective facts. This insight is of great importance to our industry. When Ferdinand Piëch drove his diesel prototype from Wolfsburg to Hamburg in 2002 and only used one liter of diesel per 100 km, it was a prime example of a highly successful PR stunt. Consequently, until "Diesel-gate" in 2015, diesel was considered economically and ecologically unbeatable: the perfect product that was about to conquer the world. Only after "Dieselgate" did Tesla have any chance of achieving significant commercial success.

A well-structured PR strategy is no trivial task, as it can pursue various objectives. For German manufacturers, a clever strategy would be to turn supposed weaknesses into strengths. For instance:

- Emphasize the tradition of German engineering ("You know what you have").
- Use the "Made in Germany" brand, which will endure long and retain a positive aura.
- Technology leadership: Despite the severe shortage of software talent among European car manufacturers, showcasing concept models that are not yet ready for mass production can provide a glimpse of cutting-edge vehicle technology and offer a vision of what the future holds for these brands (following the motto "Look what we already have").
- Environmental awareness is powerful in Germany; Germany can and should build on this.
- Emotional appeal: Skillfully engaged influencers can work wonders. PR stunts and events are highly effective for this purpose, provided they are executed credibly and supported by consistent communication before and after each event.
- Ongoing PR efforts, such as promoting "green car cities" and commitments to environmentally friendly disposal of EVs, can be effectively highlighted.

Moreover, traditional car manufacturers can rely on government support, as – unlike newcomers like Tesla – they are deeply integrated into the conventional social landscape (trade unions, works councils). This impartial assessment can be leveraged as a significant component of their PR strategy.

Overall, the PR strategy of leading German car manufacturers should be based on a balanced presentation of tradition and progress and be in line with the official agendas of governments and international bodies such as the United Nations. It should offer insights into the further development of future vehicle fleets and emphasize the commitment to environmental and climate protection. In this way, German manufacturers can strengthen their brand and present electrification as an exciting phase of transformation that will benefit customers and society.

7.7   Further Strategic Elements

In addition to the strategy factors already mentioned, there are other aspects to consider:

Table 7.6 Battery manufacturer

Manufacturer	Market share (%)	Country of origin
CATL	34	China
LG Energy Solution	14	Korea
BYD	12	China
Panasonic	10	Japan
SK On	7	Korea
Samsung SDI	5	Korea
CALB	4	China
Guoxuan	3	China
Sunvoda	2	China
SVOLT	1	China
Data source [RM-126]		

Batteries and Raw Materials

The concentration of global battery cell suppliers is a risk that needs to be addressed in the short term. More than half of all battery cells for EVs come from China (Table 7.6).

Although Chinese manufacturers are well-known for their cost efficiency, relying solely on this advantage would be unwise. The automotive industry must approach battery manufacturing with a focus on risk minimization and adopt a global strategy to avoid geopolitical and economic dependence on a limited number of battery cell suppliers.

Raw Materials Required for EV Production

In addition to lithium, the production of EVs requires other raw materials that are often not found geographically close to Germany (Fig. 7.7, reproduced with permission from Dr. Annegret Stephan/Dr. Axel Thielmann, Fraunhofer ISI, [RM-128]).

EV-relevant raw materials are distributed across regions worldwide, some of which are stable, but many come with significant risks that must be considered. (Fig. 7.8, reproduced with permission from Dr. Annegret Stephan/Dr. Axel Thielmann, Fraunhofer ISI, [RM-128]).

Notably, China is dominant in lithium-ion production and a powerhouse in other raw material categories. The country is largely self-sufficient in EV-related raw materials. However, commodity risks remain severe, as China can supply almost any resource in its pursuit to dominate the European—and, increasingly, the US—EV industry. Other EV-critical resource sources, such as Congo (cobalt), pose significant risks due to their potential political instability. Various measures can mitigate these risks. Beyond political strategies, there is always the possibility of transitioning to alternative battery cell technologies in the long term, which could further reduce or eliminate the need for materials like lithium or cobalt.

Fig. 7.7 EV-relevant raw materials in Europe

"Rightsizing"

In times of crisis, the common practice of *downsizing* must be replaced by the principle of *rightsizing*. This means it is not the *reduction* but the *adaptation* of employees' skills that is decisive. Electric vehicles require an estimated 30 to 40 percent fewer components than their counterparts with combustion engines (ICE vehicles), which leads to a long-term reduction in costs. This development is already visible in the USA, where a Tesla Model M is now more affordable than a BMW 3 Series. This trend is expected to continue due to economies of scale and technological advances that reduce production costs.

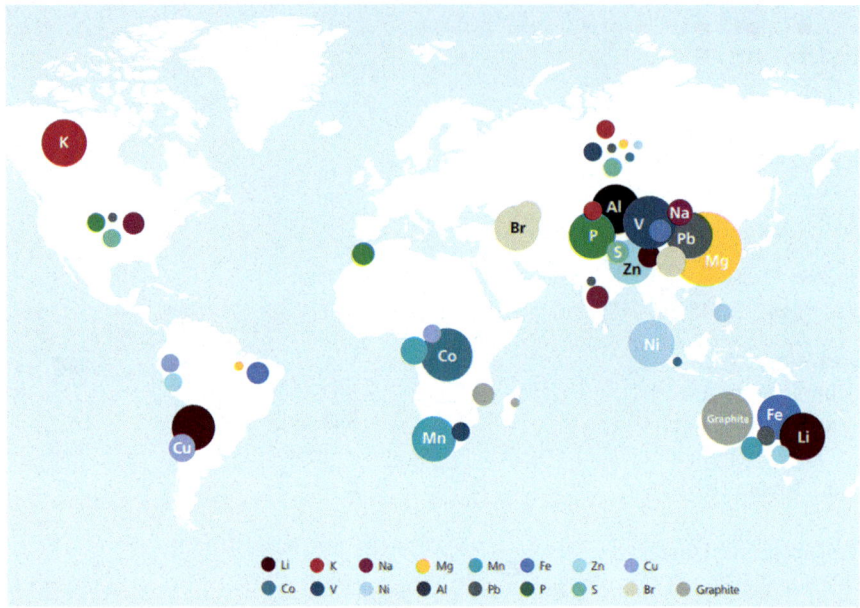

Fig. 7.8 EV-relevant raw materials (Dr. Annegret Stephan/Dr. Axel Thielmann, Fraunhofer ISI, [RM-128])

Demand is shifting towards talent in software, semiconductors, battery technologies, and artificial intelligence. Employees who previously worked in the production of combustion engines can work in new fields such as sales, project management, quality management, and developing charging networks. This shift will require more flexibility in the global workforce, and a smooth redistribution of the workforce is, therefore, of great importance. The shift to new technology brings changes that must not be overlooked.

Competitive pressure caused by new market entrants could lead to labor disputes, as seen in the US in 2023. New players could exploit these tensions to gain market share in uncertain times. Therefore, established car manufacturers must develop proactive strategies that not only prepare their employees for the future of electromobility through training and retraining but also strengthen the OEM's global market position.

Manufacturing Efficiency

Traditional car body construction occurs in multiple stages. Each part has to be cast or stamped separately and then assembled into a complete unit with a precise fit. Body production costs typically amount to around 10 percent of the total cost.

Tesla, meanwhile, has developed the innovative idea of *megacasting*. The entire body is produced in a single work step, which Tesla calls "Giga Press." This reduces costs, as the reduction in the number of body parts eliminates numerous

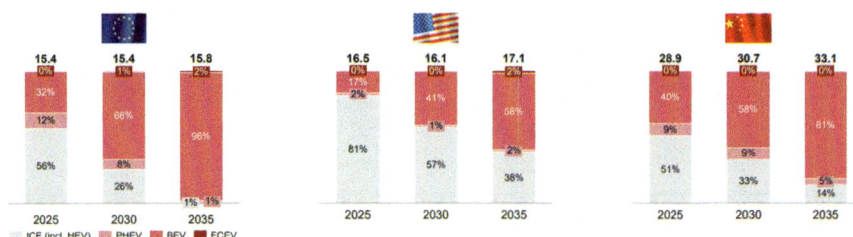

Fig. 7.9 Planned adaptation of EVs

work steps, such as welding, assembly, and the associated quality assurance. The technology promises a significantly higher production speed (time savings of up to ten hours per vehicle) and a further reduction in labor costs.

Other manufacturers are also considering or introducing this new technology (e.g., Toyota [RM-127]).

Research and Development
The adaptation of EVs is driven both by their promise to become cheaper and more attractive to buyers and by ambitious government plans, especially in the EU (see Fig. 7.9, reproduced with permission from PwC Strategy& (Germany)).

Rapid further development of EVs is therefore crucial for the competitiveness of car manufacturers. This includes not only the optimization of existing technologies for battery management and powertrains but also the development of new features that increase customer benefits and vehicle safety. Research and development (R&D) investment is the key to this progression. Car manufacturers need to invest significant resources in researching and developing future-proof technologies. This includes deploying capital to test new materials and techniques, such as advanced lithium-ion batteries with higher density or solid-state batteries, potentially improving range and reducing charging times. Equally important is the funding of pilot projects and collaborations with technology companies and academic institutions to accelerate technological progress.

It is therefore not surprising that the leading vehicle manufacturers are investing billions in R&D. Volkswagen AG, in particular, is characterized by its expenditure in research and development and is therefore at the forefront of the automotive industry (Fig. 7.10, reproduced with permission from Seeking Alpha Ltd.)

While investment in research and development is crucial, the efficiency and allocation of these research funds is equally important. With a relatively low investment volume of 3.2 billion US dollars, Tesla occupies one of the bottom places, but it is hard to deny that Tesla is a particularly innovative company. Therefore, the efficiency of R&D will certainly be scrutinized by all manufacturers in the future.

Cannibalization Versus Dividend Distributions
Electrification is a challenge that represents uncharted territory for established manufacturers. While the sale of ICE vehicles continues to be a primary source

Fig. 7.10 R and D expenditure of selected car manufacturers (in US$, source: Seeking Alpha Ltd)

of income, EV revenue cannot be produced to cover costs because of the continued low sales figures. R&D investments are required in car development and infrastructure, such as charging stations and customer service in electromobility. Investment in PR and advertising for EVs also constitutes a significant cost factor. The challenge is allocating resources effectively to achieve short-term and strategic goals.

For investors, the growing emphasis on electric vehicles—vehicles that currently do not achieve the same profit margins as traditional combustion engine cars—presents a potentially concerning scenario. Many of these investors rely heavily on dividends as a significant source of income, making dividend security a priority. If legacy OEMs fail to adapt by simplifying standards and reducing complexities, the risk of diminished returns increases, threatening the financial stability that these investors depend on (see dividends per share Fig. 7.11, reproduced with permission from Seeking Alpha Ltd.)

The challenge in the transition to EVs is that traditional car manufacturers are seen as reliable cash cows that systematically generate dividends, unlike start-ups such as Tesla or Rivian, which do not pay dividends. The necessary transformation to EVs could deliver a painful surprise to dividend-reliant investors if the substantial investments required for electrification diminish short-term returns.

Fig. 7.11 Dividends per share (in US$, source: Seeking Alpha Ltd.)

Make or buy—that is the Question

Companies are currently faced with either increasing internal development or continuing to rely on the fragmented supply chain (automotive supplier ecosystem). Despite apparent inefficiencies, some still stick with the outdated supplier-OEM model. Historically, splitting the vertical integration allowed cost savings, as components could be produced in lower-cost countries. However, this approach is becoming less efficient with the rise of EVs and their new technological requirements. Innovations such as silicon carbide inverters and efficient battery technologies are essential to remain competitive in the electric vehicle industry. These innovations could be developed more effectively and quickly in-house to secure a market advantage, provided the necessary skills are available.

Falling Margins

The politically driven electrification of the car industry seems unstoppable. Car manufacturers are in an unpleasant situation where margins will fall, at least in the long transition phase to electric vehicles. Leading EV manufacturers like Tesla are lowering prices to maintain or increase their market share. This could lead to a price war in which suppliers undercut each other, further squeezing margins.

Quality: Does "More" Equal "More"?

Chap. 4 discussed the ever-increasing number of various formal standards in the automotive industry. Most of these standards seem sensible, but in their entirety, they represent a continuous burden that costs time and money. The number and complexity of these standards and regulations do not necessarily lead to fewer car defects beyond a certain point. At the same time, EV manufacturing is increasingly dominated by American newcomers (Tesla) and Chinese manufacturers. As the popular saying says, "time is money," and the previously mentioned goal of halving development time from four to two years is unlikely to be achieved through additional quality assessments. A strategic shift from formality to effectiveness and an effective functional review—team peer reviews and validation strategies such as DRBFM—appear promising in this context.

7.8 Winds of Change

In the realms of technology and innovation, it is often the first courageous leap into the unknown, driven by bold and visionary ideas to change our world, that alters the course of history. By its very nature, however, being innovative is often tricky, and it is not uncommon for innovation to be met with skepticism, ignorance, and ridicule.

This is an inherently human reaction, but this natural resistance to innovation can sometimes be counterproductive. It's reasonable to consider the potential benefits of new ideas; otherwise, there's a risk of falling behind.

The following examples demonstrate how innovations that were initially viewed with skepticism have become global hits.

Wheeled suitcases: When Bernard Sadow invented the suitcase on wheels in 1972, it was widely ridiculed. 'Real men carrying carry their suitcases themselves' was the common refrain. At first, nobody wanted one. Today, countless wheeled suitcases roll through every airport in the world.

Computer mouse: The computer mouse was invented by American engineer Douglas Engelbart in 1968. For a long time, it was seen as having little practical use. Even when Apple introduced the first mouse in 1984, the reaction was initially mixed. The well-known *PC Magazine* columnist John C. Dvorak wrote a now-famous article in 1984 stating that he could find no meaningful use for the new type of computer peripheral that came with the Apple Macintosh. Today, almost all of us have at least one mouse on our computer desk.

Smartphone with touchscreen: The then head of Microsoft, Steve Ballmer, ridiculed the new iPhone when it was launched in 2007: 'There's no chance that the iPhone is going to get any significant market share. No chance!' he emphasized [RM-094]. Interestingly, even Steve Jobs, who is posthumously celebrated as an innovation legend, was critical of touchscreens. It was only after an Apple employee (Tony Fadell) showed him the new touchscreen prototypes that he was won over. The rest is history.

Industry-specific examples could include innovations such as EVs, hybrid vehicles, airbags, and ABS (anti-lock braking systems). Many of these innovations were ridiculed or dismissed as impractical, but now they are firmly established or, like EVs, in the process of achieving this status.

In all fairness, it should be acknowledged that 95 percent of all innovations fail [RM-093], so a certain level of skepticism toward technical innovation is justified. However, it is also well-known that companies can – and must – secure a promising future through innovation. Failure and success are two sides of the same coin. For this reason, we must not and will not shy away from new technologies in the automotive sector.

In summary, the strategic aspects to highlight include:

– Intensification of the strategic shift from mechanics to software
– Targeted global talent management with a "quality over quantity" approach
– Competence and "rightsizing" of employees
– More dynamic corporate culture, especially in project management (project orientation over traditional matrix)
– Integration of specialized chips (SoC, System-on-a-Chip)
– Development of new battery technologies
– Global risk management for EV raw materials
– Vertical integration, including supplier management
– PR and advertising strategies aligned with global economic trends
– Effective response to Chinese efforts to dominate the EV sector
– Efficient production strategies (e.g., Gigacasting)
– Proactive management of falling margin expectations
– Exploration of new opportunities in infrastructure (e.g., charging networks)

- Expansion of customer relationship management (e.g., charging stations in hotels)
- Strategic handling of the cannibalization issue
- Implementation of increasingly stringent environmental and climate protection regulations

Our industry's challenges may seem extensive and daunting, but it is built by those with courage and resilience; it's not for the faint-hearted. It's no coincidence that EV pioneers like Elon Musk have described the start of production for a new car model as "hell."

The true heroes of our industry are not the business visionaries but the dedicated and creative engineers who turn bold ideas into reality. Their collective energy and ambition will drive our industry into an exciting automotive future. The boundless energy of these brilliant engineering minds gives us confidence that everything will turn out just fine.

We turn challenges into opportunities, and the current landscape offers countless possibilities that we are committed to seizing fully. Confidence in the future is not just an option; it's a necessity in our industry. Paraphrasing Immanuel Kant, it is our *categorical imperative*. With the capability and resolve to meet these challenges head-on, we are poised to emerge even more successfully.

In Chap. 8, we will explore the strategies and actions that can help us shape the future of the automotive industry to our advantage despite the adversities we face now and those that lie ahead.

References

[RM-049] What is the history of the Takata airbag recall? https://getjustice.com/faq/takata-airbag-recall/what-is-the-history-of-the-takata-air-bag-recall, accessed 07.09.2023

[RM-055] "Dieselgate" - a timeline of the car emissions fraud scandal in Germany, https://www.cleanenergywire.org/factsheets/dieselgate-timeline-car-emissions-fraud-scandal-germany, accessed 08.09.2023

[RM-061] Tesla Growth and Production Statistics: How Many Vehicles Are Sold Across the Globe? https://www.investing.com/academy/statistics/tesla-facts, accessed 10.09.2023

[RM-062] Brilliance BS2 will undercut Golf, https://europe.autonews.com/article/20080901/ANE09/808309895/brilliance-touts-bs4-de-emphasizes-bs6, Zugrieff on 10.09.2023

[RM-064] World EV Sales Report, https://cleantechnica.com/2023/09/10/world-ev-sales-15-of-world-auto-sales, accessed on 10.09.2023

[RM-074] Internal Market, Industry, Entrepreneurship and SMEs, EU, https://single-market-economy.ec.europa.eu/sectors/automotive-industry_en, accessed on 17-Sep-2023

[RM-077] Strategic Management, H. Igor Ansoff, ISBN: 0230525482

[RM-093] Clay Christensen's Milkshake Marketing (Harvard Business Scool), https://hbswk.hbs.edu/item/clay-christensens-milkshake-marketing, accessed 30.09.2023

[RM-094] Ballmer: iPhone has "no chance" of gaining significant market share, https://arstechnica.com/information-technology/2007/04/ballmer-says-iphone-has-no-chance-to-gain-significant-market-share, accessed 30.09.2023

[RM-106] EU threatens China with tariffs on e-cars, https://www.handelsblatt.com/politik/international/subventionen-eu-droht-china-mit-zoellen-auf-e-autos/29388114.html, accessed on 30.11.2023

[RM-107] The 12 Group Initiatives of the NEW AUTO Strategy (VW), https://annualreport2021. volkswagenag.com/group-management-report/goals-and-strategies/the-12-group-initia tives-of-the-new-auto-strategy.html, accessed on 05.10.2023

[RM-109] Volkswagen Group taps new sources of income with sustainable mobility, https:// www.volkswagen-newsroom.com/de/pressemitteilungen/volkswagen-konzern-erschl iesst-neue-ertragsquellen-mit-nachhaltiger-mobilitaet-17607, accessed 05.10.2023

[RM-110] Volkswagen makes design a matter for the boss, https://www.faz.net/aktuell/wirtsc haft/auto-verkehr/volkswagen-auf-der-iaa-neue-strategie-fuer-design-und-marke-191 48968.html, accessed 05.10.2023

[RM-111] Europe's "Economic Engine" Raises a Red Flag?, https://www.pudaily.com/Home/ NewsDetails/35291, accessed 05.10.2023

[RM-112] BMW has no plans to produce its own electric vehicle batteries as it looks to ramp up offering, https://www.cnbc.com/2021/03/17/bmw-has-no-plans-to-produce-its-own-ele ctric-vehicle-batteries.html, accessed 06.10.2023

[RM-114] BMW Group - Investor presentation, https://www.bmwgroup.com/content/dam/ grpw/websites/bmwgroup_com/ir/downloads/en/2023/investor-presentation/BMW_ Investor_Presentation_2023.pdf, accessed on 06.10.2023

[RM-117] BMW looking to cut production cost per vehicle by 25% https://www.bmwblog. com/2021/06/22/bmw-looking-to-cut-production-cost-per-vehicle-by-25, accessed on 08.10.2023

[RM-118] MUSICIANS WHO SING THE MOST ABOUT CARS, https://www.goldeagle.com/ tips-tools/musicians-sing-most-about-cars, accessed 10/08/2023

[RM-119] Mercedes-Benz and Nvidia team up to develop next-generation supercomputers for cars, https://www.theverge.com/2020/6/23/21300614/mercedes-benz-nvidia-computer- orin-self-driving-adas-ota, accessed 09.10.2023

[RM-120] Mobileye Brakes After Saying It's No Longer Working With Tesla, https://www.invest ors.com/news/technology/mobileye-brakes-after-saying-its-no-longer-working-with- tesla, accessed 09.10.2023.

[RM-121] Volkswagen to Work With Mobileye on Automated Driving After Argo Exit, https:// money.usnews.com/investing/news/articles/2022-10-27/volkswagen-to-work-with- mobileye-on-automated-driving-after-argo-exit-sources, accessed 09.10.2023,

[RM-122] Qualcomm and BMW Group to Extend Their Long-Lasting Technology Collabora- tion to Automated Driving, https://www.qualcomm.com/news/releases/2021/11/qua lcomm-and-bmw-group-extend-their-long-lasting-technology-collaboration, accessed 09.10.2023

[RM-123] "Doing Our Part" (HR strategy of the Mercedes-Benz Group, https://group.mercedes- benz.com/dokumente/investoren/praesentationen/mercedes-benz-ir-esg23-presentat ion-sabine-kohleisen.pdf, accessed on 09.10.2023.

[RM-124] Global brand high-power charging network, https://group.mercedes-benz.com/innova tion/drive-systems/electric/high-power-charging-network.html, accessed 09.10.2023.

[RM-126] The Top 10 EV Battery Manufacturers in 2022, Visual Capitalist, https://www.visual capitalist.com/the-top-10-ev-battery-manufacturers-in-2022, accessed 14.10.2023

[RM-127] https://www.thedrive.com/news/toyota-will-adopt-tesla-style-cast-bodies-that-might- be-impossible-to-fix, The Drive, https://www.thedrive.com/news/toyota-will-adopt- tesla-style-cast-bodies-that-might-be-impossible-to-fix, accessed 10/14/2023

[RM-128] Alternative Battery Technologies Roadmap 2030+, Fraunhofer ISI, https://www. isi.fraunhofer.de/content/dam/isi/dokumente/cct/2023/abt-roadmap.pdf, accessed on 14.10.2023

[RM-135] Ford CEO explains why its difficult , https://www.youtube.com/watch?v=PPYb8M jJy5E, YouTube, accessed on 24.10.2023

[RM-137] The Software Defined Vehicle, IBM, https://www.ibm.com/blogs/digitale-perspektive/ 2023/06/the-software-defined-vehicle, accessed 10/30/2023

[RM-138] Digital Auto Report 2023 - Volume 2, PricewaterhouseCoopers International (PwC), Jörg Krings et al

[RM-140] Automakers Are Selling Cars Without Chips as They Struggle With A Shortage, Observer, https://observer.com/2022/03/automakers-are-selling-cars-without-chips-as-they-struggle-with-a-shortage, accessed 01-Nov-2023

[RM-147] Ex Audi manager Peter Mertens: "we have been asleep (...) it is getting bloody!" - German article about German auto industry and Tesla, translation in comments, Reddit, https://www.reddit.com/r/teslamotors/comments/hd3ukq/ex_audi_manager_peter_mertens_we_have_been_asleep, accessed on 21.12.2023

Automotive Perspectives: Resolving the Riddle

8

Abstract

This final chapter synthesizes the insights from previous chapters to propose solutions for the automotive industry's transformation. It presents nine fundamental paradigm shifts needed for success in the electric vehicle era, including agile PR strategies, increased vertical integration, and the prioritization of software development. The chapter emphasizes the importance of effective project management, sustainability, and the need for cultural transformation within automotive companies. It concludes by providing practical guidance for implementing these changes and maintaining competitiveness in a rapidly evolving industry landscape.

The giants on whose shoulders we can look to the future with hope are talented, passionate engineers who have consciously chosen their profession to shape the future of mobility. These brilliant people do not need constant injections of motivation—they are already motivated. True motivation is intrinsic and stems from the humanistic vision of being able to help others—and, perhaps, sometimes being praised for it.

At the same time, motivation is a fragile commodity. Sometimes, it takes just one wrong word, and everything falls apart. A whole legion of unmotivated experts cannot achieve what a passionate engineer can. For this reason, it's crucial that our engineers have the freedom to develop their ideas and pursue their passion without obstacles—and are empowered to do so.

The motivation to innovate is the key differentiator, setting dynamic and creative fields like the automotive industry apart from more static and conservative ones like mining, insurance, and infrastructure areas such as energy supply or waste management. Therefore, skillfully channeling this creative energy is the key to overcoming the electrification challenge. The strategic challenges discussed in

© The Author(s), under exclusive license to Springer Fachmedien Wiesbaden GmbH, part of Springer Nature 2025
R. Mildner et al., *Car IT Reloaded*, https://doi.org/10.1007/978-3-658-47691-5_8

Chap. 7 call for forward-looking answers. The following sections offer solutions based on these observations.

8.1 Concepts and Ideas

The revolutionary changes in our industry can be daunting and even paralyzing, yet they also serve as a powerful incentive for constructive change and improvement. Winston Churchill's seemingly cynical phrase, 'Never let a good crisis go to waste,' can be interpreted as a strategic approach to solving the electrification challenge. From that perspective, we can see real opportunities to revolutionize the car industry. In the following sections, we outline fundamental paradigm shifts and share ideas to help us make the most of this period of disruptive automotive transformation.

We formulate the potential within those challenges for each of the insights presented in our previous chapters.

8.1.1 Paradigm Shift 1: Agile PR Strategy

Insight
We love technology and the automotive engineering prowess. While technological excellence is indispensable, it is essential to meet customer demands. These customer trends may only sometimes seem logical and are increasingly driven by PR and marketing. For instance, diesel engines make perfect sense from a technical point of view. However, it's crucial to recognize that despite diesel's previous dominance, its expansion was derailed not by flaws in diesel technology but by the aggressive PR efforts of environmental activism, shifting the industry in a completely new direction. The lesson to be learned from the shift towards electric mobility is that the customer is being shaped less and less by conventional means of focus groups and market research but increasingly by activists, influencers, social media, and organic word of mouth. New legal requirements also shape customer preferences when purchasing cars.

Consequence
An intelligent PR approach should be the starting point for all strategic planning. A key aspect is recognizing new megatrends as early as possible and reacting to them agilely. Given that global PR machinery is omnipresent and a single company has too little global influence to set trends on its own, we must become faster at recognizing and implementing industry trends.

Therefore, Developing a New, Agile PR Strategy Must Be a Top Priority.

8.1.2 Paradigm Shift 2: Effectiveness Becomes More Important Than Mere Efficiency

Insight

In the past, efficiency was viewed as the key to success. In manufacturing, the focus has traditionally been on reducing costs and optimizing processes. However, the dynamic has shifted. While efficiency is still crucial, effectiveness has become vital to maintaining innovative leadership. Innovation, adaptability, and strategic foresight are required to be *effective*. Only in this way can disruptive ideas emerge. Traditionally valued quality aspects, such as perfectly aligned panel gaps—a hallmark of Piech's obsession—remain essential but are no longer seen as extraordinary.

Consequence

A more progressive approach to product and process quality is needed. A bureaucratic approach can hinder progress. Instead of focusing on formalities, delivering measurable results that meet customer expectations should be prioritized. Today's end customers expect high safety standards and innovative features at an acceptable price.

Pragmatic quality and effective project management must be prioritized and integrated into automotive systems development. The outcome matters more than the process. In this context, CORE SPICE has a critical role as a framework that simultaneously helps design effective development, increase quality, and drive innovation.

The Focus Must Be on Rapid Development While Maintaining High-Quality Standards.

8.1.3 Paradigm Shift 3: Project Fusion

Insight

Traditionally, collaboration between suppliers and OEMs has been marked by a latent competitive attitude, where both parties are rivals in the pursuit of cost efficiency and margins—a constant "tug of war." However, this approach is becoming a hindrance in today's fast-paced, innovation-driven market. A new and creative approach to collaboration between OEMs and suppliers is needed to meet the demand for quicker adaptability and tighter integration of innovation.

Consequence

Effective collaboration with suppliers is required to achieve innovation synergies, which are urgently needed in the age of EV disruption. It is no longer simply about getting the lowest price but achieving the best added value for everyone involved—especially the end customer.

Instead of the traditional supplier-customer relationship, we need a "project fusion"—a tightly knit collaboration between OEMs and suppliers, with streamlined communication channels and a more flexible contract structure. This approach requires greater permeability between OEMs and suppliers.

Suppliers must realize that their competence no longer lies in greater efficiency and a project offering as a "black box supplier" but in engineering-oriented talent management. It is time to recognize that OEMs will never have enough technological talent; it must be "supplied." Traditionally, suppliers' decentralization of talent management leads to friction and unexpected personnel changes within the project lifecycle. Allowing projects to source within a development cycle by including OEM and supplier engineers in a "Project Fusion" project removes those boundaries. This approach results in more vertically integrated projects where OEM experts work with talented supplier employees. This project fusion would eliminate information silos, break down information barriers and increase the speed of product development.

The risk of such close cooperation lies in the possible dependency and resulting vulnerability to market fluctuations. Nevertheless, it is a risk that should be consciously taken to avoid hindering long-term progress.

As a conceptual framework, CORE SPICE supports this "Project Fusion," serving as a bridge for new suppliers, including smaller ones, to lower the barriers to market entry. This enables them to become part of a broader, innovation-driven network, an OEM supplier community that transcends the traditional boundaries of company size and capacity.

Project Fusion Can Empower Legacy OEMs to Accelerate the Delivery of New Vehicle Platforms While Staying Competitive in a Rapidly Evolving Market.

8.1.4   Paradigm Shift 4: The Power of "Small." Integrating Independent Businesses and Tech Freelancers

Insight
The automotive industry's transformation opens opportunities for small, agile units and independent experts. Where large corporations tend to be sluggish, small teams and individual experts with outstanding skills can act quickly and flexibly. Previous structures favored large companies with extensive resources, but the digital transformation and the trend towards individualization demand greater involvement of smaller, specialized experts and tech businesses. These expert units are now globally active and permeable across countries and continents.

Consequence
Companies must encourage collaboration with freelancers, start-ups, and microenterprises to exploit this paradigm shift. These can serve as catalysts for innovation and flexibility, offering an unconventional perspective that can lead to disruptive solutions. The challenge is to create and foster a global expert ecosystem in which independent experts can act as an integral part of the value chain, with their ideas effectively integrated into the processes of larger companies. Bogus

self-employment laws, widespread in both the USA and Europe, would have to be reconsidered for this dynamic to bear fruit.

Small is Beautiful, and This Holds Especially True When Harnessing Top Talent's Expertise, Where Agility and Specialized Skills Often Lead to the Most Innovative Solutions.

8.1.5 Paradigm Shift 5: Artificial Intelligence and Automation of Business Processes

Insight

Artificial intelligence (AI) has the disruptive power to revolutionize the automotive industry at its core. This transformation extends far beyond car technology; it also has the potential to reshape the entire process of automotive system development. Manual and repetitive tasks will increasingly give way to AI-powered systems, leading to a new era of efficiency and innovation.

Consequence

Companies must invest strategically in AI-driven automation solutions to revolutionize their development processes. This includes introducing intelligent systems for car design, the supply chain, customer management, manufacturing, progressive customer service, and after-sales services. AI can help analyze complex data, optimize processes, and create personalized customer experiences. However, this technology also necessitates a reassessment of work roles, with our employees increasingly focused on creative and strategic tasks. AI also affects quality assurance, previously based on manual work using checklists and assessments. Solutions already exist that can check the entire company and every project for compliance and, for example, automatically calculate and report traceability coverage or compliance with ISO 26262 rules.

The Ultimate Goal Must Be to Achieve Automatic, AI-Driven Process Quality Assurance.

8.1.6 Paradigm Shift 6: Vertical Integration

Insight

For decades, a low level of vertical integration was considered the foundation of a successful vehicle company strategy. Deeply branched global supply chains, just-in-time approaches, and best cost strategies were in vogue.

The pandemic and electrification ended further advances in this approach. Gone are the days of unhindered globalization, which led to a low level of vertical integration. Local markets are gaining importance, which is putting the stability of extended supply chains to the test. At the same time, system development in the automotive industry is often too sluggish to compete successfully in this new balance between quality and development agility, especially when compared to newcomers in the USA and China.

Consequence
Vertical integration must increase again, as it now represents a competitive advantage. Integrated vehicle production requires OEMs to take on more responsibility, from securing raw materials for batteries to developing their own software and specialized vehicle chips (SoC—System-on-a-Chip). With no time for lengthy contract negotiations or wasteful bidding wars, in-house development is becoming desirable, placing high demands on a car company's workforce's skills. Also, make-or-buy decisions must increasingly favor in-house projects.

In-house software, chips, raw materials, sales, marketing, and charging networks are uncharted territory for established car manufacturers. However, continuing with the old approach would put the future of legacy OEMs at risk of ultimately losing the race to the electric future.

Urgent Action is Needed to Increase Vertical Integration and Secure a Competitive Edge.

8.1.7   Paradigm Shift 7: The Rise of Software

Insight
Software is often seen as an outsider at legacy OEMs, and hands-on software expertise is not yet widespread.

Consequence
Over-the-air (OTA) updating of vehicle software allows unprecedented flexibility and development speed. However, this is only one advantage of software-driven car development. Software enables new car functions, advanced ADAS features, autonomous driving, deeper customer engagement, and more. The possibilities can seem overwhelming for traditionally-minded vehicle engineers. However, it's just a question of how one *thinks* about software used in a car. The expectation is different when considering what software can do. With the right software-oriented approach, the sky truly is the limit.

However, to unlock software's full potential, we urgently need more and better software engineers who think independently and are ready to take on responsibility.

This shift requires changes in corporate structures, the ability to think across department boundaries, and a deep understanding of opportunities and how to leverage the software potential. For instance, while a mechanical engineering degree was once the best path to reach the C-level in car manufacturers, this opportunity must now be equally accessible to professionals from other disciplines, such as software engineers.

Software Expertise isn't just Another Discipline; It Must Be at the Core of Car Systems Development to Ensure Our Competitive Edge in the Industry.

8.1.8   Paradigm Shift 8: Prioritizing Projects Over Matrix

Insight

The traditional matrix organization in the automotive industry, rooted in functional hierarchies and rigid departmental boundaries, is increasingly at odds with the agility demanded by modern, software-defined car manufacturers. Once viewed as a mere nuisance, the matrix structure is now struggling to keep pace with the growing complexity of vehicle functions, ultimately hindering the speed and effectiveness of development. In contrast, the project-based approach has become essential for meeting these challenges and accelerating innovation.

Consequence

Car manufacturers must become more project-oriented, and this must also be reflected in role understanding, such as that of the project manager (or Project Lead, in the case of CORE SPICE). Strong project leadership is a recipe for more personal responsibility and safer, faster feature implementation. A clear understanding of roles and a strong focus on objectives are proven strategies for avoiding unproductive debates over methods.

To become effective, the project approach must be replicated across the organization. At the same time, overarching work must be bundled and coordinated via systematic program management.

A professional and effective career ladder must be established to ensure that working on projects is not a career dead end. In the evolving automotive industry, it's not practical for managerial roles to be the only path to higher salaries. A technical career ladder with levels such as software developer, software architect, software lead architect, system developer, and system architect must be established at the highest company level (see [RM-143]) and supported by attractive compensation.

Project Orientation and an Expert Career Path Must Be Established as Equivalent Alternatives to a Management Career to Retain Top Talent in Our Industry.

8.1.9   Paradigm Shift 9: Embedding Sustainability into the Core Business Model

Insight

Sustainability is no longer just a PR term, or a side note in corporate policy but a key factor for long-term competitiveness in the automotive industry. Customers, investors, and legislators increasingly demand ecological responsibility and transparency along the entire value chain.

Consequence

Vehicle manufacturers and suppliers must integrate sustainability into every level of their business activities. This ranges from using environmentally friendly materials to using renewable energy in production and developing circular economic concepts for the end of a vehicle's life. It requires all stakeholders to work closely to minimize environmental impact while ensuring efficiency and profitability. Sustainable innovations can help open up new markets and strengthen brand loyalty.

8.2   Car Manufacturers in a Constant State of Change

Transforming a 130-year-old industry into a collection of "T-Rex" companies (see [RM-143]) isn't something that can happen at the push of a button. The strategic elements outlined in Chap. 7 and the necessary paradigm shifts demand a true *tour de force*. This transformation requires a united effort from the entire company— from the boardroom to every employee. These changes must not only be supported but actively championed and driven forward.

Such a significant, strategic change requires careful planning. The groundbreaking book *Leading Change* by John P. Kotter, a renowned expert in leadership and change management [BF-016], presents a proven process for corporate transformation:

1. **Create a sense of urgency**: Build awareness of the need for change.
2. **From a powerful coalition**: Bring influential and committed team members together.
3. **Develop a vision and strategy**: Craft a clear vision of the future and actionable strategies.
4. **Communicate the vision**: Clearly and convincingly share the vision of change.
5. **Empower broad-based action**: Enable employees to act and remove obstacles to change.
6. **Generate short-term wins**: Achieve and celebrate quick, visible successes.
7. **Consolidate gains and produce more change**: Use early successes to drive further change.
8. **Anchor new approaches in the culture**: Integrate changes into the corporate culture.

This straightforward "change process" provides a framework for a car company to plan and achieve measurable results. It is particularly relevant for the company's shareholders that the difficult transition phase, likely to extend over several years, is communicated appropriately.

This all-encompassing change can be supported in a targeted manner, for example:

Seek expert support with expertise guiding similar transformations to a software-focused organization. Prioritize those with hands-on experience in the software industry.

Implement education and training programs focused on "software". Contrary to common belief, software is not as hard to grasp as it might seem. Successful training relies on motivation, curiosity, and effective communication of software concepts. These elements can encourage interest in software and lead to professional fulfillment. However, to ensure lasting impact, employees should receive ongoing support after the training to prevent these gains from becoming short-lived.

Actively utilize AI-powered tools. With modern AI-supported tools, teams can be highly effective. Process improvement, documentation, searching for new ideas, and even root cause analysis of tricky defects—AI-supported development tools and planning tools – make the list of AI solutions endless. AI can offer and implement training measures, act as a virtual consultant (for example, for Automotive SPICE or ISO 26262 issues), and even provide emotional support in crises, such as during extreme escalations, which are common in our industry. However, appropriate support must be offered to use these tools successfully. This is an opportunity for older, experienced employees who can, for example, develop the right "prompts" [RM-146]. This will be necessary until a general AI understands the context and can take the initiative.

Persuasion. Resistance to change is natural. Examples of the right mindset are worth their weight in gold for easing this resistance. Imagine a visionary industry leader of the caliber of Steve Jobs giving an inspiring motivational speech.

Utilize cross-functional teams. Bring teams from different company areas to generate and implement ideas and solutions. This promotes collaboration and helps to overcome silo thinking.

Maintain open communication channels and an ongoing dialog between management and employees to gather feedback, address concerns, and keep the workforce informed.

Support champions. Identify and empower critical individuals within the organization to serve as change champions who lead by example and inspire others. Establish rewards and recognition for teams and individuals to ensure that innovative "inner leadership" is cultivated and spread throughout the organization. This approach aligns closely with John Kotter's second step of management change.

Promote flexibility and adaptability. Encourage employees to take on new challenges and introduce them to the new world of software.

Pilot projects. Test new approaches and ideas on a small scale before implementing them company-wide. This idea correlates with the sixth step of John Kotter's change process.

Implement change sustainably, not just in the short term but also in the long term, integrating it into the corporate culture. This correlates with the eighth step of the Kotter process.

In all these observations and implementation ideas, the role of company management is crucial. Top management must identify and implement suitable support measures that fit the management style and the (new) corporate culture. Credible, inspiring leadership is essential for such profound change.

Such changes are currently occurring worldwide. The pace of change in the automotive sector is nothing short of breathtaking. At least in our industry, the saying 'Nothing is more constant than change' has been proven true.

8.3   Closing Words

This book has explored a wide range of topics that shape the future of the automotive industry, from the industry's early days to the challenges and innovations of autonomous driving and connected cars. It has also addressed the role of quality, effective project management, and strategic considerations.

From groundbreaking technological advances in automobiles to the revolutionary approaches to business management, the true greatness of our industry is evident.

Car IT Reloaded has taken us on a journey that reflects on the past and looks forward confidently to a future where new technologies are increasingly ubiquitous. Its topics are so diverse and captivating that we had to focus on the essentials. While we covered a narrower spectrum in our previous book, *Car IT*, and only briefly touched on many of the critical topics of digital mobility, this book has become a quasi-compendium. Each sub-chapter could easily warrant its dedicated, comprehensive book, as the automotive industry has evolved into a highly diverse field, increasingly integrating the fast-paced advancements of automotive digitalization.

Car manufacturers are profoundly transforming their brands, platforms, and corporate cultures to navigate the new era of rapid mobility change. This challenging task deserves recognition, as corporate culture is critical in determining a company's success or failure. Now is the time for vision, courage, and creativity to shape the future of mobility. We firmly believe the key to success lies in striking the right balance between honoring tradition and embracing the adaptability needed to thrive in our industry's new dynamics.

The pace of change in our industry is nothing short of astonishing. What was once a leisurely automotive Sunday stroll has become a digital race. So, let's race together into a future defined by mobile innovation and the seamless connection between man and machine. One thing is sure: we will win this race. Together, we will all emerge as winners.

May our journey through *Car IT Reloaded* inspire you to explore the new mobile world with enthusiasm and curiosity. Change remains the only constant, and the most exciting automobile innovations and adventures lie ahead.

References

[BF-016] Leading Change, John P. Kotter, ISBN: 978–1–4221–8643–5

[RM-143] The T-Rex Company, ProjectCrunch.com, https://projectcrunch.com/t-rex-company, accessed 11/06/2023

[RM-146] AI: The Golden Age of Yodas, Project Crunch, https://projectcrunch.com/ai-the-gol den-age-of-yodas, accessed 12/01/2023

The manufacturer's authorised representative in the EU is Springer
Nature Customer Service Centre GmbH, Europaplatz 3, 69115 Heidelberg,
Germany. If you have any concerns regarding our products, please
contact ProductSafety@springernature.com

Printed and bound by CPI Group (UK) Ltd, Croydon, CR0 4YY
23/04/2026
02095586-0004